FEEDBACK CONTROL SYSTEMS

JOHN VAN DE VEGTE

Department of Mechanical Engineering
University of Toronto

Prentice-Hall, Inc., Englewood Cliffs, NJ 07632

Library of Congress Cataloging in Publication Data

VAN DE VEGTE, J. (JOHN)
    Feedback control systems.

    Includes bibliographies and index.
    1. Feedback control systems.   I. Title.
TJ216.V26   1986   629.8'3   85-6567
ISBN 0-13-312950-0

Editorial/production supervision and
    interior design: *Mary Carnis and Colleen Brosnan*
Cover design: *20/20 Services, Inc.*
Manufacturing buyer: *Rhett Conklin*

© 1986 by Prentice-Hall
A Division of Simon & Schuster, Inc.
Englewood Cliffs, New Jersey 07632

Printed in the United States of America

10   9   8   7   6   5   4   3

ISBN 0-13-312950-0   01

PRENTICE-HALL INTERNATIONAL, INC., *London*
PRENTICE-HALL OF AUSTRALIA PTY. LIMITED, *Sydney*
EDITORA PRENTICE-HALL DO BRASIL, LTDA., *Rio de Janeiro*
PRENTICE-HALL CANADA INC., *Toronto*
PRENTICE-HALL HISPANOAMERICANA, S.A., *Mexico*
PRENTICE-HALL OF INDIA PRIVATE LIMITED, *New Delhi*
PRENTICE-HALL OF JAPAN, INC., *Tokyo*
PRENTICE-HALL OF SOUTHEAST ASIA PTE. LTD., *Singapore*
WHITEHALL BOOKS LIMITED, *Wellington, New Zealand*

To
  Maria

and to
  Joyce
    John
      David
        Michael

CONTENTS

14 MULTIVARIABLE SYSTEMS IN THE FREQUENCY DOMAIN

362

APPENDIX A: VECTORS, MATRICES, AND DETERMINANTS

391

APPENDIX B: COMPUTER AIDS FOR ANALYSIS AND DESIGN

396

PREFACE

This book is intended to serve as a text for a first course in control systems to third- or fourth-year engineering students. Material in later chapters beyond what can be covered in a first course is used by the author in an elective second course at the senior undergraduate/graduate level. The book has been written to be suitable also for students with a more remote or less complete educational background, and for self-study.

The text has grown out of many years of experience in teaching the subject to students in mechanical engineering, industrial engineering, engineering science, correspondence courses, and to students from industry in night courses. However, to show the generality and power of the subject, a first course should not be directed toward a particular department, and the book has been written to be suitable in all branches of engineering.

To develop insight, concepts are explained in the simplest possible mathematical framework. Thus Chapter 1 immediately identifies the two questions to be answered, that is, how dynamic systems behave and how this behavior may be changed by the use of feedback. This chapter also introduces the basic compromise between stability and accuracy which underlies all feedback system design.

To promote understanding of the subject, concepts of design (i.e., the change of dynamic behavior) should be developed in parallel with those of analysis. Thus Chapter 5 allows a focus on physical explanation of the basic actions of dynamic controllers, unencumbered by the intricacies of the root locus and frequency response techniques discussed in Chapters 6 through 8. Examples are used to motivate the study of these techniques.

Two chapters devoted to digital control systems reflect the explosive growth in this area following the introduction of microprocessors. The discussion of state space techniques has been separated into a chapter on analysis and one on design. The final chapter is concerned with frequency-domain design of multivariable systems. This alternative to the state space approach, which has long been favored in the United Kingdom, is receiving increasing attention elsewhere, and its parallel introduction is overdue.

A text must accommodate a wide spectrum of preferences on the extent to which computer aids are incorporated in a course. To achieve this, the text and the 450 problems have been made independent of such aids, and programs for interactive computer-aided analysis and design with graphics are collected in Appendix B, with examples of their use. Reference to these aids is made where appropriate. The author is pleased to acknowledge the work of graduate student Philip W. P. Cheng, who developed these programs for use in the Department of Mechanical Engineering of the University of Toronto.

Finally, the author expresses his gratitude to his wife, who, starting from a two-finger level of expertise, typed every word, and every equation, of the manuscript and valiantly resisted momentary urges to heave a brick at the word processor.

John Van de Vegte
Toronto, Canada

INTRODUCTION AND LINEARIZED DYNAMIC MODELS

1.1 INTRODUCTION

In the first part of this chapter, after a general introduction, the concepts of open-loop and closed-loop control are discussed in the context of a water level control system. This example is then used to introduce fundamental considerations in control system analysis and design.

In the second part of the chapter, Laplace transforms are discussed and used to define the transfer function of a system. This is a linearized model of the dynamic behavior of the system that will serve as the basis for system analysis and design in most of this book. Block diagram reduction is used to obtain the transfer function of a system consisting of interconnected subsystems from those of the subsystems. This completes the framework necessary for Chapter 2, in which transfer functions are derived for a variety of physical (sub)systems.

1.2 EXAMPLES AND CLASSIFICATIONS OF CONTROL SYSTEMS

Control systems exist in a virtually infinite variety, both in type of application and level of sophistication. The heating system and the water heater in a house are systems in which only the sign of the difference between desired and actual temperatures is used for control. If the temperature drops below a set value, a constant heat source is switched on, to be switched off again when the temperature rises above a set maximum. Variations of such relay or on-off control systems, sometimes quite sophisticated, are very common in practice because of their relatively low cost.

In the nature of such control systems, the controlled variable will oscillate continuously between maximum and minimum limits. For many applications this control is not sufficiently smooth or accurate. In the power steering of a car, the controlled variable or system output is the angle of the front wheels. It must follow the system input, the angle of the steering wheel, as closely as possible but at a much higher power level.

In the process industries, including refineries and chemical plants, there are many temperatures and levels to be held to usually constant values in the presence of various disturbances. Of an electrical power generation plant, controlled values of voltage and frequency are outputs, but inside such a plant there are again many temperatures, levels, pressures, and other variables to be controlled.

In aerospace, the control of aircraft, missiles, and satellites is an area of often very advanced systems.

*****

One classification of control systems is the following:

1. *Process control or regulator systems:* The controlled variable, or output, must be held as close as possible to a usually constant desired value, or input, despite any disturbances.
2. *Servomechanisms:* The input varies and the output must be made to follow it as closely as possible.

Power steering is one example of the second class, equivalent to systems for positioning control surfaces on aircraft. Automated manufacturing machinery, such as numerically controlled machine tools, uses servos extensively for the control of positions or speeds.

This last example brings to mind the distinction between continuous and discrete systems. The latter are inherent in the use of digital computers for control.

The classification into linear and nonlinear control systems should also be mentioned at this point. Analysis and design are in general much simpler for the former, to which most of this book is devoted. Yet most systems become nonlinear if the variables move over wide enough ranges. The importance in practice of linear techniques relies on linearization based on the assumption that the variables stay close enough to a given *operating point*.

1.3 OPEN-LOOP CONTROL AND CLOSED-LOOP CONTROL

To introduce the subject, it is useful to consider an example. In Fig. 1.1, let it be desired to maintain the actual water level c in the tank as close as possible to a desired level r. The desired level will be called the system *input,* and the actual level the *controlled variable* or system *output*. Water flows from the tank via a valve V_o and enters the tank from a supply via a *control valve* V_c. The control valve is adjustable, either manually or by some type of *actuator*. This

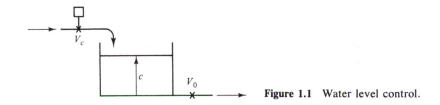

Figure 1.1 Water level control.

may be an electric motor or a hydraulic or pneumatic cylinder. Very often it would be a pneumatic diaphragm actuator, indicated in Fig. 1.2. Increasing the pneumatic pressure above the diaphragm pushes it down against a spring and increases valve opening.

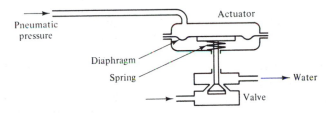

Figure 1.2 Pneumatically actuated valve.

Open-Loop Control

In this form of control, the valve is adjusted to make output c equal to input r, but not readjusted continually to keep the two equal. Open-loop control, with certain safeguards added, is very common, for example, in the context of sequence control, that is, guiding a process through a sequence of predetermined steps. However, for systems such as the one at hand, this form of control will normally not yield high performance. A difference between input and output, a system *error $e = r - c$* would be expected to develop, due to two major effects:

1. *Disturbances* acting on the system
2. *Parameter variations* of the system

These are prime motivations for the use of feedback control. For the example, pressure variations upstream of V_c and downstream of V_o can be important disturbances affecting inflow and outflow, and hence level. In a steel rolling mill, very large disturbance torques on the drive motors of the rolls when steel slabs enter or leave affect speeds.

For the water level example, a sudden or gradual change of flow resistance of the valves due to foreign matter or valve deposits represents a system parameter variation. In a broader context, not only are the values of the parameters of a process often not precisely known, but they may also change greatly with operating condition.

For an aircraft or a rocket, the effectiveness of control surfaces changes rapidly as the device rises through the atmosphere. In an electrical power plant, parameter values are different at 20% and 100% of full power. In a valve, the relation between pressure drop and flow rate is often nonlinear, and as a result the resistance parameter of the valve changes with flow rate. Even if all parameter variations were known precisely, it would be complex, say in the case of the level example, to schedule the valve opening to follow time-varying desired levels.

Closed-Loop Control or Feedback Control

To improve performance, the operator could continuously readjust the valve based on observation of the system error e. A *feedback control system* in effect automates this action, as follows:

> The output c is measured continuously and fed back to be compared with the input r. The error $e = r - c$ is used to adjust the control valve by means of an actuator.

The feedback loop causes the system to take corrective action if output c (actual level) deviates from input r (desired level), whatever the reason.

A broad class of systems can be represented by the block diagram shown in Fig. 1.3. The *sensor* in Fig. 1.3 measures the output c and, depending on type, represents it by an electrical, pneumatic, or mechanical signal. The input r is represented by a signal in the same form. The *summing junction* or *error junction* is a device that combines the inputs to it according to the signs associated with the arrows: $e = r - c$.

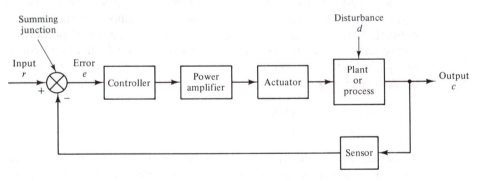

Figure 1.3　System block diagram.

It is important to recognize that if the control system is any good, the error e will usually be small, ideally zero. Therefore, it is quite inadequate to operate an actuator. A task of the *controller* is to amplify the error signal. The controller output, however, will still be at a low power level. That is, voltage or pressure have been raised but current or airflow are still small. The *power amplifier* raises power to the levels needed for the actuator.

The *plant* or *process* has been taken to include the valve characteristics as well as the tank. In part this is related to the identification of a *disturbance d* in Fig. 1.3 as an additional input to the block diagram. For the level control, *d* could represent supply pressure variations upstream of the control valve.

1.4 CONTROL SYSTEM ANALYSIS AND DESIGN

Control system analysis and design can be summarized in terms of the following two questions:

1. *Analysis:* What is the performance of a given system in response to changes of inputs or disturbances?
2. *Design:* If the performance is unsatisfactory, how can it be improved without changing the process, actuator, and power amplifier blocks?

It is particularly important to note the constraints imposed on the designer. The blocks indicated generally represent relatively, or very, expensive equipment, and must be considered as a fixed part of the system. The power of design techniques that will permit large changes in performance to be achieved by changing only the controller should be appreciated.

*****

The term *performance* is used to summarize several aspects of the behavior. Assume that in Fig. 1.3 a sudden change of input is applied, to a new constant value. A certain period of time will be required for transient response terms to decay and for the output to level off at the new value. One key feature of this transient period is that it should be sufficiently short. Another is that the transient response should not be excessively oscillatory or severely overshoot the final level.

The *steady-state* response, after the transients have decayed, is an equally important aspect of the performance. Any steady-state errors between *r* and *c* must be satisfactorily small. To a disturbance input, the output should ideally not respond at all, and in any case the steady-state value of this output should be acceptably small.

The performance of a design is also measured by its success in reducing the dynamic and steady-state effects of parameter variations in the plant on the output.

Disturbances and parameter variations were given as motivations for feedback control. However, the transient response and steady-state error characteristics can also be improved by the use of feedback, and the motivations for feedback can be listed as follows:

1. Reducing the effects of parameter variations
2. Reducing the effects of disturbance inputs
3. Improving transient response characteristics
4. Reducing steady-state errors

In fact, improvements in the first two items are usually achieved in the course of design procedures aimed at the last two.

An intuitive idea concerning these can be obtained by assuming the controller in Fig. 1.3 to be an amplifier with *gain K*; that is, the output of this block is *K* times its input. Larger *K* means greater amplification of the error signal *e*. Therefore, the errors for given output values are smaller. Hence large gains are desirable to reduce errors, that is, to improve accuracy. Also, larger gain means a larger change of valve opening for a certain change of error. This suggests a faster change of the output, and greater speed of response of the system.

On the other hand, these faster changes of output intuitively suggest increasing danger of severe overshoot and oscillations of the output following a sudden change of the input. Figure 1.4 shows the large effect on the response to a step change of *r* which can easily result if the gain is increased from a rather low to a rather high value. In fact, with a further increase of gain the oscillations may grow instead of decay. The system is then *unstable*.

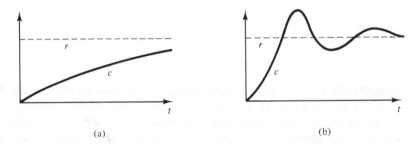

Figure 1.4   Step responses for low (a) and high (b) gains.

Stability is always the primary concern in feedback control design. But to be useful a system must also possess adequate *relative stability;* that is, the overshoot of a step response must be acceptably small, and this response must not be unduly oscillatory during the transient period.

Relative stability considerations usually impose an upper limit on gain, and hence on accuracy and speed of response. Much of control system design can be summarized as being concerned with achieving a satisfactory compromise between these features. If this is not possible with only a gain *K*, controller complexity is increased.

The remainder of this chapter provides a basis for the tools needed to move beyond this intuitive discussion and to answer the questions it raises.

1.5 LINEARIZED DYNAMIC MODELS

The concept of a transfer function will be developed to describe individual blocks and their interconnections in a block diagram. This is a linear model and requires a linearized description of system dynamic behavior.

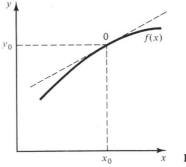

Figure 1.5 Linearization.

Figure 1.5 shows a nonlinear relation $y = f(x)$ between the input x and the output y of a block in a block diagram. It could represent the force–deflection relation of a rubber spring, or the flow versus valve opening characteristic of a control valve. If the variations of x about x_0 are small enough, the nonlinear curve can be approximated by its tangent at 0. The model is then said to be linearized about the steady-state operating point 0. In Fig. 1.5, if $\Delta x = x - x_0$, $\Delta y = y - y_0$, the linearized model of $f(x)$ about x_0 is

$$\Delta y = \left(\frac{df}{dx}\right)_{x_0} \Delta x \qquad (1.1)$$

where $(df/dx)_{x_0}$ is the slope of the tangent at 0.

It may also be that the output y of a block is a nonlinear function of two or more inputs. For example, valve flow is a function of valve opening and the pressure drop across the valve. If $y = f(x, z)$, the linearized model for small variations Δx and Δz about an operating point (x_0, z_0) is

$$\Delta y = \left(\frac{\partial f}{\partial x}\right)_{x_0, z_0} \Delta x + \left(\frac{\partial f}{\partial z}\right)_{x_0, z_0} \Delta z \qquad (1.2)$$

where $(\partial f/\partial x)_{x_0, z_0}$ and $(\partial f/\partial z)_{x_0, z_0}$ are partial derivatives. Note that the variables in these linearized models are not the actual values of the variables, but the deviations from those at the steady-state operating point.

It is a common practice in these linearized models to redefine the variables to represent the variations. Thus the models (1.1) and (1.2) will often be written as

$$y = Kx \qquad y = K_1 x + K_2 z \qquad (1.3)$$

*****

For a broad class of systems the dynamic behavior of the variations about operating-point values can be approximated by an nth-order linear differential equation

$$\frac{d^n c}{dt^n} + a_{n-1}\frac{d^{n-1}c}{dt^{n-1}} + \cdots + a_1\frac{dc}{dt} + a_0 c = b_m\frac{d^m r}{dt^m} + \cdots + b_1\frac{dr}{dt} + b_0 r \qquad (1.4)$$

where the variations $c(t)$ and $r(t)$ of output and input are functions of time t, and a_i and b_i are real constants.

Frequently, it is sufficient to determine the dynamic behavior of the output variations $c(t)$. If the actual variables are required, the steady-state solutions at the operating point must be added, corresponding to x_0 and y_0 in Fig. 1.5. These can be found from the actual nonlinear equations by setting derivatives with respect to time to zero.

Example 1.5.1   Spring–Mass–Damper System

The linear case of the classical system in Fig. 1.6 is considered in Example 2.2.1. The differential equation (2.2) that describes the position $x(t)$ of the mass m in response to an external force $f(t)$ is

$$m\ddot{x} + c\dot{x} + kx = f$$

As discussed in Section 2.2, $-kx$ is the spring force on m, with spring constant k, and $-c\dot{x}$ the damping force, with damping coefficient c. In the linear case k and c are constant, and weight mg does not occur in the equation if $x = 0$ is chosen at the position of static equilibrium, where the weight is counterbalanced by a spring force.

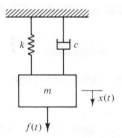

Figure 1.6   Spring–mass–damper system.

But often spring force does not change proportionally with x, and a nonlinear differential equation model is necessary, for example:

$$m\ddot{x} + c\dot{x} + kx^3 = f + mg$$

Weight mg is added to the downward force $f(t)$ and $x = 0$ at the position of zero spring force.

At a steady-state operating point where $f = f_0 = $ constant, the solution $x = x_0 = $ constant, so that $kx_0^3 = f_0 + mg$, or $x_0 = [(f_0 + mg)/k]^{1/3}$. To linearize the equation for small variations $\Delta x = x - x_0$, $\Delta f = f - f_0$ about the operating point, the nonlinear term $f(x) = kx^3$ is linearized according to (1.1) as

$$kx_0^3 + \left(\frac{df}{dx}\right)_{x_0} \Delta x = kx_0^3 + 3kx_0^2 \Delta x$$

Then, since $(\Delta\dot{x}) = \dot{x}$, $(\Delta\ddot{x}) = \ddot{x}$, the linear model is

$$m(\Delta\ddot{x}) + c(\Delta\dot{x}) + 3kx_0^2(\Delta x) + kx_0^3 = f_0 + \Delta f + mg$$

Since $kx_0^3 = f_0 + mg$, redefining x and f to represent the variations about x_0 and f_0 gives

$$m\ddot{x} + c\dot{x} + (3kx_0^2)x = f$$

If, say to determine physical clearances, the actual positions are needed, x_0 must be added to the solution of this linear differential equation.

For linear systems, the equations for actual variables and deviations are the same. For example, if $\ddot{x} + c\dot{x} + dx = y$, then substituting $x = x_0 + \Delta x$, $y = y_0 + \Delta y$, where x_0 and y_0 are constant, yields

$$\Delta\ddot{x} + c\,\Delta\dot{x} + d\,\Delta x + dx_0 = \Delta y + y_0$$

But $dx_0 = y_0$ is the steady-state solution, so the equation in terms of Δx is the same as that for x.

In Chapter 12 the state space model will be introduced. It is a description of dynamic systems in terms of a set of first-order differential equations, written compactly in matrix form. This form of model is very powerful for the study of even very large systems. But in classical control theory, systems are commonly described by means of transfer functions. These will be defined by the use of Laplace transforms.

1.6 LAPLACE TRANSFORMS

Laplace transform theory is quite extensive, and it is therefore fortunate that only a small and isolated part of it is needed. The *Laplace transform F(s)* of a function $f(t)$ is defined by

$$F(s) = L[f(t)] = \int_0^\infty f(t)e^{-st}\,dt \qquad (1.5)$$

For the present it suffices to consider the *Laplace variable s* simply as a complex variable with real part σ and imaginary part ω:

$$s = \sigma + j\omega \qquad (1.6)$$

From (1.5), the transform changes a function of time into a function of this new variable s. The advantage will be found to be that differentiation and integration are changed into algebraic operations.

Important examples and theorems are discussed below. The solutions of the integrals, which need not be evaluated in the use that will be made of all these results, may be verified, if desired, from tables of integrals.

1. *Unit step function u(t)*: Shown in Fig. 1.7(a), this is a common test input to evaluate the performance of control systems. Its Laplace transform, from substitution of $f(t) = 1$ into (1.5), is

$$L[u(t)] = \int_0^\infty (1)e^{-st}\,dt = \frac{1}{s} \qquad (1.7)$$

For a step of magnitude A, $Au(t)$, the transform is A/s.

2. *Ramp function At*: This is also a common test input, and is shown in Fig. 1.7(b). With $f(t) = At$, integration by parts in equation (1.5) yields

$$L[At] = \int_0^\infty At e^{-st} \, dt = \frac{A}{s^2} \tag{1.8}$$

3. *Decaying exponential $Ae^{-\alpha t}$*: In many physical systems, the transient response that follows a change of input or a disturbance decays according to this characteristic, which is shown in Fig. 1.7(c).

$$L[Ae^{-\alpha t}] = \int_0^\infty (Ae^{-\alpha t})e^{-st} \, dt = \frac{A}{s + \alpha} \tag{1.9}$$

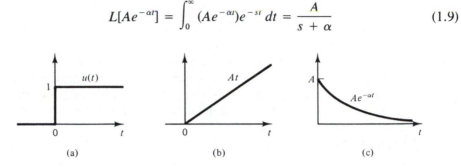

(a)   (b)   (c)

Figure 1.7   (a) Unit step; (b) ramp; (c) decaying exponential.

4. *Derivatives $d^n f(t)/dt^n$ if $F(s) = L[f(t)]$*:

$$L\left[\frac{df(t)}{dt}\right] = \int_0^\infty \left[\frac{df(t)}{dt}\right]e^{-st} \, dt \tag{1.10}$$

$$= e^{-st}f(t)\Big|_0^\infty + s\int_0^\infty f(t)e^{-st} \, dt = sF(s) - f(0)$$

Repeating this procedure yields for the second derivative:

$$L\left[\frac{d^2 f(t)}{dt^2}\right] = s^2 F(s) - sf(0) - \frac{df(0)}{dt} \tag{1.11}$$

In general, for the nth derivative, if the initial conditions of $f(t)$ and its derivatives are zero,

$$L\left[\frac{d^n f(t)}{dt^n}\right] = s^n F(s) \tag{1.12}$$

From this it is useful to remember that taking a derivative of a function is equivalent to multiplying its transform by s.

5. *Integrals of $f(t)$ if $F(s) = L[f(t)]$*: By using integration by parts in the definition (1.5), it can be shown that

$$L\left[\int_{-\infty}^t f(\zeta) \, d\zeta\right] = \frac{1}{s}F(s) + \frac{1}{s}\left[\int_{-\infty}^t f(\zeta) \, d\zeta\right]_{t=0} \tag{1.13}$$

where $[\int'_{-\infty} f(\zeta)\, d\zeta]_{t=0}$ is the initial value of the integral. For zero initial conditions

$$L[n\text{th integral of } f(t)] = \frac{F(s)}{s^n} \tag{1.14}$$

6. The product $e^{-at}f(t)$ if $F(s) = L[f(t)]$:

$$L[e^{-at}f(t)] = \int_0^\infty e^{-at}f(t)e^{-st}\, dt$$

$$= \int_0^\infty f(t)e^{-(s+a)t}\, dt = F(s + a) \tag{1.15}$$

since the last integral is the definition of the transform, with s replaced by $(s + a)$. This result will be used in the following context: If $F(s + a) = 1/(s + a)^2$, then $F(s) = 1/s^2$, which from (1.8) is known to be the transform of t. Hence

$$L[te^{-at}] = \frac{1}{(s + a)^2} \tag{1.16}$$

7. *The translated function $f(t - t_d)$ if $F(s) = L[f(t)]$:* As shown in Fig. 1.8, $f(t - t_d)$ equals $f(t)$ translated by t_d along the t-axis, and is zero for $t < t_d$.

$$L[f(t - t_d)] = \int_0^\infty f(t - t_d)e^{-st}\, dt$$

$$= \int_{t_d}^\infty f(t - t_d)e^{-st}\, dt$$

$$= e^{-st_d} \int_{t_d}^\infty f(\tau)e^{-s\tau}\, dt \tag{1.17}$$

$$= e^{-st_d} \int_0^\infty f(\tau)e^{-s\tau}\, d\tau = e^{-st_d}F(s)$$

where $\tau = t - t_d$.

Thus translation over t_d is equivalent to multiplying the transform by e^{-st_d}.

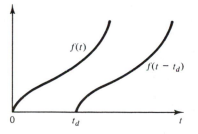

$f(t)$

$f(t - t_d)$

0   t_d t **Figure 1.8** Translated function.

8. *The linearity theorem:* If $F_1(s) = L[f_1(t)]$, $F_2(s) = L[f_2(t)]$, and c_1 and c_2 are independent of t and s, then

$$L[c_1f_1(t) + c_2f_2(t)] = c_1F_1(s) + c_2F_2(s) \qquad (1.18)$$

9. *The final value theorem:*

$$\lim_{t \to \infty} f(t) = \lim_{s \to 0} sF(s) \qquad (1.19)$$

This allows the final or steady-state value of $f(t)$, that is, its value as $t \to \infty$, to be found from $F(s)$. The theorem is valid only if the limit exists.

Table 1.6.1 is a table of *Laplace transform pairs* which includes the preceding examples. The additional entries will be considered later. The transform of a unit impulse can be obtained by a limit process in which the width of a rectangular pulse of unit area is allowed to approach zero. The left column shows the inverse Laplace transforms of the expressions on the right. The entries in the table will be used to determine the inverse transforms of more complex expressions by expanding these expressions in terms of their partial fractions.

However, it is desirable to define the concept of a transfer function first, to show how such transforms arise in the study of control systems.

1.7 TRANSFER FUNCTIONS AND SYSTEM RESPONSE

Figure 1.9(a) shows a block representing a system with input r and output c, which could also be a subsystem in a block diagram. The general linearized differential equation (1.4) will be taken to relate r and c. Based on the linearity theorem in Section 1.6, its Laplace transform can be written simply by transforming each term in turn. If the initial conditions are assumed to be zero, the result is, using (1.12) or Table 1.6.1,

$$(s^n + a_{n-1}s^{n-1} + \cdots + a_1s + a_0)C(s) = (b_ms^m + \cdots + b_1s + b_0)R(s) \qquad (1.20)$$

where

$$C(s) = L[c(t)] \qquad R(s) = L[r(t)] \qquad (1.21)$$

Equation (1.20) gives immediately the common system representation of classical control theory.

Definition.   The *transfer function* of a (sub)system is the ratio of the Laplace transforms of its output and input, assuming zero initial conditions:

$$G(s) = \frac{C(s)}{R(s)} \qquad (1.22)$$

From (1.20), the transfer function is given by

$$G(s) = \frac{b_ms^m + b_{m-1}s^{m-1} + \cdots + b_1s + b_0}{s^n + a_{n-1}s^{n-1} + \cdots + a_1s + a_0} \qquad (1.23)$$

TABLE 1.6.1 LAPLACE TRANSFORM PAIRS

$f(t)$	$F(s)$
$u(t)$—unit step	$\dfrac{1}{s}$
At; At^n	$\dfrac{A}{s^2}$; $\dfrac{An!}{s^{n+1}}$
$Ae^{-\alpha t}$	$\dfrac{A}{s+\alpha}$
$\dfrac{df(t)}{dt}$	$sF(s) - f(0)$
$\dfrac{d^2f(t)}{dt^2}$	$s^2F(s) - sf(0) - \dfrac{df(0)}{dt}$
$\dfrac{d^nf(t)}{dt^n}$ (0 initial conditions)	$s^nF(s)$
$\displaystyle\int_{-\infty}^{t} f(\zeta)\,d\zeta \equiv f^{(-1)}(t)$	$\dfrac{1}{s}F(s) + \dfrac{1}{s}\left[\displaystyle\int_{-\infty}^{t} f(\zeta)\,d\zeta\right]_{t=0}$
$f^{(-n)}(t)$ (0 initial conditions)	$\dfrac{F(s)}{s^n}$
$e^{-at}f(t)$	$F(s+a)$
te^{-at}; t^ne^{-at}	$\dfrac{1}{(s+a)^2}$; $\dfrac{n!}{(s+a)^{n+1}}$
$f(t - t_d), t > t_d; 0, t < t_d$	$e^{-st_d}F(s)$
$c_1f_1(t) + c_2f_2(t)$	$c_1F_1(s) + c_2F_2(s)$
$\delta(t)$—unit impulse	1
$A \sin \omega t$	$\dfrac{A\omega}{s^2 + \omega^2}$
$e^{-\zeta\omega_n t} \sin[\omega_n(1 - \zeta^2)^{1/2}t]$; $\zeta < 1$	$\dfrac{\omega_n(1 - \zeta^2)^{1/2}}{s^2 + 2\zeta\omega_n s + \omega_n^2}$
$1 - e^{-\zeta\omega_n t}(1 - \zeta^2)^{-1/2}$ $\quad\times \sin[\omega_n(1 - \zeta^2)^{1/2}t + \phi]$ $\left(\zeta < 1; \phi = \tan^{-1}\left[\dfrac{(1 - \zeta^2)^{1/2}}{\zeta}\right]\right)$	$\dfrac{\omega_n^2}{s(s^2 + 2\zeta\omega_n s + \omega_n^2)}$

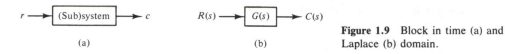

(a)   (b)

Figure 1.9 Block in time (a) and Laplace (b) domain.

The representation of Fig 1.9(a) is now replaced by that of Fig. 1.9(b), in terms of transforms and the transfer function. The following interpretation will often be applied:

$$C(s) = G(s)R(s) \qquad (1.24)$$

In words, the transform $C(s)$ of the output equals the transfer function $G(s)$ times the transform $R(s)$ of the input.

*****

The assumption of zero initial conditions is very common in system analysis and design. It is natural for linearized equations, in which the variables are variations about operating point values. Furthermore, for linear systems the nature of the transient response is independent of the initial conditions: that is, whether or not the response is oscillatory, and if it is, whether the oscillations decay sufficiently fast. This is the type of information usually of most interest to the designer. At the same time, it is well to point out that the assumption of zero initial conditions in the definition of $G(s)$ is not a constraint. For linear systems the principle of superposition applies, which means that the total response is the sum of those to the input and to the initial conditions applied separately. Thus the initial conditions can be set to zero in determining the response to input $r(t)$, and $r(t)$ set to zero when calculating the response to initial conditions.

*****

The above in effect provides the scenario for calculating the response $c(t)$ of a system to a given input $r(t)$ with specified initial conditions. For example, using the dot notation for derivatives, if

$$\ddot{c} + a_1\dot{c} + a_0c = r \qquad c(0) = c_0, \quad \dot{c}(0) = \dot{c}_0 \qquad (1.25)$$

then the transfer function is

$$G(s) = \frac{C_1(s)}{R(s)} = \frac{1}{s^2 + a_1s + a_0} \qquad (1.26)$$

and for a unit step input $r(t)$ [$R(s) = 1/s$] the transform of the output due to $r(t)$ is

$$C_1(s) = \frac{1}{s(s^2 + a_1s + a_0)} \qquad (1.27)$$

For the response to initial conditions, $r = 0$ in (1.25) and this equation is transformed, using (1.10) and (1.11) or Table 1.6.1, to

$$[s^2C_2(s) - sc_0 - \dot{c}_0] + a_1[sC_2(s) - c_0] + a_0C_2(s) = 0$$

Rearranging yields the output transform

$$C_2(s) = \frac{(s + a_1)c_0 + \dot{c}_0}{s^2 + a_1 s + a_0}$$ (1.28)

Hence the transform of the total output is

$$C(s) = C_1(s) + C_2(s)$$ (1.29)

*****

Inverse Laplace transformation of these transforms will give the corresponding responses in the time domain. For very simple cases these inverses may be available directly in a table of Laplace transform pairs such as Table 1.6.1, but it is now evident that techniques are needed to invert more complex expressions. This will be considered in detail and in a broader context in Chapter 3. But, in part to help motivate the derivation of transfer functions for physical systems in Chapter 2 by showing at least one of the uses to which they can be put, the simplest case of the technique to be used is introduced here.

Inverse Transformation by Partial Fraction Expansion

[Roots of denominator of $C(s)$ assumed real and distinct.] Examples will be used to demonstrate the technique.

Example 1.7.1

Determine the response to initial conditions $c(0) = 1$, $\dot{c}(0) = -2$ of the system

$$\ddot{c} + 7\dot{c} + 6c = r$$ (1.30)

For the response to initial conditions, the input r can be taken to be zero, and the transform is

$$[s^2 C - sc(0) - \dot{c}(0)] + 7[sC - c(0)] + 6C = 0$$

or

$$C = \frac{(s + 7)c(0) + \dot{c}(0)}{s^2 + 7s + 6} = \frac{s + 9}{s^2 + 7s + 6}$$

For inverse transformation, the denominator is factored as $(s + 1)(s + 6)$, and $C(s)$ is expanded into partial fractions according to

$$C(s) = \frac{s + 9}{(s + 1)(s + 6)} = \frac{K_1}{s + 1} + \frac{K_2}{s + 6}$$

with a separate term for each real root factor in the denominator of $C(s)$. These terms are of the form $A/(s + \alpha)$, for which Table 1.6.1 gives the inverse $Ae^{-\alpha t}$. The constants K_i are the *residues* at the corresponding roots. To find K_1, multiply both sides of the equation by its denominator $(s + 1)$:

$$(s + 1)C(s) = \frac{s + 9}{s + 6} = K_1 + \frac{K_2(s + 1)}{s + 6}$$

If s is permitted to approach the root -1 of $(s + 1)$, the second term on the right disappears and the residue K_1 is

$$K_1 = [(s + 1)C(s)]_{s=-1} = \left(\frac{s + 9}{s + 6}\right)_{s=-1} = 1.6$$

Residue K_2 at root -6 is found in the same manner by multiplying through by $(s + 6)$:

$$K_2 = [(s + 6)C(s)]_{s=-6} = \left(\frac{s + 9}{s + 1}\right)_{s=-6} = -0.6$$

The partial fraction expansion is now

$$C(s) = \frac{1.6}{s + 1} - \frac{0.6}{s + 6}$$

By the linearity theorem, the inverse is the sum of the inverses of the terms, so that the transient response $c(t)$ is the sum of two decaying exponentials:

$$c(t) = 1.6e^{-t} - 0.6e^{-6t}$$

Example 1.7.2

Find the response $c(t)$ of a system with transfer function

$$G(s) = \frac{2(s + 3)}{(s + 1)(s + 6)} \tag{1.31}$$

to a decaying exponential input $r(t) = e^{-2t}$.

From Table 1.6.1, $R(s) = 1/(s + 2)$, so that $C(s)$ and its partial fraction expansion can be written as follows:

$$C(s) = G(s)R(s) = \frac{2(s + 3)}{(s + 2)(s + 1)(s + 6)}$$

$$= \frac{K_1}{s + 2} + \frac{K_2}{s + 1} + \frac{K_3}{s + 6}$$

The residues are calculated next:

$$K_1 = [(s + 2)C(s)]_{s=-2} = \left[\frac{2(s + 3)}{(s + 1)(s + 6)}\right]_{s=-2} = -0.5$$

$$K_2 = [(s + 1)C(s)]_{s=-1} = \left[\frac{2(s + 3)}{(s + 2)(s + 6)}\right]_{s=-1} = +0.8$$

$$K_3 = [(s + 6)C(s)]_{s=-6} = \left[\frac{2(s + 3)}{(s + 2)(s + 1)}\right]_{s=-6} = -0.3$$

The partial fraction expansion is

$$C(s) = -\frac{0.5}{s + 2} + \frac{0.8}{s + 1} - \frac{0.3}{s + 6}$$

so that the response sought is

$$c(t) = -0.5e^{-2t} + 0.8e^{-t} - 0.3e^{-6t}$$

The foregoing technique for calculating residues also applies when the roots of the denominator of $C(s)$ are distinct but include complex conjugate pairs. But then the residues also include complex conjugate pairs. A vector-based method is preferable for these cases and for more complicated transforms generally. This is discussed in Chapter 3, where repeated roots are considered as well.

1.8 BLOCK DIAGRAM REDUCTION

The discussion of Section 1.7 appears to imply that if the transfer function relating input r and output c in a block diagram such as Fig. 1.3 is desired, a differential equation relating these two variables must be obtained first. Fortunately, this is not necessary. The transfer function can be derived instead by certain algebraic manipulations of those of the subsystems or blocks. Some examples will show this block diagram reduction technique and provide some useful results.

Example 1.8.1

Reduce the *cascade* or series connection of two blocks in Fig. 1.10 to a single block G as in Fig. 1.9(b). By definition,

$$C = G_2M \qquad M = G_1R$$

Hence, substituting the second into the first yields

$$C = GR \qquad G = G_1G_2 \qquad\qquad (1.32)$$

By direct extension it follows that

The overall transfer function of a series of blocks equals the product of the individual transfer functions.

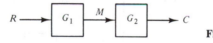

Figure 1.10   Two-block cascade.

Example 1.8.2

The configuration in Fig. 1.11, equivalent to that in Fig. 1.3, is extremely common. By definition

$$C = G_2M \qquad M = G_1E \qquad E = R - B \qquad B = HC$$

Combining each pair yields

$$C = G_1G_2E \qquad E = R - HC$$

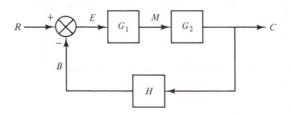

Figure 1.11   Standard feedback loop.

Eliminating E gives

$$C = G_1 G_2 R - G_1 G_2 H C$$

and rearranging this, the

$$\text{closed-loop transfer function } \frac{C}{R} = \frac{G_1 G_2}{1 + G_1 G_2 H} \tag{1.33}$$

In words, and in somewhat generalized form, this important result states that

> The *closed-loop transfer function* of the standard loop equals the product of the transfer functions in the forward path divided by the sum of 1 and the loop gain function. The *loop gain function* is defined as the product of the transfer functions around the loop.

Note further that, since $C = G_1 G_2 E$,

$$\frac{E}{R} = \frac{1}{1 + G_1 G_2 H} \tag{1.34}$$

If $H = 1$, then $E = R - C$ is the system error, as in Fig. 1.3, and E/R is the input-to-error transfer function. It will permit the error response for a given input $r(t)$ to be found directly. The transfer function relating the input to any variable of interest can be found similarly.

Example 1.8.3

The configuration in Fig. 1.12(a), which includes a *minor feedback loop*, is very common in servomechanisms. Derivation of C/R by the approach of Example 1.8.2

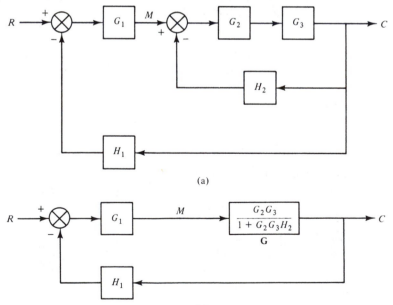

(a)

(b)

Figure 1.12 Minor loop feedback.

would be laborious, but becomes simple if its result in (1.33) is used. It is applied first to reduce the minor feedback loop C/M to a single block, as shown in Fig. 1.12(b). But (1.33) applies again to this new loop and now yields the closed-loop transfer function

$$\frac{C}{R} = \frac{G_1 G}{1 + G_1 G H_1} = \frac{G_1 G_2 G_3}{1 + G_2 G_3 H_2 + G_1 G_2 G_3 H_1} \tag{1.35}$$

Example 1.8.4

A two-input system is shown in Fig 1.13. The additional input D often represents a disturbance, such as a supply pressure variation in the level control example in Section 1.3. With the additional block L, the diagram models the effect of the disturbance on the system.

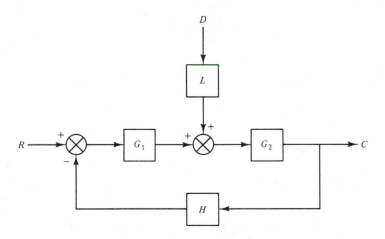

Figure 1.13 Two-input system.

For linear systems the principle of superposition applies, and the total output is the sum of the outputs due to each input separately. Thus the output due to R is found as before, and while finding that due to D, R is put equal to zero.

The rule of Example 1.8.2 applies when finding the response to D, but note that the product of the transfer functions in the forward path consists, aside from L, only of G_2. Note also that for $R = 0$ the minus sign for the feedback at R can be moved to the summing junction for D. Inspection now yields

$$\frac{C}{D} = \frac{G_2 L}{1 + G_1 G_2 H} \tag{1.36}$$

Example 1.8.5

In Fig. 1.14 the two feedback loops interfere with each other. The rearrangements (a) and (b) are alternative first steps to make the result (1.33) again applicable. Verify that neither changes the system, and that applying (1.33) twice to (a) or (b) yields the closed-loop transfer function

$$\frac{C}{R} = \frac{G_1 G_2}{1 + G_1 H_2 + G_2 H_1} \tag{1.37}$$

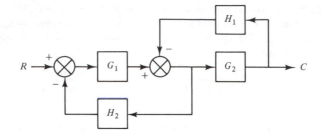

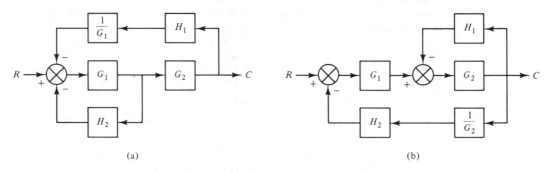

(a)   (b)

Figure 1.14   Example 1.8.5.

Signal Flow Graphs

Signal flow graphs are an alternative to block diagrams. Figure 1.15 shows the flow graphs equivalent to the block diagrams in Figs. 1.10 to 1.13. As may be seen from the equivalent of Fig. 1.10 in Fig. 1.15, variables are shown as nodes instead of line segments, and line segments with associated arrows and transfer functions represent the relations between variables.

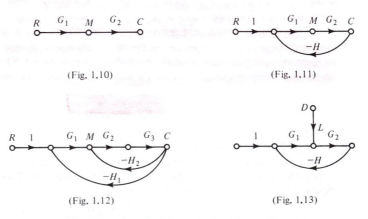

Figure 1.15   Signal flow graphs for Figs. 1.10 to 1.13.

For complex systems, the block diagram reduction procedure can become laborious. For signal flow graphs a general equation is available from which overall transfer functions can be obtained without graphical reductions. For a first course, the block diagram is preferable because it is closer to the physical reality of the system. Block diagrams are used in this book.

1.9 CONCLUSION

In this chapter a general introduction has been given first, including physical discussion of some fundamental features of control system behavior. A level control example led to a common block diagram configuration.

Laplace transforms led to the transfer function description of dynamic behavior, and block diagram reduction to the description of an interconnected system of blocks.

The application of transfer functions and transforms to calculation of the response $c(t)$ to an input $r(t)$ and initial conditions has been demonstrated for cases where the roots of the denominator of the transform $C(s)$ are real and distinct.

This provides a framework and motivation for study of the next chapter, and a basis for detailed discussion of transient response in Chapter 3. It also allows for an introductory examination of some of the effects of feedback in the problems below.

A final example is given which combines the techniques of this chapter with the topic of the next one.

Example 1.9.1   Temperature Control System

In Example 2.5.2 the differential equation and transfer function of a space heating system are derived. If $T(t)$ is the difference with a constant ambient temperature, it makes sense to take the heat loss to ambient as being proportional to T and model it as T/R, where R is constant. If $q_i(t)$ is the heat inflow rate from a space heater, the net heat inflow rate is $(q_i - T/R)$. Clearly, T will rise faster if this is larger, and it is reasonable that $\dot{T}(t)$ should be proportional to net inflow. This gives $C\dot{T} = q_i - T/R$, with C a constant, and on rearranging the differential equation

$$RC\dot{T}(t) + T(t) = Rq_i(t)$$

The process transfer function in the system block diagram of Fig. 1.3, and the transfer function $G_2(s)$ in the standard loop of Fig. 1.11, is then

$$G_2(s) = \frac{T(s)}{Q_i(s)} = \frac{R}{RCs + 1}$$

This so-called *simple lag* form of transfer function will be found to approximate the behavior of many physical systems. In fact, the sensor in Fig. 1.3 used to obtain a measure of the temperature for use in feedback control may respond rather slowly, and is often modeled as a simple lag. $H(s)$ in Fig. 1.11 is therefore taken to be

$$H(s) = \frac{1}{\tau s + 1}$$

If it is assumed that the power amplifier and actuator in Fig. 1.3 respond relatively fast, and if the controller is simply an amplifier, then $G_1(s)$ in Fig. 1.11 is, with K constant:

$$G_1(s) = K$$

Using (1.33), the closed-loop transfer function $C/R = G_1G_2/(1 + G_1G_2H)$ becomes

$$\frac{C(s)}{R(s)} = \frac{KR(\tau s + 1)}{(\tau s + 1)(RCs + 1) + KR}$$

For a unit step input, $R(s) = 1/s$, and the transform of the system output is

$$C(s) = \frac{KR(\tau s + 1)}{s[\tau RCs^2 + (\tau + RC)s + 1 + KR]}$$

If, in a consistent system of units,

$$\tau = 2 \qquad C = 80 \qquad R = \tfrac{1}{4}$$

then

$$C(s) = \frac{0.00625K(2s + 1)}{s(s^2 + 0.55s + 0.025 + 0.00625K)}$$

For $K = 4$, Laplace transform inversion by partial fraction expansion is found to yield

$$c(t) = 0.5 - 0.523e^{-0.115t} + 0.023e^{-0.435t}$$

This unit step response is plotted in Fig. 1.16 and is decidedly unattractive since in the steady state ($t \to \infty$) the output changes by only 0.5° for a 1° change of desired temperature.

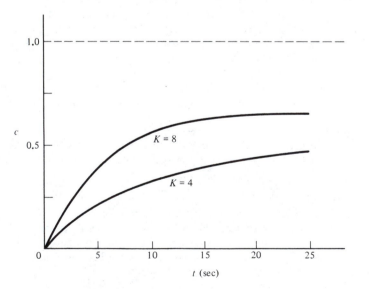

Figure 1.16   Step response, Example 1.9.1.

Doubling the gain to $K = 8$ yields the solution

$$c(t) = 0.667 - 2e^{-0.25t} + 1.333e^{-0.30t}$$

and the response, while the steady-state error is still quite large, is certainly much improved, also in that the response reaches steady state sooner; that is, the speed of response is increased.

It is noted, finally, that the steady-state value c_{ss} of the output could also have been found directly from $C(s)$ and the final value theorem:

$$c_{ss} = \lim_{t \to \infty} c(t) = \lim_{s \to 0} sC(s) = \frac{KR}{1 + KR}$$

PROBLEMS

1.1. Set up a basic block diagram for a ship's autopilot system. Describe the blocks in the diagram, indicating what hardware might be used to indicate the desired heading, and how the actual heading might be made to agree with it.

1.2. Set up a basic block diagram for a position control system in which an electric motor must control angular shaft position. A potentiometer is used to measure the position.

1.3. The output y and input x of a block in a block diagram are related by $y = 2x + 0.5x^3$.
   (a) Find the values of the output for steady-state operation at the operating points
$$\text{(i) } x_0 = 0 \quad \text{(ii) } x_0 = 1 \quad \text{(iii) } x_0 = 2$$
   (b) Obtain linearized models for small variations about these operating points. For these models, redefine x and y to represent variations about operating-point values.

1.4. The output z of a block in a block diagram is a function of two inputs, x and y, according to

$$z = 5x - 3y + x^3 + y^3 + x^2y$$

Derive a linearized model for small variations about the operating point $x_0 = 1$, $y_0 = 2$. For this model, redefine x, y, and z to represent only the variations about operating-point values.

1.5. Find the Laplace transform $R(s)$ of a signal $r(t)$ if $r(t) = 2t, 0 \leq t < 1; r(t) = 2, t \geq 1$.

1.6. In a pipeline of length L with fluid velocity v, the temperature T_i of the inflow undergoes a step change of magnitude A at $t = 0$. Neglecting heat loss and with variables representing only changes:
   (a) Describe the outflow temperature T_o and its transform.
   (b) Give the transfer function relating T_i and T_o.

1.7. Use the final value theorem to determine the steady-state value $c_{ss} = \lim_{t \to \infty} c(t)$ of $c(t)$ for the following forms of the transform $C(s)$ of $c(t)$:
   (a) $C(s) = \dfrac{s + 11}{(s + 2)(s + 5)}$ **(c)** $C(s) = \dfrac{4s^2 + 5s + 6}{s(2s^3 + 7s^2 + 13s + 2)}$
   (b) $C(s) = \dfrac{2}{s(s + 3)}$ **(d)** $C(s) = \dfrac{2s^2 + 5s + 1}{s^3 + 2s^2 + 3s + 4}$

1.8. For the following differential equations and initial conditions, write the transforms of the solutions:

(a) $\ddot{c} + 6\dot{c} + 13c = 5u(t)$, $c_0 = 1$, $\dot{c}_0 = 4$

(b) $\ddot{c} + 3\dot{c} + 4c = 7u(t) + 2t$, $c_0 = 1$, $\dot{c}_0 = 2$

(c) $\ddot{c} + 3\dot{c} + 4c = 6 \sin \omega t$, $c_0 = 4$, $\dot{c}_0 = 5$

(d) $\dot{c} + 2c = u(t)$, $c_0 = 1$

1.9. Calculate the unit step response of a system with the transfer function

$$G(s) = \frac{5(1 - 0.4s)}{(s + 1)(0.2s + 1)}$$

1.10. Calculate the response to a decaying exponential input $r(t) = e^{-t}$ of a system with the transfer function

$$G(s) = 3 \frac{s^2 + 9s + 18}{s^2 + 6s + 8}$$

1.11. For the systems described by the differential equations with initial conditions

(i) $\dot{c} + 6c = r$, $c(0) = 1$

(ii) $\ddot{c} + 7\dot{c} + 10c = r$, $c(0) = 1$, $\dot{c}(0) = 4$

determine:

(a) The transfer functions C/R.

(b) The transient responses to the initial conditions.

1.12. For the systems shown in Fig. P1.12:

(a) Find the closed-loop transfer functions.

(b) Calculate the responses to unit step inputs.

The types of transfer functions given here and in other problems approximate the behavior of many physical systems.

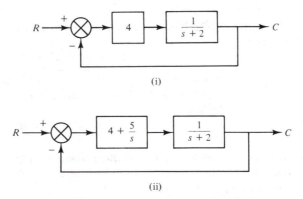

(i)

(ii)

Figure P1.12

1.13. Calculate the unit step responses of the systems shown in Fig. P1.13.

1.14. For the systems in Fig. P1.13, determine the transient responses to the initial conditions:

(i) $c(0) = 1$ (ii) $c(0) = 1$ $\dot{c}(0) = 1$

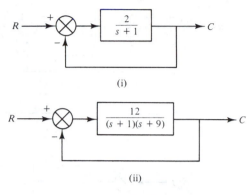

(i)

(ii)

Figure P1.13

1.15. For the systems in Fig. P1.15:
  (a) Determine the transfer functions E/R.

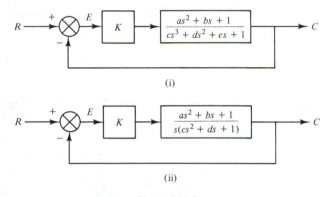

(i)

(ii)

Figure P1.15

  (b) Use the final value theorem to find the steady-state values of the system error $e = r - c$ in response to unit step inputs.
  (c) Observe how this evidently important aspect of system behavior depends on the value of K and the type of system.

1.16. For the system in Fig. P1.16, express the transfer function E/R and find the steady-state values of the error following unit step and unit ramp inputs.

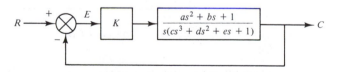

Figure P1.16

1.17. Reduce the block diagrams in Fig. P1.17 and find the closed-loop transfer functions.

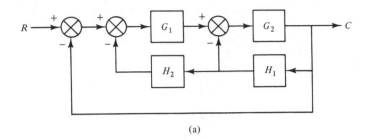

(a)

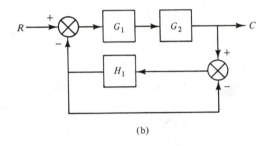

(b)

Figure P1.17

1.18. Find the transfer functions C/R and C/D for the block diagram shown in Fig. P1.18.

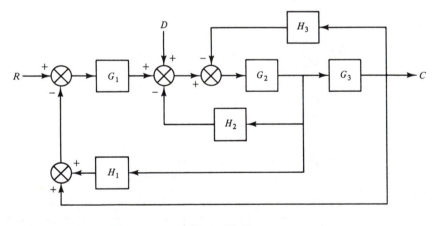

Figure P1.18

1.19. For the two-input/two-output system shown in Fig. P1.19, determine the transfer functions C_1/R_1 and C_1/R_2 relating C_1 to R_1 and R_2, respectively, and express C_1 when both inputs are present.

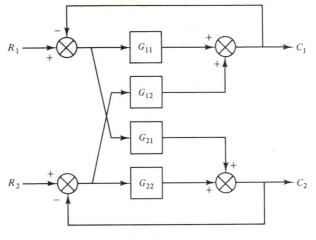

Figure P1.19

1.20. In Fig. P1.20, R is the input which the output C should follow as closely as possible, and D is a disturbance input to which the output should ideally not respond at all. The output of a block K_i equals K_i times its input. The total output is the sum of the outputs due to each input separately.

(a) Calculate C/R (with $D = 0$) and C/D (with $R = 0$).

(b) Find K_1 and K_2 for two sets of specifications:
       (i) $C/D = 0.1$, $C/R = 0.9$ (ii) $C/D = 0.01$, $C/R = 0.99$

(c) Note the effect of the more severe specifications (ii) on the values of the gains and on their distribution over the two blocks.

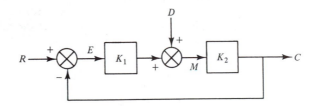

Figure P1.20

1.21. To show the use of feedback in reducing the effects of parameter variations, express C/R in Fig. P1.20, with $D = 0$, both with and without feedback.

(a) In the open-loop case, what is the effect on C/R if the value of K_2 is halved?

(b) Determine the value of K_1K_2 for which when K_2 is halved, the closed-loop ratio C/R does not fall below 0.9 of its original value.

1.22. To show the use of feedback in reducing the effect of disturbance inputs:

(a) Express C/D in Fig. P1.20, with $R = 0$, both with and without the feedback.

(b) Give the condition for reducing C/D to 10% of its value without feedback.

1.23. It is desired to examine how adding a feedback loop and a gain K as shown in Fig. P1.23 (ii) to the block in part (i) can be used to reduce the effect of variations of the system parameter a.

(a) Calculate the unit step responses in Fig. P1.23 (i) and (ii).

(b) Use the final value theorem to verify the values of $\lim_{t \to \infty} c(t)$ found in part (a).

(c) If a is nominally 1, so that the desired steady-state output is 1, compare the open-loop and closed-loop systems for $K = 10$ and $K = 100$ if the parameter a doubles in value.

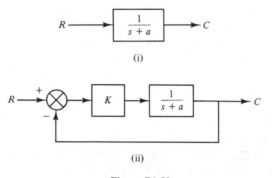

(i)

(ii)

Figure P1.23

1.24. In Fig. P1.24, R is the input that output C should follow as closely as possible, and D is a disturbance input to which ideally C should not respond at all.

(a) Express the transfer functions C/R and C/D and hence determine the ratio of the outputs due to unit step inputs of D and R, and the choice of K that will reduce this ratio.

(b) Calculate the response to a unit step R, both with and without a unit step D occurring simultaneously, if $K = 80$. Note the effect of D relative to that of R.

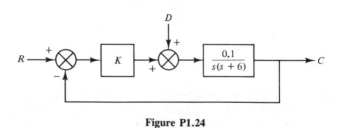

Figure P1.24

1.25. In Fig. P1.25, the controller in part (i) is called *proportional control,* and that in part (ii) is *proportional plus integral control.* (In Table 1.6.1, $1/s$ is the transform of integration.) For $K_c = 4$, $K_i = 5$:

(a) Find the transfer functions E/R and E/D.

(b) Find the steady-state errors $e = r - c$ for unit step inputs of R and D in turn.

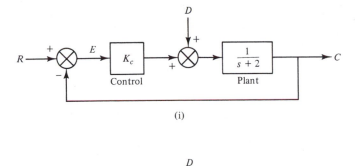

(i)

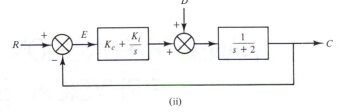

(ii)

Figure P1.25

(c) Compare the results for parts (i) and (ii) and comment on the effects on the steady-state performance of adding the integral control.

1.26. In Fig. P1.25, determine the steady-state values of the system error e if the input r is a unit ramp signal, for $K_c = 4$, $K_i = 5$. Which of parts (i) and (ii) is in effect not capable of following a unit ramp input?

2

TRANSFER FUNCTION MODELS
OF PHYSICAL SYSTEMS

2.1 INTRODUCTION

In this chapter differential equations are derived to describe the dynamic behavior of mechanical, electrical, thermal, and fluid systems. These are used to obtain transfer functions between selected variables. The same differential equations can also be formulated into state-space models. This alternative, mentioned in Section 1.5, is discussed in Chapter 12.

The chapter will concentrate on subsystem blocks, with attention to the restrictions that must be observed when separating a system into blocks. Block diagram reduction can then provide overall system transfer functions. For many systems that include feedback control, the division into blocks is far from obvious. Such systems, in which all system equations and the block diagram are derived directly, are discussed in Chapter 4. Frequently, the precise nature of the feedback may be evident only from this block diagram.

It is useful to note that a model should not be expected to be evident "by inspection." Rather, it usually will emerge gradually from equations written for parts of the system.

2.2 MECHANICAL SYSTEMS

Figure 2.1 shows common elements of mechanical systems with linear and rotational motion, together with the equations used to describe them. It is noted that in Fig. 2.1 and diagrams to follow, the arrows identify the positive directions of the associated variables.

30

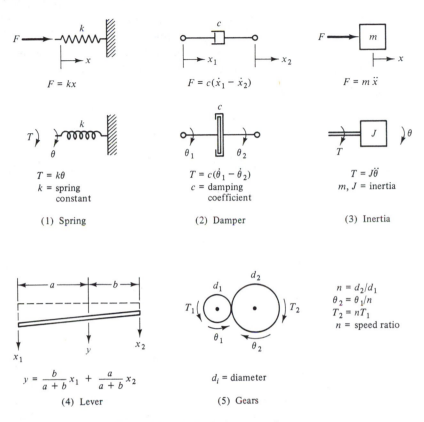

Figure 2.1 Mechanical system elements.

1. *Springs:* These occur in many design configurations and materials. For linearized models the force F (torque T) is taken to be proportional to linear deflection x (angular deflection θ), but for larger variations of deflections the behavior can sometimes be very nonlinear. Note that in linearized models F and x represent the variations of force and deflection about the operating point values, and that the spring constant k for nonlinear springs changes with the operating point.

2. *Dampers or dashpots:* These generate a damping force F (torque T) proportional to the difference $\dot{x}_1 - \dot{x}_2$ ($\dot{\theta}_1 - \dot{\theta}_2$) of the velocities across the damper, and in the opposite direction. In practice, the friction in mechanical systems may differ greatly from the viscous friction of this linear damper model. For dry friction, or Coulomb friction, the force or torque is opposite the velocity difference but independent of its magnitude. The approximate linearization of such behavior is a subject in the study of nonlinear systems.

3. *Mass and inertia:* By Newton's law, force F (torque T) equals mass m (inertia J) times acceleration $\ddot{x}$ ($\ddot{\theta}$).

4. *Lever mechanism:* For small enough angles from horizontal, the total motion y equals the sum of the motion due to x_1 with $x_2 = 0$ and that due to x_2 with $x_1 = 0$. It is useful to observe that the lever is a mechanical implementation of a summing junction in a block diagram, and is in fact often used for this purpose. If $a = b$, then $y = 0.5\,(x_1 - x_2)$ if the direction of x_2 is reversed. If input and output are available in mechanical form and are applied to x_1 and x_2, respectively, the linkage with $a = b$ and x_2 opposite implements the feedback loop as well as providing the system error $e = x_1 - x_2$.

5. *Gears:* This very common element is often identified in terms of its gear ratio n.

$$n = \frac{\text{speed of driving gear}}{\text{speed of driven gear}} = \frac{\omega_1}{\omega_2} = \frac{\dot{\theta}_1}{\dot{\theta}_2} \left(= \frac{\theta_1}{\theta_2} = \frac{\ddot{\theta}_1}{\ddot{\theta}_2} \right) \tag{2.1}$$

where $\omega_i = \dot{\theta}_i$ is the angular velocity (rad/sec) of the gear with diameter d_i. The relation $T_2 = nT_1$ between the torques arises because the two gears have a common contact force, and the torque equals this force times the gear radius.

Examples incorporating these elements are considered next.

Example 2.2.1 Spring–Mass–Damper System

By Newton's law, $m\ddot{x}$ equals the resultant of all external forces on m in Fig. 2.2 in the downward direction. To help in determining the signs of the terms in such problems, it is useful to make any assumption concerning the motion: for example, that the mass is moving downward from $x = 0$. In that case the spring is stretched, so spring force kx is upward, and hence opposes downward acceleration. It therefore receives a minus sign on the right side of the equation for $m\ddot{x}$. Since the mass moves down, the damping force $c\dot{x}$ is upward, and this term must also have a minus sign. The external force $f(t)$ helps downward acceleration, and therefore has a plus sign. The resulting equation is

$$m\ddot{x} = -kx - c\dot{x} + f(t)$$

Rearranging gives the differential equation of motion in the usual form:

$$m\ddot{x} + c\dot{x} + kx = f(t) \tag{2.2}$$

It may have been observed that the effects of gravity do not appear, so that turning the system upside down will not affect the equation. This is done by choosing

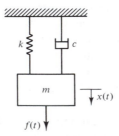

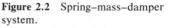

Figure 2.2 Spring–mass–damper system.

$x = 0$ at the position of static equilibrium, where the weight mg is counterbalanced by a spring force.

For a transfer function model, assume that the response $x(t)$ to $f(t)$ is desired. With zero initial conditions, the transform of (2.2) is

$$(ms^2 + cs + k)X(s) = F(s)$$

and the transfer function of interest is

$$\frac{X(s)}{F(s)} = \frac{1}{ms^2 + cs + k} \qquad (2.3)$$

Example 2.2.2   Two-Mass System

The system in Fig. 2.3 can represent a dynamic absorber, where a relatively small mass m_1 is attached to a main mass m via spring k_1 and damper c_1 to reduce vibrations x due to force f. Assume, say, that m and m_1 both move to the right from the zero positions, m farther and faster than m_1. Then spring force $k_1 (x - x_1)$ "opposes" m and "helps" m_1, and damper force $c_1 (\dot{x} - \dot{x}_1)$ has the same effect. Hence the differential equations of motion become

$$m\ddot{x} = -kx - k_1(x - x_1) - c_1(\dot{x} - \dot{x}_1) + f$$
$$m_1\ddot{x}_1 = k_1(x - x_1) + c_1(\dot{x} - \dot{x}_1)$$

or

$$m\ddot{x} + c_1\dot{x} + (k + k_1)x = c_1\dot{x}_1 + k_1x_1 + f \qquad (2.4)$$
$$m_1\ddot{x}_1 + c_1\dot{x}_1 + k_1x_1 = c_1\dot{x} + k_1x$$

For a transfer function model, the effect of f on x would be of interest. This requires the elimination of x_1, an algebraic operation if (2.4) are first transformed:

$$(ms^2 + c_1s + k + k_1)X(s) = (c_1s + k_1)X_1(s) + F(s) \qquad (2.5)$$
$$(m_1s^2 + c_1s + k_1)X_1(s) = (c_1s + k_1)X(s)$$

Solving $X_1(s)$ from the second equation, substituting it into the first, and rearranging would give the transfer function X/F.

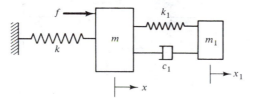

Figure 2.3   Dynamic absorber.

Example 2.2.3   Rotating Drive System

Figure 2.4 indicates a drive system, with c representing a friction coupling, and torsion spring k the twisting of a long shaft due to torque. Angle θ_1 is taken to be the input and θ_3 the output. Any other variables may be introduced to facilitate

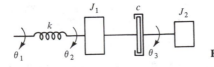

Figure 2.4   Rotating drive system.

the writing of the equations, such as θ_2, the angle to the right as well as the left of J_1.

The approach of the preceding examples could be used, but an alternative is often convenient. Equations are written in order, starting at the input. To help visualize signs, assume that $\theta_1 > \theta_2$, $\dot{\theta}_2 > \dot{\theta}_3$. The shaft torque $k(\theta_1 - \theta_2)$ accelerates inertia J_1 and supplies the damping torque:

$$k(\theta_1 - \theta_2) = J_1\ddot{\theta}_2 + c(\dot{\theta}_2 - \dot{\theta}_3)$$

It is the damping torque which in turn accelerates inertia J_2:

$$c(\dot{\theta}_2 - \dot{\theta}_3) = J_2\ddot{\theta}_3$$

Rearranging yields the differential equations

$$J_1\ddot{\theta}_2 + c\dot{\theta}_2 + k\theta_2 = c\dot{\theta}_3 + k\theta_1 \tag{2.6}$$

$$J_2\ddot{\theta}_3 + c\dot{\theta}_3 = c\dot{\theta}_2$$

and the transformed equations

$$(J_1s^2 + cs + k)\theta_2(s) = cs\theta_3(s) + k\theta_1(s)$$

$$(J_2s^2 + cs)\theta_3(s) = cs\theta_2(s)$$

where $\theta_i(s)$ is the transform of $\theta_i(t)$. From the second equation,

$$\theta_2(s) = \left(\frac{J_2}{c}s + 1\right)\theta_3(s)$$

Substituting this into the first and rearranging gives the transfer function

$$\frac{\theta_3(s)}{\theta_1(s)} = \frac{k}{(J_1J_2/c)s^3 + (J_1 + J_2)s^2 + (J_2k/c)s + k} \tag{2.7}$$

Example 2.2.4   Systems with Gears

Although not necessary, it is often convenient for analysis to replace a system with gears by a dynamically equivalent system without gears. Figure 2.5 shows the

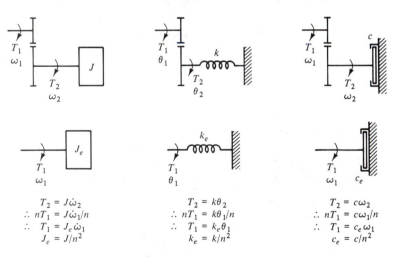

Figure 2.5 Equivalent systems without gears.

derivation of the equivalent inertia, spring, and damping, identified as J_e, k_e, and c_e, on the first shaft which can replace J, k, and c on the second shaft. In addition to the relation $T_2 = nT_1$ from Fig. 2.1, (2.1) is used, and the fact that $\ddot{\theta}_i = \omega_i$. *Note that in each case the equivalent element is obtained by dividing the original element by the square of the speed ratio n.*

Example 2.2.5   Nongeared Equivalent System

Using Fig. 2.5, the system in Fig. 2.6(a) can be replaced immediately by its nongeared equivalent in Fig. 2.6(b), where J_e, k_e, and c_e are as given in Fig. 2.5.

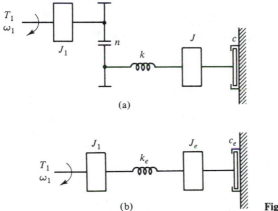

(a)

(b)   **Figure 2.6**   Example 2.2.5.

2.3 ELECTRICAL SYSTEMS: CIRCUITS

Figure 2.7 summarizes some important results for the modeling of electrical circuits. Figure 2.7(a) gives the voltage–current relations of the basic elements in the time domain and the Laplace domain, assuming zero initial conditions. In the general transformed relation $V = IZ$, Z is the impedance. In an *ideal* voltage source, the voltage difference across the terminals is independent of the current through them. In a *real* voltage source, the voltage v_s across the terminals decreases with increasing current due to the voltage drop across the internal impedance of the source, which may be a resistance R.

The use of these results and those in Fig. 2.7(b) and (c) will be illustrated by application to the examples shown in Fig. 2.8. The transfer functions $E_o(s)/E_i(s)$ between inputs $e_i(t)$ and outputs $e_o(t)$ are desired. The current through the output terminals must be assumed to be negligibly small, because otherwise its value would affect e_o, and E_o/E_i would not be uniquely defined. In such a case a transfer function can be derived from differential equations written for the combination of the circuit and its load. This illustrates a key condition when subdividing a system into subsystem blocks:

> The division of systems into blocks must be such that each block does not *load*, that is, affect the output of, the preceding block.

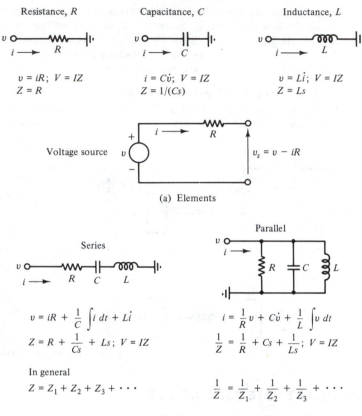

(a) Elements

$v = iR + \dfrac{1}{C} \displaystyle\int i\, dt + L\dot{i}$

$Z = R + \dfrac{1}{Cs} + Ls; \quad V = IZ$

In general

$Z = Z_1 + Z_2 + Z_3 + \cdots$

$i = \dfrac{1}{R} v + C\dot{v} + \dfrac{1}{L} \displaystyle\int v\, dt$

$\dfrac{1}{Z} = \dfrac{1}{R} + Cs + \dfrac{1}{Ls}; \quad V = IZ$

$\dfrac{1}{Z} = \dfrac{1}{Z_1} + \dfrac{1}{Z_2} + \dfrac{1}{Z_3} + \cdots$

(b) Circuits

Loop method
  Voltage equations:
Sum of voltage drops
around a closed loop = 0

Node method
  Current equations:
Sum of currents at a circuit
node or junction = 0

(c) Kirchoff's laws

Figure 2.7 Electrical circuit fundamentals.

Example 2.3.1 Common Electrical Circuits [Fig. 2.8(a) to (e)]
   These circuits are used extensively as controllers to improve the performance of feedback control systems. The transfer functions E_o/E_i can be found by use of the results in Fig. 2.7(b). With no current through the output terminals, all are in effect voltage dividers, in which e_o is a fraction of the voltage e_i, determined by the current i through the input terminals caused by e_i. Consider each in turn:

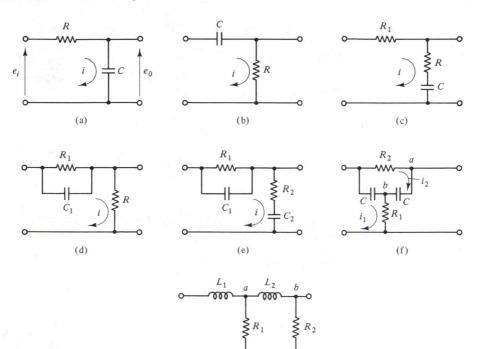

Figure 2.8 Electrical networks: (a) simple lag; (b) transient lead; (c) phase lag; (d) phase lead; (e) lag–lead (notch); (f) bridged-T network; (g) RL ladder network.

Figure 2.8(a):

$$E_i = IZ = I(Z_r + Z_c) \qquad Z_r = R \qquad Z_c = \frac{1}{Cs}$$

$$E_i = I\left(R + \frac{1}{Cs}\right)$$

$$E_o = I\frac{1}{Cs}$$

$$\frac{E_o}{E_i} = \frac{1}{RCs + 1} \qquad (2.8a)$$

Figure 2.8(b):

$$E_i = I\left(R + \frac{1}{Cs}\right)$$

$$E_o = IR$$

$$\frac{E_o}{E_i} = \frac{RCs}{RCs + 1} \qquad (2.8b)$$

Figure 2.8(c):

$$E_i = I\left(R_1 + R + \frac{1}{Cs}\right)$$

$$E_o = I\left(R + \frac{1}{Cs}\right)$$

$$\frac{E_o}{E_i} = \frac{\tau s + 1}{(\tau/\alpha)s + 1} \qquad (2.8c)$$

where $\tau = RC \qquad \alpha = \dfrac{R}{R_1 + R}$

Figure 2.8(d):

$$E_i = I\left(\frac{1}{(1/R_1) + C_1 s} + R\right)$$

since from Fig. 2.7(b) the equivalent impedance Z_1 of the parallel impedances $Z_r = R_1$ and $Z_c = 1/(C_1 s)$ is given by

$$\frac{1}{Z_1} = \frac{1}{Z_r} + \frac{1}{Z_c} \qquad \text{or} \qquad Z_1 = 1 \left/ \left(\frac{1}{Z_r} + \frac{1}{Z_c}\right)\right.$$

Hence

$$E_i = I\left(\frac{R_1 + R + RR_1 C_1 s}{1 + R_1 C_1 s}\right)$$

$$E_o = IR$$

$$\frac{E_o}{E_i} = \alpha \frac{\tau s + 1}{\alpha \tau s + 1} \qquad (2.8d)$$

where $\tau = R_1 C_1 \qquad \alpha = \dfrac{R}{R_1 + R}$

Figure 2.8(e):

$$E_o = I\left(\frac{1}{(1/R_1) + C_1 s} + R_2 + \frac{1}{C_2 s}\right)$$

$$E_o = I\left(R_2 + \frac{1}{C_2 s}\right)$$

$$\frac{E_o}{E_i} = \frac{(\tau_1 s + 1)(\tau_2 s + 1)}{\tau_1 \tau_2 s^2 + (\tau_1 + \tau_2 + \tau_{12})s + 1} \qquad (2.8e)$$

where $\tau_1 = R_1 C_1 \qquad \tau_2 = R_2 C_2 \qquad \tau_{12} = R_1 C_2$

Example 2.3.2   Bridged-T Network [Fig. 2.8(f)]

This network is often used in ac control systems, that is, systems in which signals are represented by modulation of an ac carrier. With e_o occurring inside a loop, it may be seen that the voltage-divider approach used above does not apply in equally straightforward fashion. But E_o/E_i can be obtained by the use of Kirchhoff's laws in Fig. 2.7(c), which could also have been used in Example 2.3.1. E_o/E_i will be derived by both the loop and node methods.

(a) *Loop method:* Kirchhoff voltage equations are written for each closed loop in terms of loop current variables:

$$E_i = \left(\frac{1}{Cs} + R_1\right)I_1 - \frac{1}{Cs}I_2 \qquad \text{(loop 1)}$$

$$0 = -\frac{1}{Cs}I_1 + \left(R_2 + \frac{2}{Cs}\right)I_2 \qquad \text{(loop 2)}$$

Since $E_o = E_i - I_2 R_2$, this set is solved for I_2 by use of *Cramer's rule:*

$$I_2 = \begin{vmatrix} \dfrac{1}{Cs} + R_1 & E_i \\[2mm] -\dfrac{1}{Cs} & 0 \end{vmatrix} \Bigg/ \begin{vmatrix} \dfrac{1}{Cs} + R_1 & -\dfrac{1}{Cs} \\[2mm] -\dfrac{1}{Cs} & \dfrac{2}{Cs} + R_2 \end{vmatrix} = \frac{E_i Cs}{1 + (2R_1 + R_2)Cs + R_1 R_2 C^2 s^2}$$

Then, from $E_o = E_i - I_2 R_2$,

$$\frac{E_o}{E_i} = \frac{1 + 2R_1 Cs + R_1 R_2 C^2 s^2}{1 + (2R_1 + R_2)Cs + R_1 R_2 C^2 s^2} \qquad (2.8f)$$

(b) *Node method:* Here the Kirchhoff current equations are written in terms of voltage variables at each of the circuit nodes. In Fig. 2.8(f) the unknown node voltages are E_a ($=E_o$) and E_b, and the equations are

$$\frac{E_a - E_i}{R_2} + \frac{E_a - E_b}{1/(Cs)} = 0 \qquad \text{(node } a)$$

$$\frac{E_b - E_a}{1/(Cs)} + \frac{E_b - E_i}{1/(Cs)} + \frac{E_b}{R_1} = 0 \qquad \text{(node } b)$$

Rearranging yields

$$\left(\frac{1}{R_2} + Cs\right)E_a - CsE_b = \frac{1}{R_2}E_i$$

$$- CsE_a + \left(\frac{1}{R_1} + 2Cs\right)E_b = CsE_i$$

Solution for E_a ($=E_o$) by Cramer's rule yields E_o/E_i as in part (a), and in the case of this example in a somewhat more direct fashion.

Example 2.3.3 Ladder Network [Fig. 2.8(g)]

Using the node method, current equations are written for the circuit nodes a and b:

$$\frac{E_a - E_i}{L_1 s} + \frac{E_a}{R_1} + \frac{E_a - E_b}{L_2 s} = 0 \qquad \frac{E_b - E_a}{L_2 s} + \frac{E_b}{R_2} = 0$$

Solution for $E_b = E_o$ gives

$$\frac{E_o}{E_i} = \frac{R_1 R_2}{L_1 L_2 s^2 + (L_1 R_1 + L_1 R_2 + L_2 R_1)s + R_1 R_2}$$

2.4 ELECTROMECHANICAL SYSTEMS: TRANSFER FUNCTIONS OF MOTORS AND GENERATORS

Schematic diagrams of several arrangements of motors and generators are shown in Fig. 2.9. In all cases the motor load is assumed to consist of an inertia J and a damper with damping constant B. Motor shaft position θ and developed motor torque T are then related by

$$T(t) = J\ddot{\theta}(t) + B\dot{\theta}(t) \qquad T(s) = s(Js + B)\theta(s) \qquad (2.9)$$

For convenience, the same variables are used in the time and Laplace domains. The identifier (t) or (s) will generally be omitted if it is evident from the context. Each of the systems in Fig. 2.9 will be considered in turn.

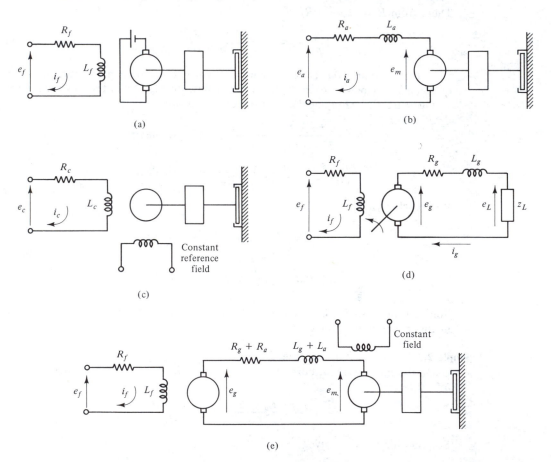

(a)

(b)

(c)

(d)

(e)

Figure 2.9 Motors and generators: (a) field-controlled dc motor; (b) armature-controlled dc motor; (c) two-phase ac servomotor; (d) dc generator; (e) motor–generator set.

Example 2.4.1   Field-Controlled DC Motor [Fig. 2.9(a)]

The equation for the field loop is

$$e_f = R_f i_f + L_f \dot{i}_f \qquad E_f = (R_f + L_f s) I_f$$

With constant armature voltage, the developed motor torque T in (2.9) can be taken to be proportional to field current:

$$T = K_t i_f \qquad T = K_t I_f \qquad K_t = \text{motor torque constant}$$

Eliminating I_f and T between these transformed equations and (2.9) yields the desired transfer function between applied field voltage e_f and shaft position θ:

$$\frac{\theta}{E_f} = \frac{K_t/(R_f B)}{s(T_m s + 1)(T_f s + 1)} \tag{2.10}$$

where $T_m = J/B$ = motor time constant

$T_f = L_f/R_f$ = field time constant

Often $T_f \ll T_m$, and a satisfactory approximation in the operating range of interest is

$$\frac{\theta}{E_f} = \frac{K_t/(R_f B)}{s(T_m s + 1)} \tag{2.11}$$

Note that the transfer function was derived for the combination of the motor and its load. This load affects motor speed (i.e., it loads the motor), so a series connection of two blocks with individual transfer functions would be incorrect.

The factor s in the denominator of (2.10) and (2.11) should be noted. From Table 1.6.1, dividing $F(s)$ by s is equivalent to integrating $f(t)$. Thus the factor s represents the fact that a motor is basically an integrator: For a constant input e_f it has a shaft angle θ which increases at a constant rate, so θ is proportional to the integral of e_f.

Example 2.4.2   Armature-Controlled DC Motor [Fig. 2.9(b)]

The armature loop is described by

$$e_a = R_a i_a + L_a \dot{i}_a + e_m \qquad E_a = (R_a + L_a s) I_a + E_m$$

Here the counter emf (electromotive force) voltage can be taken to be proportional to shaft speed,

$$e_m = K_e \dot{\theta} \qquad E_m = K_e s \theta$$

and the developed torque proportional to current i_a,

$$T = K_t i_a \qquad T = K_t I_a$$

Eliminating I_a, E_m, and T between these equations and (2.9) permits the desired transfer function to be arranged in the common form

$$\frac{\theta}{E_a} = \frac{1/K_e}{s(T_a T_m s^2 + (T_m + \gamma T_a)s + \gamma + 1)} \tag{2.12}$$

where $T_m = J R_a/(K_e K_t)$ = motor time constant

$T_a = L_a/R_a$ = armature time constant

$\gamma = B R_a/(K_e K_t)$ = damping factor

Example 2.4.3 Two-Phase AC Servomotor [Fig. 2.9(c)]

Fixed and variable magnitude ac voltages are applied to the reference and control fields, respectively. A 90° phase shift arranged between these voltages is made positive or negative depending on the desired direction of rotation. The control field is described by

$$e_c = R_c i_c + L_c \dot{i}_c \qquad E_c = (R_c + L_c s)I_c \tag{2.13}$$

The developed motor torque T can be taken to be proportional to i_c and to decrease proportionally with increasing speed, and is described by

$$T = K_c i_c - K_\omega \dot{\theta} \qquad T = K_c I_c - K_\omega s\theta \tag{2.14}$$

This dependence on speed is assumed to be the same under dynamic conditions as for the steady-state torque–speed motor characteristic curves indicated in Fig. 2.10. Eliminating T and I_c between these equations and (2.9) and rearranging gives the transfer function

$$\frac{\theta}{E_c} = \frac{K}{s(T_m s + 1)(T_c s + 1)} \tag{2.15}$$

where $T_m = J/(B + K_\omega)$ = motor time constant

$T_c = L_c/R_c$ = electrical time constant

$K = K_c/[R_c(B + K_\omega)]$ = motor constant

Often simplification to the form (2.11) is again satisfactory: $K/[s(T_m s + 1)]$.

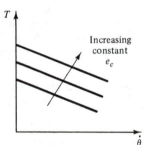

Figure 2.10 Ac motor characteristics.

Example 2.4.4 DC Generator [Fig 2.9(d)]

The field loop equation is

$$e_f = R_f i_f + L_f \dot{i}_f \qquad E_f = (R_f + L_f s)I_f$$

The developed generator voltage e_g can be assumed to be proportional to field current:

$$e_g = K_g i_f \qquad E_g = K_g I_f$$

The voltage e_L across the load is given by

$$E_L = Z_L I_g \qquad Z_L = \text{load impedance}$$

and the generator loop is described by

$$e_g = R_g i_g + L_g \dot{i}_g + e_L \qquad E_g = (L_g s + R_g + Z_L)I_g$$

Hence

$$\frac{E_g}{E_f} = \frac{K_g}{L_f s + R_f} \qquad \frac{E_L}{E_g} = \frac{Z_L}{L_g s + R_g + Z_L} \qquad \frac{E_L}{E_f} = \frac{E_L}{E_g}\frac{E_g}{E_f} \qquad (2.16)$$

Example 2.4.5   Motor–Generator Set [Fig. 2.9(e)]

The generator serves as a rotating power amplifier. $\theta/E_f = (\theta/E_g)(E_g/E_f)$ is obtained directly by appropriate substitutions in (2.12) and (2.16).

2.5 THERMAL SYSTEMS

As in the preceding sections, thermal system elements are discussed first.

Thermal resistance [Fig. 2.11(a)].   A wall of area A separates regions with temperatures T_1 and T_2. The heat flow rate q, in units of heat per unit of time, say Btu/sec, is proportional to the temperature difference $T_1 - T_2$ and to the area A, and flows toward the lowest temperature. The constant of proportionality is the heat transfer coefficient h. In the case illustrated, this is an effective coefficient which combines the effects of heat convection at the surfaces and heat conduction through the wall. The equation would also represent heat convection across a single surface.

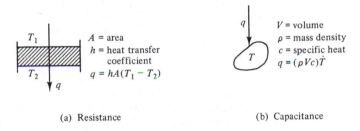

(a) Resistance   (b) Capacitance

Figure 2.11   (a) Thermal resistance; (b) thermal capacitance.

For heat conduction through a wall with surface temperatures T_1 and T_2 and thickness d, the heat flow is proportional to the temperature gradient $(T_1 - T_2)/d$:

$$q = \frac{kA(T_1 - T_2)}{d} = \frac{k}{d}A(T_1 - T_2) \qquad (2.17)$$

where k is the thermal conductivity. Hence the equivalent heat transfer coefficient is k/d.

To identify these relations in terms of a thermal equivalent of an electrical resistance, the equation in Fig. 2.11(a) is written as

$$T_1 - T_2 = qR_t \qquad R_t = \frac{1}{hA} \qquad (2.18)$$

With $T_1 - T_2$ analogous to voltage drop v, and q to current i, R_t becomes the thermal resistance.

Thermal capacitance [Fig. 2.11(b)].   Let q be the net heat flow rate into a volume V of a material with mass density ρ and specific heat c (= heat required to raise the temperature of a unit mass by 1°). This net inflow q of heat per second must equal the change per second (i.e., the rate of change) of heat stored in V. Since the mass is ρV, the heat required for a 1° rise of temperature is ρVc, and hence the heat stored at a temperature T is ρVcT. Assuming ρ, V, and c to be constant, its rate of change is $\rho Vc\dot{T}$: hence the equation shown in Fig. 2.11(b).

From the equation $i = C\dot{v}$ in Fig. 2.7 for an electrical capacitance follows immediately the equivalent notion of a thermal capacitance C_t:

$$q = C_t\dot{T} \qquad C_t = \rho Vc \tag{2.19}$$

It is seen that C_t is the heat required for a 1° temperature rise.

Some examples of thermal systems follow.

Example 2.5.1   Process Control

Figure 2.12 shows a tank of volume V filled with an incompressible fluid of mass density ρ and specific heat c. Volume flow rates entering and leaving are f_i and f_o. T_i is the temperature of the inflow. It is assumed that the tank is well stirred, so that the outlet temperature equals the tank temperature T.

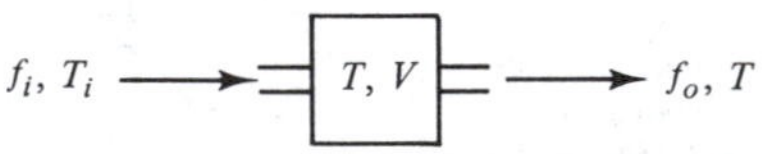

$$f_i, T_i \longrightarrow \boxed{T, V} \longrightarrow f_o, T$$

Figure 2.12   Process flow.

The tank is filled and the fluid incompressible, so $f_i = f_o$, and the mass flow rate entering and leaving is $f_i\rho$. Hence the heat inflow rate is $f_i\rho cT_i$, the outflow rate $f_i\rho cT$, and the net inflow rate $f_i\rho c(T_i - T)$. As explained, this must equal $V\rho c\dot{T}$, the rate of change of heat $V\rho cT$ stored in the tank, $V\rho c\dot{T} = f_i\rho c(T_i - T)$, or

$$\frac{V}{f_i}\dot{T} + T = T_i \tag{2.20}$$

The transform is $[(V/f_i)s + 1]T(s) = T_i(s)$, so that the following *simple lag* transfer function relates T_i and T:

$$\frac{T(s)}{T_i(s)} = \frac{1}{(V/f_i)s + 1} \tag{2.21}$$

Example 2.5.2   Space Heating

In Fig. 2.13, let T be the difference with a constant ambient temperature. By (2.18), the heat loss q_o to ambient can be modeled by $q_o = T/R_t$, where R_t is the thermal resistance. If q_i is the heat inflow rate from an electrical heater, the net inflow $(q_i - T/R_t)$ must equal $C_t\dot{T}$, where C_t is the thermal capacitance. Hence the behavior is modeled by the differential equation

$$R_tC_t\dot{T} + T = R_tq_i \tag{2.22}$$

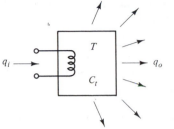

Figure 2.13   Space heating.

Therefore, the effect of heat flow q_i on the temperature T is approximated by the transfer function

$$\frac{T(s)}{Q_i(s)} = \frac{R_t}{R_t C_t s + 1}$$ (2.23)

As in Example 2.5.1 and the electrical RC circuit in Example 2.3.1, the behavior is described by a simple lag transfer function.

Example 2.5.3   Three-Capacitance System

Extending Fig. 2.12, in Fig. 2.14 allowance is made for heat loss to the ambient temperature T_a via surface area A_o with heat transfer coefficient h_o and for the heating up, via surface A_m with heat transfer coefficient h_m, of an internal space or material with capacitance C_m and uniform temperature T_m. This temperature is measured by a sensor with a significant capacitance C_s, heated via a surface of area A_s and heat transfer coefficient h_s.

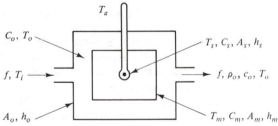

Figure 2.14   Three-capacitance system.

The most convenient approach is to equate the net heat flow rate to the rate of change of heat for each capacitance in turn:

$$C_o \dot{T}_o = f\rho_o c_o(T_i - T_o) - A_o h_o(T_o - T_a) - A_m h_m(T_o - T_m)$$

$$C_m \dot{T}_m = A_m h_m(T_o - T_m)$$ (2.24)

$$C_s \dot{T}_s = A_s h_s(T_m - T_s)$$

The second of these equations does not include the heat loss term $-A_s h_s(T_m - T_s)$ to the sensor, on the assumption that it is relatively negligible. The last equation immediately gives the transfer function relating T_m and its value as measured by the sensor:

$$\frac{T_s(s)}{T_m(s)} = \frac{1}{\tau_s s + 1} \qquad \tau_s = \frac{C_s}{A_s h_s}$$ (2.25)

This is again a simple lag transfer function, as is that relating T_o and T_m from the second of equations (2.24):

$$\frac{T_m(s)}{T_o(s)} = \frac{1}{\tau_m s + 1} \qquad \tau_m = \frac{C_m}{A_m h_m} \tag{2.26}$$

Substituting this for T_m in the first of equations (2.24), its transform, on bringing all terms for $T_o(s)$ to one side, is

$$\left[C_o s + (f\rho_o c_o + A_o h_o) + A_m h_m \left(1 - \frac{1}{\tau_m s + 1} \right) \right] T_o(s) = f\rho_o c_o T_i(s) + A_o h_o T_a(s)$$

and some algebraic manipulation then yields

$$T_o(s) = \frac{(\tau_m s + 1)[f\rho_o c_o T_i(s) + A_o h_o T_a(s)]}{C_o \tau_m s^2 + [(f\rho_o c_o + A_o h_o)\tau_m + C_o + C_m]s + f\rho_o c_o + A_o h_o} \tag{2.27}$$

This transform reflects the condition that variations of ambient temperature T_a act as a disturbance input on the system. Thus the system has two inputs, as shown in Fig. 2.15, where G is the transfer function implied by (2.27). As discussed in Section 1.8, the total output is the superposition of the outputs for each input separately with the other put equal to zero.

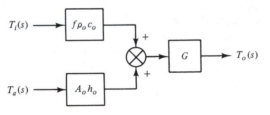

Figure 2.15 Two-input system.

It should also be observed, however, that if ambient temperature variations are known to be small or slow, $T_a = 0$ can be assumed in the original equations (2.24). This is because in a linearized model the variables represent variations from operating point values, so that constant variables are zero.

For $T_a = 0$, the overall transfer function $T_s(s)/T_i(s)$ can now be written from (2.25) to (2.27), or represented by the series connection of blocks in Fig. 2.16, which also identifies the responses $T_o(s)$ and $T_m(s)$.

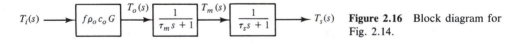

Figure 2.16 Block diagram for Fig. 2.14.

2.6 FLUID SYSTEMS

Fluid system elements are defined in Fig. 2.17, again in terms of their electrical equivalents.

1. *Fluid resistance R_f*: This exists in flow orifices, valves, and fluid lines. With pressure drop $p_1 - p_2$ equivalent to voltage drop and flow rate q to current, R_f is equivalent to electrical resistance. Commonly, the actual

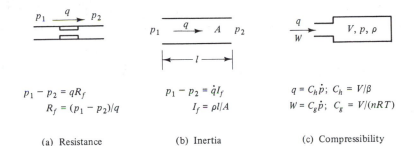

$$p_1 - p_2 = qR_f$$
$$R_f = (p_1 - p_2)/q$$

$$p_1 - p_2 = \dot{q}I_f$$
$$I_f = \rho l/A$$

$$q = C_h\dot{p}; \quad C_h = V/\beta$$
$$W = C_g\dot{p}; \quad C_g = V/(nRT)$$

(a) Resistance   (b) Inertia   (c) Compressibility

Figure 2.17 Fluid system elements.

relation is nonlinear, and Fig. 2.17(a) shows a linearized model, with p_1, p_2, and q being variations about the values at an operating point used to calculate R_f. R_f may be obtained by calculation or experiment, and the units may yield either volume or mass flow rates.

2. *Fluid inertia I_f:* The mass of fluid of mass density ρ in a line of length l and cross-sectional area A is $\rho A l$. The pressure drop $p_1 - p_2$ generates a force $(p_1 - p_2)A$ to accelerate this mass. Fluid velocity v and volumetric flow rate q are related by $q = Av$, so the acceleration $\dot{v}$ can be expressed as $\dot{v} = \dot{q}/A$. Newton's law then yields $(p_1 - p_2)A = \rho A l(\dot{q}/A)$, which reduces to the equation in Fig. 2.17(b). It is equivalent to $v = Li$ for an electrical inductance.

3. *Fluid compressibility C_h, C_g:* In Fig. 2.17(c), pressure and mass density in volume V are p and ρ. The mass in V is ρV, and its rate of change $d(\rho V)/dt$ must clearly be equal to the mass flow rate W entering V. With V constant, therefore,

$$W = V\dot{\rho} \tag{2.28}$$

Liquids and gases are considered in turn.

Liquids. In high-performance hydraulic systems it is necessary to include the effect of oil compressibility. At constant temperature, near an operating point p_0, ρ_0, a bulk modulus β is defined by

$$\rho - \rho_0 = \frac{\rho_0}{\beta}(p - p_0) \qquad \dot{\rho} = \frac{\rho_0}{\beta}\dot{p} \tag{2.29}$$

A value $\beta \approx 200,000$ psi is possible theoretically, but air entrainment usually makes 100,000 psi more realistic. Substituting (2.29) into (2.28) gives

$$W = \frac{V\rho_0}{\beta}\dot{p} \tag{2.30}$$

In hydraulics, volume flow rate $q = W/\rho_0$ is more commonly used:

$$q = C_h\dot{p} \qquad \text{capacitance } C_h = \frac{V}{\beta} \tag{2.31}$$

This is equivalent to $i = C\dot{v}$ for an electrical capacitance. The flow rate q causes a larger rate of change of pressure if the volume V is smaller or the oil stiffer (i.e., β larger).

Gases. For a polytropic process in a gas described by the ideal gas law $p = \rho RT$ (T = absolute temperature; R = gas constant) the p–ρ relation is

$$p = C\rho^n \qquad (\ln p = \ln C + n \ln \rho) \tag{2.32}$$

where

$$n = \begin{cases} 1 & \text{for isothermal processes} \\ k = \dfrac{c_p}{c_v} & \text{for adiabatic frictionless processes} \end{cases}$$

For the latter, where c_p and c_v are the specific heat values at constant pressure and constant volume, there is no heat exchange with the environment. Taking the derivative of (2.32) gives

$$\frac{dp}{p} = \frac{n\,d\rho}{\rho} \qquad \text{or} \qquad \dot{\rho} = \frac{\rho}{np}\dot{p}$$

and substituting into (2.28) yields

$$W = \frac{V\rho}{np}\dot{p} \tag{2.33}$$

By comparison with (2.30), this shows that

$$\beta = np = \text{bulk modulus for gases} \tag{2.34}$$

The form more common for gases

$$W = C_g\dot{p} \qquad \text{capacitance } C_g = \frac{V}{nRT} \tag{2.35}$$

($n = 1$: isothermal; $n = k = c_p/c_v$: adiabatic) results by substitution of $p = \rho RT$.

Example 2.6.1   Hydraulic Tank (Fig. 2.18)

This was the process in the level control system of Section 1.3. The net volumetric inflow rate is $(q_i - q_o)$. This is the volume entering per unit time, so must equal the change per unit time (i.e., the rate of change) of the volume Ah in the tank:

$$A\dot{h} = q_i - q_o \tag{2.36}$$

The outflow q_o depends on the pressure drop across the valve. The pressure p at depth h below the water level is the force per unit area due to the weight of water, so equals the weight of a column of water of unit area cross section and height h: $p = \rho gh$. Here ρ is mass density and g the acceleration due to gravity. The valve resistance R is often expressed in terms of the head h instead of the pressure p.

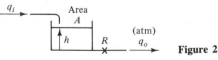

Figure 2.18   Tank level.

The actual relation between h and q_o is nonlinear, but it is approximated by the linearized model

$$h = q_o R \qquad (2.37)$$

with the variables representing the variations about operating-point values. R is determined as in Fig. 1.5 from the slope of the nonlinear characteristic of h versus q_o at the operating point. Transforming these equations and substituting the second into the first yields

$$(ARs + 1)H(s) = RQ_i(s)$$

and the transfer function

$$\frac{H(s)}{Q_i(s)} = \frac{R}{ARs + 1} \qquad (2.38)$$

Note that this again has the form of a simple lag transfer function, already encountered in electrical and thermal systems.

Example 2.6.2   Two-Tank System with Control Valve (Fig. 2.19)

A control valve V_c with valve opening x controls flow rate q_{i1} into the first tank from a supply with constant pressure P_s. From (2.38), the following transfer functions can be written immediately:

$$\frac{H_1(s)}{Q_{i1}(s)} = \frac{R_1}{A_1 R_1 s + 1} \qquad \frac{H_2(s)}{Q_{i2}(s)} = \frac{R_2}{A_2 R_2 s + 1} \qquad (2.39)$$

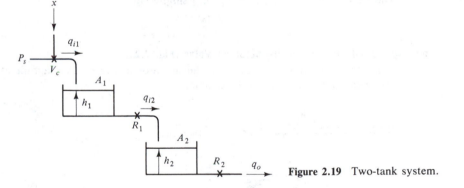

Figure 2.19   Two-tank system.

To express q_{i1} and q_{i2}, different linearized models may be used for R_1 and the control valve:

$$h_1 = q_{i2} R_1 \qquad q_{i1} = K_v x \qquad (2.40)$$

The input to the valve model is the valve opening x, which in effect controls the valve resistance parameter. The transfer functions corresponding to (2.40) are

$$\frac{Q_{i1}(s)}{X(s)} = K_v \qquad \frac{Q_{i2}(s)}{H_1(s)} = \frac{1}{R_1} \qquad (2.41)$$

Transfer functions (2.39) and (2.41) can be combined into the block diagram shown in Fig. 2.20. It should be noted that this subdivision into blocks would not apply if in Fig. 2.19 the outflow of the first tank fed into the bottom of the second tank. The net head on R_1 would then be $h_1 - h_2$, so the second tank would affect the

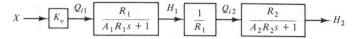

Figure 2.20   Block diagram for system of Fig. 2.19.

output of the first. As discussed earlier, this means that the second tank loads the first, and in this case an overall transfer function must be derived directly from the equations for the combined system.

Example 2.6.3   Pneumatic Tank (Fig. 2.21)

The linearized model for subsonic flow through R can be written as

$$p_i - p_o = WR \tag{2.42}$$

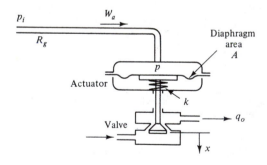

Figure 2.21   Pneumatic tank.

From Fig. 2.17(c), this flow raises tank pressure p_o according to

$$W = C_g \dot{p_o} \qquad C_g = \text{capacitance} \tag{2.43}$$

Transforming these equations gives

$$P_i(s) - P_o(s) = W(s)R \qquad W(s) = C_g s P_o(s)$$

The transfer function P_o/P_i is again a simple lag:

$$\frac{P_o(s)}{P_i(s)} = \frac{1}{RC_g s + 1} \tag{2.44}$$

Example 2.6.4   Pneumatically Actuated Valve (Fig. 2.22)

This extremely common device is the actuator in the block diagram of Fig. 1.3 assumed for the level control example.

The pneumatic line that connects control pressure p_i to pressure p above the diaphragm is represented by the linearized resistance R_g:

$$p_i - p = W_a R_g \qquad P_i(s) - P(s) = W_a(s)R_g \tag{2.45}$$

Diaphragm motion x is so small that the capacitance C_g of the space above it is about constant. Then W_a raises p according to

$$W_a = C_g \dot{p} \qquad W_a(s) = C_g s P(s) \tag{2.46}$$

A very simplified model, by no means always acceptable, will be used for diaphragm and valve poppet motion x. The mass and friction forces of the moving parts and the fluid flow forces on the poppet will be neglected. The downward pressure force pA on the diaphragm must then be counterbalanced by the spring force kx:

$$Ap = kx \qquad AP(s) = kX(s) \tag{2.47}$$

Finally, valve flow q_o is modeled by

$$q_o = K_v x \qquad Q_o(s) = K_v X(s) \tag{2.48}$$

Eliminating W_a between (2.45) and (2.46) gives

$$\frac{P(s)}{P_i(s)} = \frac{1}{R_g C_g s + 1} \tag{2.49}$$

and this with (2.47) and (2.48) gives the block diagram in Fig. 2.23. The overall transfer function is again a simple lag:

$$\frac{Q_o(s)}{P_i(s)} = \frac{AK_v/k}{R_g C_g s + 1} \tag{2.50}$$

Note how the overall model gradually emerged from equations written for the parts of the system, and not from some form of grand view of the total system.

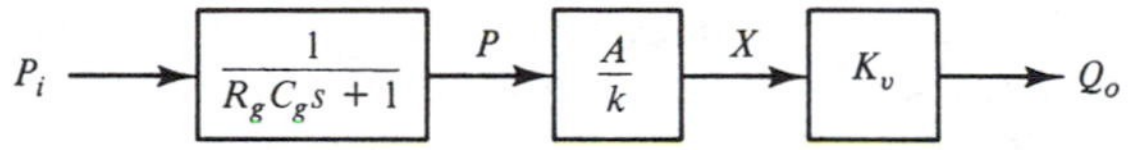

Figure 2.23   Block diagram for Fig. 2.22.

It is also useful to remark on the signs in (2.47) and (2.48). Remember that the variables represent changes from operating-point values. Figure 2.22 shows that a positive change of p causes a positive change of x, hence the positive signs in (2.47). Also, a positive change of x causes a positive change of q_o, so the signs in (2.48) must also be positive. If x had been defined as positive in upward direction, the signs in both equations would be negative on the right or left sides.

2.7 FLUID POWER CONTROL ELEMENTS

In this section the modeling of a number of very common subsystems of fluid power servos and controls is discussed by means of examples.

Example 2.7.1   Control Valves

Two common types of valves are shown in Fig. 2.24. The supply pressure is p_s and the output pressure p. For $x = 0$ the output port is just blocked off. The valve model used earlier was $q = K_v x$, where q is the volumetric flow rate. The level control example in Chapter 1 serves to show that this model is incomplete. Supply pressure variations were a major potential disturbance and a prime reason why feedback control was needed in the first place, yet their effect on valve flow is not incorporated. Valve flow increases with both x and valve pressure drop $(p_s - p)$, and an appropriate linearized valve model is $q = K_x x + K_p(p_s - p)$. In linearized models the variables represent variations about the operating point. In fluid power systems, if the supply pressure is constant, the valve model reduces to that shown in Fig. 2.25:

$$q = K_x x - K_p p \tag{2.51}$$

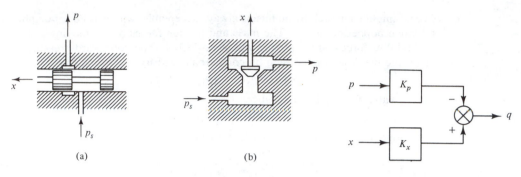

Figure 2.24 Control valves: (a) spool; (b) poppet.   Figure 2.25 Linearized valve model.

The constants K_x and K_p are found from the slopes of the steady-state valve characteristics at the operating point (x_o, p_o).

Example 2.7.2 Hydraulic Cylinder Control

The spool-valve-controlled actuator in Fig. 2.26 is used extensively. For $x = 0$ the valve spool is centered, and the lands on this spool exactly block the ports of fluid lines to the ends of the cylinder, so that the piston is stopped. If the valve spool is moved slightly to the left, the ports are partially unblocked. The left side of the cylinder is now connected to the supply and the right side to a low-pressure reservoir. Thus the piston can move to the right.

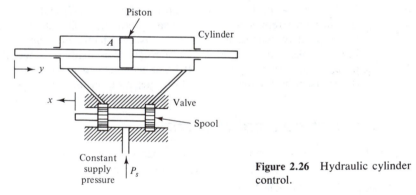

Figure 2.26 Hydraulic cylinder control.

Consider first the simplest possible model, in which the load connected to the piston is very small and pressure variations are negligible. Oil compressibility can then be ignored, and if the effective area A on both sides of the piston is the same, the flow rate q through both valve ports is also the same and can be modeled as

$$q = K_v x \qquad Q = K_v X \qquad (2.52)$$

The change $A\dot{y}$ of volume on each side of the piston per second must equal q:

$$q = A\dot{y} \qquad Q = AsY \qquad (2.53)$$

This is modeled in Fig. 2.27, and the transfer function is

$$\frac{Y(s)}{X(s)} = \frac{K_v}{As} \qquad (2.54)$$

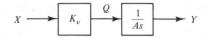

Figure 2.27 Simple model of Fig. 2.26.

The factor s in the denominator represents the fact that, like the electric motor, the cylinder is an integrator, since for a constant flow q the output y increases linearly.

Example 2.7.3 Loaded Hydraulic Cylinder Control

Let the load on the hydraulic cylinder in Fig. 2.26 consist of mass m and damping b. Then if the net pressure on the piston is p, the force balance equation is

$$pA = m\ddot{y} + b\dot{y} \qquad AP(s) = (ms + b)sY(s) \tag{2.55}$$

The valve flow is modeled by (2.51):

$$q = K_x x - K_p p \qquad Q(s) = K_x X(s) - K_p P(s) \tag{2.56}$$

Equation (2.53) must be modified to include the effect of oil compressibility. From Fig. 2.17(c), near a piston position where the volume under pressure p is V, the compressibility flow associated with pressure variations is $(V/\beta)\dot{p}$. The flow q to the cylinder supplies this compressibility flow as well as the flow $A\dot{y}$ corresponding to piston velocity:

$$q = A\dot{y} + \frac{V}{\beta}\dot{p} \qquad Q(s) = s[AY(s) + \frac{V}{\beta}P(s)] \tag{2.57}$$

From (2.56) and (2.57),

$$K_x X - K_p P = AsY + \frac{V}{\beta}sP \qquad K_x X - AsY = \left(\frac{V}{\beta}s + K_p\right)P$$

Then substituting for P from (2.55) and rearranging yields the following improvement of the model (2.54):

$$\frac{Y(s)}{X(s)} = \frac{K_x}{s\left[\dfrac{mV}{\beta A}s^2 + \dfrac{1}{A}\left(K_p m + \dfrac{V}{\beta}b\right)s + \dfrac{K_p b}{A} + A\right]} \tag{2.58}$$

Example 2.7.4 Hydraulic Motor and Hydrostatic Transmission

Figure 2.28 shows a schematic diagram. A constant-speed hydraulic pump supplies flow to a hydraulic motor. Motor speed can be changed by adjusting pump flow per revolution via a setting ϕ_p. Delivered pump flow is proportional to ϕ_p:

$$q_p = K_p \phi_p = q_l + q_c + q_m \tag{2.59}$$

Of this flow rate, a part q_l is lost in internal leakages, q_c is compressibility flow, and only part q_m causes motor rotation. Let

p = pressure drop across the motor

V = volume of oil under high pressure p

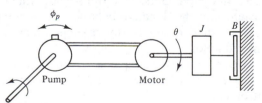

Figure 2.28 Hydrostatic transmission.

D_m = motor displacement, the volume of oil needed for 1 rad motor rotation

Then

$$q_l = Lp: \text{leakage flow proportional to } p$$

$$q_c = \frac{V}{\beta}\dot{p}: \text{compressibility flow [Fig. 2.17(c)]}$$

$$q_m = D_m\dot{\theta}: \text{motor flow } (\dot{\theta} = \text{motor speed, rad/sec})$$

This gives the flow equation:

$$K_p\phi_p = D_m\dot{\theta} + Lp + \frac{V}{\beta}\dot{p} \qquad K_p\phi_p(s) = D_m s\theta(s) + \left(L + \frac{V}{\beta}s\right)P(s) \qquad (2.60)$$

To obtain the load equation, for 100% motor efficiency its mechanical output power equals its hydraulic input power. If the developed motor torque is T, the mechanical output power is $T\dot{\theta}$. The hydraulic input power is $q_m p$. This may be verified by thinking of the motor as a cylinder of area A. With flow q_m, the piston velocity is then q_m/A, and piston force pA. The power is their product, pq_m. Thus $T\dot{\theta} = q_m p = D_m\dot{\theta}p$, so that

$$T = D_m p \qquad (2.61)$$

This torque accelerates inertia J and overcomes damping B, so that the load equation becomes

$$T = D_m p = J\ddot{\theta} + B\dot{\theta} \qquad D_m P(s) = s(Js + B)\theta(s) \qquad (2.62)$$

Substituting $P(s)$ from this into (2.60) and rearranging yields the transfer function:

$$\frac{\theta(s)}{\phi_p(s)} = \frac{K_p D_m}{s\left[\frac{VJ}{\beta}s^2 + \left(\frac{VB}{\beta} + LJ\right)s + BL + D_m^2\right]} \qquad (2.63)$$

2.8 CONCLUSION

In this chapter transfer functions were derived for a variety of physical subsystem blocks. If feedback systems consist of a connection of such blocks, overall transfer functions can now be obtained by block diagram reduction. But often such a separation into blocks and the precise nature of the feedback may become evident only when all equations for the system have been written. The modeling of such systems, and their representation by block diagrams that clarify system behavior and exhibit feedback effects, are considered in Chapters 4 and 5. The operational amplifier and some of its many uses are also discussed in Chapter 5.

First, however, it is desirable to obtain a better understanding of the nature of system transient responses. In Chapter 3 the powerful s-plane will be introduced, to provide fundamental insight into the dynamic behavior of systems described by given transfer functions, whether these describe a single block or have been obtained by block diagram reduction.

As an introduction to this, a final example is given to demonstrate the application of techniques and results of Chapters 1 and 2 to the transient analysis of a simple feedback control system.

Example 2.8.1   DC Motor Position Servo

From (2.11), let the transfer function from field voltage to shaft position of a field-controlled dc motor be

$$G(s) = \frac{0.5}{s(0.25s + 1)}$$

The output of the sensor measuring shaft position is compared with a signal in the same form to obtain the error signal E in the block diagram in Fig. 2.29. Assuming a fast response power amplifier and a constant-gain controller, the loop can be considered to be closed as shown, by an amplifier block of gain K. The closed-loop transfer function, using (1.33), is

$$\frac{C(s)}{R(s)} = \frac{2K}{s^2 + 4s + 2K}$$

and for a unit step input $[R(s) = 1/s]$, the transform of the shaft position is

$$C(s) = \frac{2K}{s(s^2 + 4s + 2K)}$$

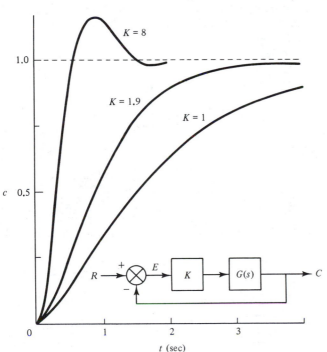

Figure 2.29   Position servo, Example 2.8.1.

For $K = 1$, partial fraction expansion yields the solution

$$c(t) = 1 - 1.207e^{-0.586t} + 0.207e^{-3.414t}$$

This is plotted in Fig. 2.29. As discussed physically in Chapter 1, for larger K the motor input changes more for a given change of error and a faster response can be expected. This is confirmed by the plot for $K = 1.9$, for which the solution is

$$c(t) = 1 - 2.737e^{-1.553t} + 1.737e^{-2.447t}$$

The physical discussion also suggested a danger of overcorrection and possibly severe oscillations if K is raised too high. The plot for $K = 8$ exemplifies this trend. Although the speed of response is improved, a maximum overshoot of about 16% exists. In many cases this is acceptable, but not, for example, for the tool slide on an automatically controlled machine tool. It is noted that the inverse transform for this case is obtained from the corresponding entry in Table 1.6.1 with $\omega_n = 4$, $\zeta = 0.5$:

$$c(t) = 1 - 1.1547e^{-2t} \sin\left(3.4641t + \frac{\pi}{3}\right)$$

The steady-state or final value of the error is seen to be zero for all three values of K, as is also evident from the final value theorem:

$$c_{ss} = \lim_{t \to \infty} c(t) = \lim_{s \to 0} sC(s) = 1$$

For servos, the steady-state error in following a unit ramp input $[R(s) = 1/s^2]$ is also an important measure of performance. For a machine tool slide moving at constant velocity, it can cause errors in part dimensions. Here application of the final value theorem to $C(s)$ yields the expected but useless result that $c_{ss} \to \infty$. The theorem is therefore applied to the error $E(s)$, as is in fact the usual practice also for step inputs. Using (1.34) yields

$$\frac{E(s)}{R(s)} = \frac{s(0.25s + 1)}{0.25s^2 + s + 0.5K}$$

so that for $R(s) = 1/s^2$,

$$e_{ss} = \lim_{t \to \infty} e(t) = \lim_{s \to 0} sE(s) = \frac{1}{0.5K}$$

As expected, this is reduced by increasing K.

PROBLEMS

2.1. For the systems shown in Fig. P2.1, write the differential equations and obtain the transfer functions indicated.

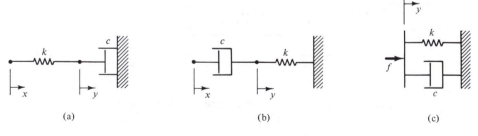

(a)   (b)   (c)

Figure P2.1   (a) Y/X; (b) Y/X; (c) Y/F.

2.2. Figure P2.2 shows a dynamic vibration absorber, often used for the control of mechanical vibrations. A relatively small mass m_2 is attached to the main mass m_1

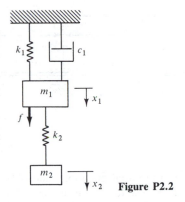

Figure P2.2

via spring k_2. For a sinusoidal force f of constant frequency, m_2 and k_2 can be chosen so that the main mass m_1 will not vibrate. Write the system differential equations and obtain the transfer function X_1/F.

2.3. The accelerometer in Fig. P2.3 is mounted on the machine of which the acceleration is to be measured. Under certain conditions, the displacement $(x - y)$ of m relative to the housing is a measure of acceleration. Write the differential equation and the transfer function $(X - Y)/X$.

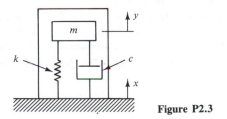

Figure P2.3

2.4. The suspension system of a car is illustrated in Fig. P2.4. On a per-wheel basis, the vehicle mass is m_1 and the mass moving with the wheel m_2. The suspension spring and tire are represented by spring constants k_1 and k_2, and the shock absorber by damping constant c. Write the differential equations and obtain the transfer function Y_1/X, which represents the vehicle response to road-surface irregularities.

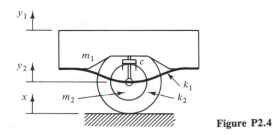

Figure P2.4

2.5. For the system shown in Fig. P2.5, derive the differential equation and obtain the transfer function X/F relating small mass motions to force f.

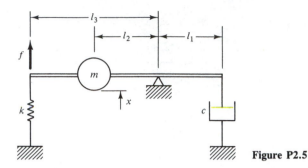

Figure P2.5

2.6. Write the differential equations and obtain the transfer function relating θ_i and θ_o for the drive system shown in Fig. P2.6, where the springs represent long shafts and damping effects due to bearings and shaft couplings are present.

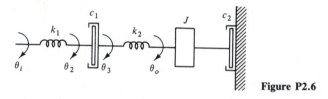

Figure P2.6

2.7. For the drive system in Fig. P2.7:
   (a) Obtain the differential equations for T_1 (torque) as input and θ_o as output and express the corresponding transfer function.
   (b) Replace the system by its nongeared equivalent to obtain the transfer function relating T_1 and θ_1 of the input shaft, and verify the result via the equations in part (a).

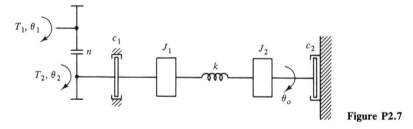

Figure P2.7

2.8. Derive the transfer functions E_o/E_i for the *RLC* circuits shown in Fig. P2.8.

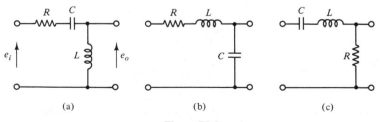

(a)   (b)   (c)

Figure P2.8

2.9. Obtain the transfer function E_o/E_i for the circuit shown in Fig. P2.9.

2.10. Obtain the transfer function E_o/E_i for the circuit shown in Fig. P2.10.

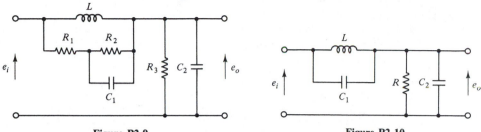

Figure P2.9   Figure P2.10

2.11. Derive E_o/E_i for the RC ladder network in Fig. P2.11.

2.12. Derive E_o/E_i for the circuit shown in Fig. P2.12.

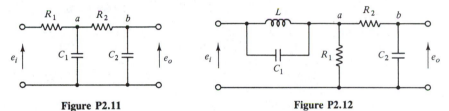

Figure P2.11   Figure P2.12

2.13. Show that the transfer function θ/E_f of the motor–generator set in Fig. 2.9(e) can be expressed as

$$\frac{\theta}{E_f} = \frac{K_g/K_e}{s(L_f s + R_f)(T_a T_m s^2 + (T_m + \gamma T_a)s + \gamma + 1)}$$

where
$$T_m = \frac{J(R_g + R_a)}{K_e K_t}$$

$$\gamma = \frac{B(R_g + R_a)}{K_e K_t}$$

$$T_a = \frac{L_g + L_a}{R_g + R_a}$$

2.14. A flexibly supported ring as shown in Fig. P2.14 is mounted on the load inertia J

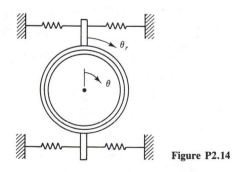

Figure P2.14

of the field-controlled dc motor. The rotational spring constant of the support is k_r and the rotational damping constant between the inertia and the ring is B_r. Determine the transfer function θ/E_f, where θ is the angle of the motor shaft.

2.15. Ac motors often run at high speed and are connected to the load via a gear reduction of ratio $n > 1$. In Fig. P2.15, J represents motor and driving gear inertia, B motor bearing damping, and J_L and B_L the inertia and damping on the driven shaft. Obtain θ_L/E_c if the electrical time constant of the ac servomotor may be neglected.

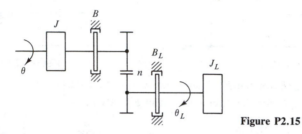

Figure P2.15

2.16. A mass M of material of specific heat c and temperature T_i is placed inside an oven at time $t = 0$, and so quickly that the constant oven temperature T_h can be considered as a step input to M. The surface area of M and its coefficient of heat transfer are A and h. Express the transform $T(s)$ of the temperature $T(t)$ of M and solve for this response $T(t)$.

2.17. A constant mass flow rate w of a liquid of specific heat c flows through a tank that contains a mass W of the liquid, and in which a resistance heater adds heat to the flow at rate q_h. As indicated in Fig. P2.17, the inflow temperature is T_i and the outflow (and tank) temperature is T_o. Obtain the transform of T_o which reflects both the effects of varying T_i and varying q_h. What is the transfer function relating q_h and T_o?

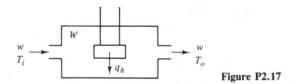

Figure P2.17

2.18. For the mercury thermometer shown in Fig. P2.18, the temperature and the thermal capacitance of the glass and mercury are T_g, C_g and T_m, C_m, respectively. The thermal resistance between ambient temperature T_a and T_g is R_g, and that between the glass and mercury is R_m.
(a) Obtain the transfer function $T_m(s)/T_a(s)$.
(b) Also obtain $T_m(s)/T_a(s)$ for a commonly used simplification in which C_g is neglected

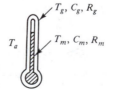

Figure P2.18

and an effective thermal resistance R_t is used. What type of transfer function is this?

2.19. In the heat exchanger in Fig. P2.19, the temperature T_h in the outer chamber can be taken to be constant, because of high flow rate through it. The mass flow rate q through the inner chamber is constant. The volume of this chamber is V and its surface area A. Inflow and outflow (and tank) temperatures are T_i and T_o. The density of the fluid is ρ and its specific heat c. The surface coefficient of heat transfer is h. Express $T_o(s)$, and also show the transfer function T_o/T_i for constant T_h.

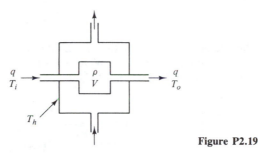

Figure P2.19

2.20. In Fig. P2.20 a mechanical brake block, of mass M, is pressed against the drum with force F. The coefficient of friction is μ, so that the friction force is μF. The surface velocity between the block and the drum is V. The friction power is converted into heat. The conversion factor that changes mechanical power into heat power is H. Determine the transfer function relating F and the temperature T of the block if all heat power is assumed to enter the block, of which the specific heat is c. The block loses heat to the ambient T_a through a surface area A with heat transfer coefficient h. Note again what type of transfer function results.

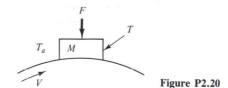

Figure P2.20

2.21. For the system in Fig. P2.21, write the linearized differential equations and obtain the transfer function relating volumetric flow rate q_i and level h_2 in the second tank.

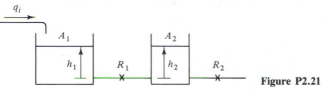

Figure P2.21

2.22. Part of a pneumatic controller is shown in Fig. P2.22. The opposing bellows are spring-centered, and the very small displacement x of the center plate may be taken to be proportional to the difference of the pressures in the bellows. Obtain the transfer function from input pressure p_i to x.

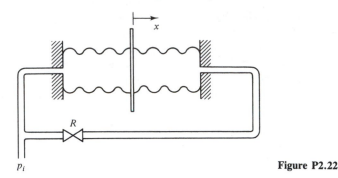

Figure P2.22

2.23. Write all linearized equations for the pneumatic system in Fig. P2.23. The R_i are resistances relating mass flow rates W_i to pressure drops, and C_{g3} and C_{g4} are the tank capacitances according to Fig. 2.17(c).

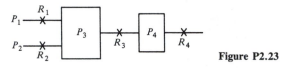

Figure P2.23

2.24. In Problem 2.23, obtain the transfer function $P_4(s)/P_1(s)$.

2.25. Figure P2.25 shows a flapper–nozzle valve, very common in pneumatic and hydraulic systems. A linearized model is analogous to that of other control valves, with flow depending on both the change x of flapper-to-nozzle distance and the change of pressure drop across the nozzle. Obtain the transfer function $P(s)/X(s)$, introducing parameters as needed, if the supply pressure p_s and the drain pressure are constant.

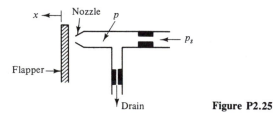

Figure P2.25

2.26. In Fig. P2.26 the flapper–nozzle valve of Problem 2.25 is used to control piston position y. Piston mass, damping, and spring constants are m, b, and k, and piston area is A. Write the linearized equations and obtain $Y(s)/X(s)$ if the volume under pressure p is V and the bulk modulus β.

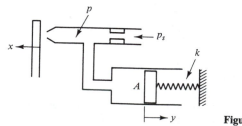

Figure P2.26

2.27. Obtain the transfer function $Y(s)/X(s)$ in Fig. P2.27 if piston mass and damping are negligible and if the volume of oil under pressure p is V, with bulk modulus β.

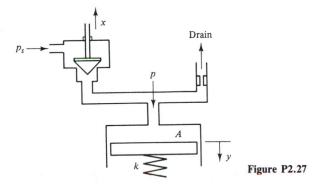

Figure P2.27

2.28. The system in Fig. P2.28 with the simple lag plant $G(s) = 1/(s + 5)$ could model, among other possibilities, a temperature, pressure, level, or speed control system. **(a)** Calculate and plot unit step responses for $K = 10$ and $K = 45$.

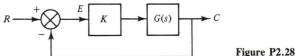

Figure P2.28

(b) Compare the responses for speed of response and steady-state error (i.e., the difference between r and c for $t \to \infty$).

2.29. In Fig. P2.28, the plant transfer function $G(s) = 20/[(s + 5)(s + 20)]$ could be a more precise model for the same systems as in Problem 2.28.
   (a) Calculate and plot unit step responses for the same values of K as in Problem 2.28(a).
   (b) Note the effect of gain on steady-state error, and note the difference with Problem 2.28 with respect to the possibility of making this error quite small.

3

TRANSIENT PERFORMANCE
AND THE s-PLANE

3.1 INTRODUCTION

In this chapter the transient behavior of systems described by given transfer functions is considered, whether these describe a single block or have been obtained by block diagram reduction. The purpose is to establish correlations and to specify requirements that ensure satisfactory performance. The use of feedback to satisfy such requirements will be considered later.

The correlations between transfer functions and response characteristics are developed in terms of the positions of the system poles and zeros in the s-plane. These are powerful concepts which will also be used in the design of feedback to improve unsatisfactory behavior.

Inverse transformation of $C(s) = G(s)R(s)$ was used in Section 1.7 to find the response $c(t)$ to an input $r(t)$, for the case where the roots of the denominator of $C(s)$ are real and distinct. The cases of repeated roots and complex conjugate pairs will now also be considered, and the s-plane will provide the basis for a graphical alternative to the analytical method of calculating the residues. This alternative enhances insight and is often preferred for more complicated transforms.

3.2 THE s-PLANE, POLE–ZERO PATTERNS, AND RESIDUE CALCULATION

In the input–output relation $C(s) = G(s)R(s)$, each of C, G, and R is in general a ratio of polynomials in s. Definitions:

Zeros of C, G, and R are the roots of their numerator polynomials.

Poles of C, G, and R are the roots of their denominator polynomials.

System zeros and *system poles* are those of the system transfer function $G(s)$.

System characteristic polynomial is the name often used for the denominator polynomial of $G(s)$.

System characteristic equation identifies the result if the system characteristic polynomial is equated to zero. Evidently, its roots are the system poles.

*****

Since the polynomials have real coefficients, poles and zeros are either real or occur in complex conjugate pairs. They are values of s and can be plotted on a complex plane called the *s-plane*. Because $s = \sigma + j\omega$, the real axis of this plane is the σ-axis, and frequencies ω are plotted along the imaginary $j\omega$-axis. It is important to note that angles in the *s*-plane have great significance, and therefore the scales on both axes must be identical.

Figure 3.1(a) shows this plane and the system poles and zero of the transfer function

$$G(s) = \frac{2K}{3} \frac{0.5s + 1}{(\frac{1}{3})s^2 + (\frac{4}{3})s + 1} = \frac{K(s + 2)}{(s + 1)(s + 3)} \tag{3.1}$$

This is the *system pole–zero pattern*. The potential power of representing all dynamic characteristics by the positions of a number of points in this manner may be appreciated.

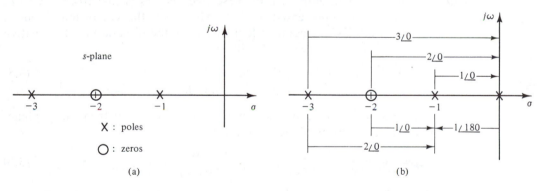

Figure 3.1   Pole–zero patterns of $G(s)$ (a) and $C(s)$ (b).

For a unit step input $R(s) = 1/s$, the pole–zero pattern of $R(s)$ consists of just a pole at the origin of the *s*-plane. Since $C(s) = G(s)R(s)$, the pole–zero pattern of $C(s)$ in Fig. 3.1(b) is simply the superposition of those of $G(s)$ and of $R(s)$:

$$C(s) = G(s)R(s) = \frac{K(s + 2)}{s(s + 1)(s + 3)} \tag{3.2}$$

Graphical Determination of Residues

To derive this important alternative to the analytical method in Section 1.7, it is recalled that in the partial fraction expansion

$$C(s) = \frac{K(s + 2)}{s(s + 1)(s + 3)} = \frac{K_1}{s} + \frac{K_2}{s + 1} + \frac{K_3}{s + 3} \tag{3.3}$$

of the output transform, the residues are

$$K_1 = \frac{K(s + 2)}{(s + 1)(s + 3)}\bigg|_{s=0} \quad K_2 = \frac{K(s + 2)}{s(s + 3)}\bigg|_{s=-1} \quad K_3 = \frac{K(s + 2)}{s(s + 1)}\bigg|_{s=-3} \tag{3.4}$$

Figure 3.2 helps to interpret the typical factor $(s + a)\,|_{s=b}$ in these expressions in terms of a vector in the s-plane. The factor is

$$(s + a)\,|_{s=b} = b + a = b - (-a)$$

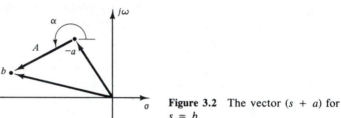

Figure 3.2 The vector $(s + a)$ for $s = b$.

Here b is a vector from the origin to point b in the s-plane, and $-a$ from the origin to point $-a$. The difference of these two vectors is that from $-a$ to b, and this vector can be represented by $Ae^{j\alpha}$, where A is the vector length and α the angle, measured positive counterclockwise from the direction of the positive real axis:

$$(s + a)\,|_{s=b} = b + a = b - (-a) \tag{3.5}$$
$$= (\text{vector from } -a \text{ to } b) = Ae^{j\alpha} = A\,\underline{/\alpha}$$

Here A and α may be found by measurement or calculation from the s-plane. Now a residue of the form of the K_i in (3.4) is

$$K_i = \frac{K(s + a)}{(s + b)(s + c)}\bigg|_{s=-d} = \frac{K(A\,\underline{/\alpha})}{(B\,\underline{/\beta})(C\,\underline{/\gamma})} = \frac{KA}{BC}\,\underline{/\alpha - \beta - \gamma} \tag{3.6}$$

To find K_1 in (3.4), vectors as shown in the upper half of Fig. 3.1(b) are drawn to the pole $s = 0$ of $C(s)$ from its zero at -2 and its other two poles at -1 and -3, and these give

$$K_1 = \frac{K(2\,\underline{/0})}{(1\,\underline{/0})(3\,\underline{/0})} = \tfrac{2}{3}K\,\underline{/0} = \tfrac{2}{3}K$$

The residue K_2 at the pole -2 is found from the vectors to this point, in the bottom half of Fig. 3.1(b). The vector from the pole at $s = 0$ to that at -1 is in the direction of the negative real axis, so has angle $\pm 180°$.

$$K_2 = \frac{K(1 \ \underline{/0})}{(1 \ \underline{/180})(2 \ \underline{/0})} = \tfrac{1}{2} K \ \underline{/-180} = -\tfrac{1}{2} K$$

Similarly,

$$K_3 = \frac{K(1 \ \underline{/180})}{(3 \ \underline{/180})(2 \ \underline{/180})} = \tfrac{1}{6} K \ \underline{/180 - 360} = -\tfrac{1}{6} K$$

These values of K_1, K_2, and K_3 may be verified using the analytical method, and give

$$c(t) = K(\tfrac{2}{3} - \tfrac{1}{2} e^{-t} - \tfrac{1}{6} e^{-3t}) \tag{3.7}$$

The gain factor K in (3.1) to (3.3) will be called the root locus gain:

> The *root locus gain* factor of a transform or a transfer function is that which results if the coefficients of the highest powers of s in numerator and denominator are made equal to unity.

For, say,

$$C(s) = 2 \frac{0.5s + 1}{(s + 3)(0.1s + 1)} = \frac{2(0.5)}{0.1} \frac{s + 2}{(s + 3)(s + 10)}$$

the root locus gain is $2(0.5/0.1) = 10$.

Reviewing the expressions for K_1, K_2, and K_3 reveals the following general rule:

> The residue K_i at the pole $-p_i$ of $C(s)$ equals the root locus gain times the product of the vectors from all zeros of $C(s)$ to $-p_i$ divided by the product of the vectors from all other poles of $C(s)$ to $-p_i$.

Example 3.2.1

Figure 3.3(a) models a level control system of which the process consists of the two-tank system with control valve in Figs. 2.19 and 2.20 and the level in the second

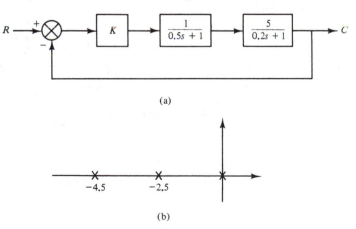

(a)

(b)

Figure 3.3 Example 3.2.1.

tank is to be controlled. Find the unit step response of the system for $K = 0.025$. The closed-loop transfer function is

$$\frac{C}{R} = \frac{(0.025)5}{(0.5s + 1)(0.2s + 1) + (0.025)5} = \frac{1.25}{s^2 + 7s + 11.25} = \frac{1.25}{(s + 2.5)(s + 4.5)}$$

and with $R(s) = 1/s$,

$$C(s) = \frac{1.25}{s(s + 2.5)(s + 4.5)} = \frac{K_1}{s} + \frac{K_2}{s + 2.5} + \frac{K_3}{s + 4.5}$$

The pole–zero pattern of $C(s)$ in Fig. 3.3(b) and the graphical residue rule now yield

$$K_1 = \frac{1.25}{(2.5 \ \underline{/0})(4.5 \ \underline{/0})} = \frac{1}{9}$$

$$K_2 = \frac{1.25}{(2.5 \ \underline{/180})(2 \ \underline{/0})} = -\frac{1}{4}$$

$$K_3 = \frac{1.25}{(4.5 \ \underline{/180})(2 \ \underline{/180})} = \frac{1.25}{9}$$

(3.8)

$$c(t) = \frac{1}{9} - \frac{1}{4} e^{-2.5t} + \frac{1.25}{9} e^{-4.5t}$$

3.3 TRANSIENT RESPONSE, INCLUDING REPEATED AND COMPLEX POLES

In this section, transient responses are calculated for the cases of repeated real poles, a complex conjugate pair of poles, and combinations of distinct real poles and complex conjugate pairs.

Repeated Real Poles

Suppose that $C(s)$ has m poles at $-p_1$:

$$C(s) = \frac{Z(s)}{(s + p_1)^m(s + p_2) \cdots (s + p_n)}$$

(3.9)

Then it can be shown that the partial fraction expansion must be written as follows:

$$C(s) = \frac{K_1}{(s + p_1)^m} + \frac{K_2}{(s + p_1)^{m-1}} + \cdots + \frac{K_m}{s + p_1} + \frac{K_{m+1}}{s + p_2} + \cdots$$

$$+ \cdots + \frac{K_{m+n-1}}{s + p_n}$$

(3.10)

The residues $K_{m+1}, \ldots, K_{m+n-1}$ are found as before, but for the repeated root the equation is

$$K_i = \frac{1}{(i - 1)!} \frac{d^{i-1}}{ds^{i-1}} [C(s)(s + p_1)^m] \bigg|_{s = -p_1} \qquad i = 1, \ldots, m$$

(3.11)

Example 3.3.1

Find the response to a unit ramp input, $R(s) = 1/s^2$, of the system

$$G(s) = \frac{1}{Ts + 1} \quad \text{(simple lag)} \tag{3.12}$$

$$C(s) = G(s)R(s) = \frac{1}{s^2(Ts + 1)} = \frac{K_1}{s^2} + \frac{K_2}{s} + \frac{K_3}{s + 1/T}$$

K_1 and K_3 can be found in the usual way:

$$K_3 = \left[\left(s + \frac{1}{T}\right)\frac{1}{s^2(Ts + 1)}\right]\Bigg|_{s=-1/T} = \frac{1}{Ts^2}\Bigg|_{s=-1/T} = T$$

$$K_1 = \left[s^2 \frac{1}{s^2(Ts + 1)}\right] = \frac{1}{Ts + 1}\Bigg|_{s=0} = 1$$

But if to determine K_2 the equation is multiplied by its denominator s, the first term becomes K_1/s and tends to infinity for $s \to 0$. Equation (3.11) arises by first multiplying both sides of the equation by s^2:

$$s^2 C(s) = K_1 + K_2 s + \frac{K_3 s^2}{s + 1/T}$$

K_1 is eliminated by differentiating both sides:

$$\frac{d}{ds}[s^2 C(s)] = K_2 + K_3 s \frac{2(s + 1/T) - s}{(s + 1/T)^2}$$

If now $s \to 0$, only K_2 remains on the right, so that, as in (3.11),

$$K_2 = \frac{d}{ds}(s^2 C)\Bigg|_{s=0} = \frac{d}{ds}\left(\frac{1}{Ts + 1}\right)\Bigg|_{s=0} = -T$$

Hence

$$C(s) = \frac{1}{s^2} - \frac{T}{s} + \frac{T}{s + 1/T}$$

Using Table 1.6.1, the transient response is

$$c(t) = t - T + Te^{-t/T}$$

Since the simple lag is so common, this response has been sketched in Fig. 3.4. The value T of the steady-state ($t \to \infty$) difference between r and c could have been found more directly by the final value theorem:

$$\lim_{t\to\infty}[r(t) - c(t)] = \lim_{s\to0} s[R(s) - C(s)] = \lim_{s\to0} s\left[\frac{T}{s(Ts + 1)}\right] = T$$

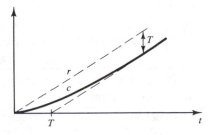

Figure 3.4   Simple lag ramp response.

Example 3.3.2   Figure 3.3 for $K = 0.045$

The response for $K = 0.025$ in Example 3.2.1, with a steady-state output of $\frac{1}{9}$ for a unit step input, so a steady-state error of $\frac{8}{9}$, is really quite poor. For $K = 0.045$ it is found that

$$\frac{C}{R} = \frac{2.25}{s^2 + 7s + 12.25} = \frac{2.25}{(s + 3.5)^2}$$

and for a unit step input

$$C = \frac{2.25}{s(s + 3.5)^2} = \frac{K_1}{s} + \frac{K_2}{(s + 3.5)^2} + \frac{K_3}{s + 3.5}$$

$$K_1 = [sC(s)]\bigg|_{s=0} = \frac{9}{49} = 0.1837$$

$$K_2 = [(s + 3.5)^2 C(s)]\bigg|_{s=-3.5} = \frac{2.25}{s}\bigg|_{s=-3.5} = -0.7429$$

$$K_3 = \frac{d}{ds}\left(\frac{2.25}{s}\right)\bigg|_{s=-3.5} = \frac{-2.25}{s^2}\bigg|_{s=-3.5} = -0.1837$$

Hence

$$c(t) = 0.1837 - 0.7429te^{-3.5t} - 0.1837e^{-3.5t}$$

The steady-state error $1 - 0.1837 = 0.8163$ has been reduced, as expected with an increase of K, but is still quite large.

A Complex Conjugate Pair of Poles

The technique for distinct real poles applies also for distinct complex poles, but the residues are now complex, and the graphical approach tends to be simpler than the analytical one.

Example 3.3.3

From Table 1.6.1

$$C(s) = \frac{\omega_n^2}{s(s^2 + 2\zeta\omega_n s + \omega_n^2)} \tag{3.13}$$

This could be the unit impulse response of a system $G(s) = C(s)$ or the unit step response of

$$G(s) = \frac{\omega_n^2}{s^2 + 2\zeta\omega_n s + \omega_n^2}$$

For $\zeta < 1$, the poles of $C(s)$ are

$$s_1 = 0 \qquad s_2 = -\zeta\omega_n + j\omega_n\sqrt{1 - \zeta^2} \qquad s_3 = -\zeta\omega_n - j\omega_n\sqrt{1 - \zeta^2} \tag{3.14}$$

The pole–zero pattern of $C(s)$ is shown in Fig. 3.5, and its partial fraction expansion

$$C(s) = \frac{K_1}{s} + \frac{K_2}{s + \zeta\omega_n - j\omega_n\sqrt{1 - \zeta^2}} + \frac{K_3}{s + \zeta\omega_n + j\omega_n\sqrt{1 - \zeta^2}}$$

shows that the response will be of the form

$$c(t) = K_1 + e^{-\zeta\omega_n t}[K_2 \exp(j\omega_n\sqrt{1 - \zeta^2}t) + K_3 \exp(-j\omega_n\sqrt{1 - \zeta^2}t)]$$

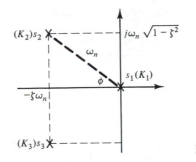

Figure 3.5 $C(s)$ of (3.13).

To determine the K_i, it is readily verified that the distance of s_2 and s_3 to the origin equals ω_n, and the angle ϕ in Fig. 3.5 is given by

$$\tan \phi = \frac{\sqrt{1 - \zeta^2}}{\zeta} \tag{3.15}$$

The root locus gain is ω_n^2, and the graphical residue rule gives the following, where it is noted that angles clockwise from the positive real-axis direction are negative:

$$K_1 = \frac{\omega_n^2}{(\omega_n \ \angle -\phi)(\omega_n \ \angle +\phi)} = 1$$

$$K_2 = \frac{\omega_n^2}{(\omega_n \ \angle \pi - \phi)(2\omega_n\sqrt{1 - \zeta^2} \ \angle \pi/2)} = \frac{1}{2\sqrt{1 - \zeta^2}} \ \angle \phi - 3\pi/2$$

$$= \frac{1}{2\sqrt{1 - \zeta^2}} \ \angle \phi + \pi/2$$

$$K_3 = \frac{\omega_n^2}{(\omega_n \ \angle -\pi + \phi)(2\omega_n\sqrt{1 - \zeta^2} \ \angle -\pi/2)} = \frac{1}{2\sqrt{1 - \zeta^2}} \ \angle -\phi - \pi/2$$

Using $e^{j\pi/2} = j = -1/j$ and $e^{-j\pi/2} = -j = 1/j$, substitution into $c(t)$ yields

$$c(t) = 1 - \frac{1}{\sqrt{1 - \zeta^2}} e^{-\zeta\omega_n t} \frac{1}{2j} \{\exp [j(\omega_n\sqrt{1 - \zeta^2}t + \phi)]$$

$$- \exp [-j(\omega_n\sqrt{1 - \zeta^2}t + \phi)]\} \tag{3.16}$$

$$= 1 - \frac{1}{\sqrt{1 - \zeta^2}} e^{-\zeta\omega_n t} \sin (\omega_n\sqrt{1 - \zeta^2}t + \phi)$$

This verifies the entry in Table 1.6.1.

An alternative form will actually prove much more useful when the extension to systems with more poles is considered. It is obtained by noting that

$$K_2 = \frac{1}{2\sqrt{1 - \zeta^2}} \ \angle \theta \qquad K_3 = \frac{1}{2\sqrt{1 - \zeta^2}} \ \angle -\theta \qquad \theta = \phi + \frac{\pi}{2} \tag{3.17}$$

where θ is the phase angle of residue K_2. The solution is then seen to be

$$c(t) = 1 + \frac{1}{\sqrt{1 - \zeta^2}} e^{-\zeta\omega_n t} \cos (\omega_n\sqrt{1 - \zeta^2}t + \theta) \tag{3.18}$$

Distinct Real Poles and Complex Conjugate Pairs

For distinct poles, whether real or complex, the partial fraction expansion of $C(s) = G(s)R(s)$ and the corresponding solution $c(t)$ are

$$C(s) = \frac{K_1}{s + p_1} + \frac{K_2}{s + p_2} + \cdots + \frac{K_n}{s + p_n}$$

$$c(t) = K_1 \exp(-p_1 t) + K_2 \exp(-p_2 t) + \cdots + K_n \exp(-p_n t)$$

(3.19)

All residues can be found by the graphical rule. Those corresponding to real poles will be real. For complex conjugate pairs, if $-p_i$ and $-p_{i+1}$ form such a pair, K_i and K_{i+1} are also complex conjugate, as in (3.17), because all poles and zeros are real or occur in pairs which are complex conjugate. Hence let

$$p_i = \zeta_i \omega_{ni} - j\omega_{ni}\sqrt{1 - \zeta_i^2} \qquad K_i = \mathbf{K}_i e^{j\theta_i} = \mathbf{K}_i \,\underline{/\theta_i}$$

(3.20)

where $\mathbf{K}_i$ is the magnitude and θ_i the phase. Then

$$p_{i+1} = \zeta_i \omega_{ni} + j\omega_{ni}\sqrt{1 - \zeta_i^2} \qquad K_{i+1} = \mathbf{K}_i \,\underline{/-\theta_i}$$

and the corresponding terms in the inverse are:

$$K_i \exp(-p_i t) + K_{i+1} \exp(-p_{i+1} t)$$

$$= \mathbf{K}_i \exp(-\zeta_i \omega_{ni} t)\{\exp[j(\omega_{ni}\sqrt{1 - \zeta_i^2}\,t + \theta_i)]$$

$$+ \exp[-j(\omega_{ni}\sqrt{1 - \zeta_i^2}\,t + \theta_i)]\}$$

$$= 2\mathbf{K}_i \exp(-\zeta_i \omega_{ni} t)\cos(\omega_{ni}\sqrt{1 - \zeta_i^2}\,t + \theta_i)$$

(3.21)

Thus, in general, if $C(s)$ has k real poles $-p_i$ and m complex conjugate pairs of the form of (3.20), the total response is

$$c(t) = \sum_{i=1}^{k} K_{ri} \exp(-p_i t)$$

$$+ 2\sum_{i=1}^{m} \mathbf{K}_i \exp(-\zeta_i \omega_{ni} t)\cos(\omega_{ni}\sqrt{1 - \zeta_i^2}\,t + \theta_i)$$

(3.22)

Example 3.3.4

Find the unit step response of a system with transfer function

$$G(s) = \frac{0.89}{(0.5s + 1)(s^2 + s + 0.89)}$$

(3.23)

The roots of the quadratic are $(-0.5 \pm 0.8j)$, and Fig. 3.6 shows the pole–zero pattern of

$$C(s) = \frac{1.78}{s(s + 2)(s + 0.5 - 0.8j)(s + 0.5 + 0.8j)}$$

The root locus gain is 1.78, and Fig. 3.6 shows significant vector lengths and angles for determination of the residues. The residues corresponding to the poles indicated are, from the graphical rule,

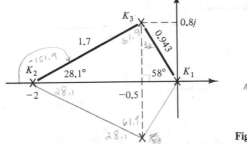

Figure 3.6   Example 3.3.4.

clockwise are ⊖

$$K_1 = \frac{1.78}{(2 \angle 0)(0.943 \angle -58°)(0.943 \angle 58°)} = 1$$

$$K_2 = \frac{1.78}{(2 \angle 180)(1.7 \angle 208.1)(1.7 \angle 151.9)} = 0.307 \angle 180° = -0.307$$

$$K_3 = \frac{1.78}{(1.7 \angle 28.1)(0.943 \angle 122)(1.6 \angle 90)} = 0.693 \angle 120$$

From (3.22), then, the response is

$$c(t) = 1 - 0.307e^{-2t} + 1.386e^{-0.5t} \cos(0.8t + 120°) \qquad (3.24)$$

since, using Fig. 3.5,

× 2 because 2 poles

$$-\zeta\omega_n = -0.5 = \text{real part of complex poles}$$

$$\omega_n\sqrt{1 - \zeta^2} = 0.8 = \text{imaginary part}$$

Note that $c(0) = 1 - 0.307 + 1.386 \cos 120° = 0$.

The denominator of $C(s) = G(s)R(s)$ and its partial fraction expansion contain terms due to the poles of input $R(s)$ and those of the system $G(s)$. The terms due to $R(s)$ yield the forced solution, corresponding to the particular integral solution of a differential equation, and the system poles give the transient solution. These parts of the solution may be identified in all examples thus far.

The system poles are real or occur in complex pairs, so the transient solution, which must decay to zero for the system to be useful, is the sum of the responses for these two types, called simple lag and quadratic lag.

These two basic types of systems will now be considered in turn, with emphasis on the correlations between the nature of the response and the pole positions in the s-plane.

3.4 SIMPLE LAG: FIRST-ORDER SYSTEMS

Chapter 2 has shown that this system, shown with its pole–zero pattern in Fig. 3.7, is very common. For a step input $R(s) = 1/s$,

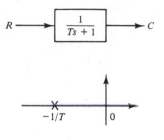

Figure 3.7 Simple lag.

$$C(s) = \frac{1/T}{s(s + 1/T)} = \frac{K_1}{s} + \frac{K_2}{s + 1/T}$$

$$K_1 = \frac{1/T}{s + 1/T}\bigg|_{s=0} = 1 \qquad K_2 = \frac{1/T}{s}\bigg|_{s=-1/T} = -1$$

Hence the transient response is

$$c(t) = 1 - e^{-t/T} \tag{3.25}$$

The first term is the forced solution, due to the input, and the second the transient solution, due to the system pole. Figure 3.8 shows this transient as well as $c(t)$. The transient is seen to be a decaying exponential. If it takes long to decay, the system response is slow, so the speed of decay is of key importance. The commonly used measure of this speed of decay is the time constant:

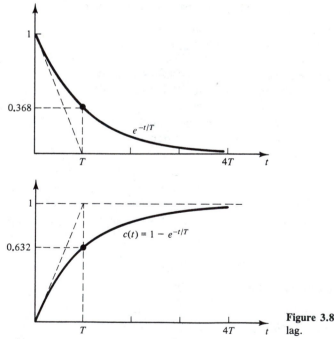

Figure 3.8 Step response of simple lag.

The *time constant* is the time in seconds for the decaying exponential transient to be reduced to $e^{-1} = 0.368$ of its initial value.

Since $e^{-t/T} = e^{-1}$ when $t = T$, it is seen that

> The time constant for a simple lag $1/(Ts + 1)$ is T seconds. This is, in fact, the reason a simple lag transfer function is often written in this form. The coefficient of s then immediately indicates the speed of decay. It takes $4T$ seconds for the transient to decay to 1.8% of its initial value. At $t = T$, $c(T) = 1 - 0.368 = 0.632$.

The values at $t = T$ provide one point for sketching the curves in Fig. 3.8. Since

$$\frac{d}{dt}(e^{-t/T})\bigg|_{t=0} = -\frac{1}{T}e^{-t/T}\bigg|_{t=0} = -\frac{1}{T}$$

it follows also that the curves are initially tangent to the dashed lines in Fig. 3.8. These two facts provide a good sketch of the response.

*****

Now consider the correlation between this response and the pole position at $s = -1/T$ in Fig. 3.7. The purpose of developing such insight is that it will permit the nature of the transient response of a system to be judged by inspection of its pole–zero pattern.

For the simple lag, two features are important:

1. *Stability:* If $-1/T$ is positive, the pole lies in the right half of the s-plane. The transient $e^{-t/T}$ then grows instead of decays as t increases. The system is unstable and useless. Hence the most important rule for design:

   > For system stability, the system pole(s) must lie in the left half of the s-plane.

2. *Speed of response:* To speed up the response of the system (i.e., to reduce its time constant T), the pole $-1/T$ must be moved left.

How such movement is to be achieved is a problem of design, considered later. Example 2.8.1 and comparison of Examples 3.2.1 and 3.3.2 suggest, however, the use of feedback around the simple lag plant, together with an adjustable gain.

3.5 QUADRATIC LAG: SECOND-ORDER SYSTEMS

This very common transfer function can always be reduced to the standard form

$$G(s) = \frac{\omega_n^2}{s^2 + 2\zeta\omega_n s + \omega_n^2} \tag{3.26}$$

where ω_n = undamped natural frequency
  ζ = damping ratio

The significance of these parameters will be discussed. For a unit step input $R(s) = 1/s$, the transform of the output is

$$C(s) = \frac{\omega_n^2}{s(s^2 + 2\zeta\omega_n s + \omega_n^2)} \tag{3.27}$$

For $\zeta < 1$, this is an entry in Table 1.6.1, verified in Example 3.3.3. However, there are three possibilities, depending on the roots of the system characteristic equation

$$s^2 + 2\zeta\omega_n s + \omega_n^2 = 0 \tag{3.28}$$

These system poles depend on ζ:

$$\zeta > 1: \quad \text{overdamped:} \ s_{1,2} = -\zeta\omega_n \pm \omega_n \sqrt{\zeta^2 - 1}$$

$$\zeta = 1: \quad \text{critically damped:} \ s_{1,2} = -\omega_n \tag{3.29}$$

$$\zeta < 1: \quad \text{underdamped:} \ s_{1,2} = -\zeta\omega_n \pm j\omega_n \sqrt{1 - \zeta^2}$$

Figure 3.9 shows the s-plane for plotting the pole positions. For $\zeta > 1$ these are on the negative real axis, on both sides of $-\omega_n$. For $\zeta = 1$, both poles coincide at $-\omega_n$. For $\zeta < 1$, the poles move along a circle of radius ω_n centered at the origin, as may be seen from the following expression for the distance of the poles to the origin:

$$|s_{1,2}| = [(\zeta\omega_n)^2 + (\omega_n\sqrt{1 - \zeta^2})^2]^{1/2} = \omega_n$$

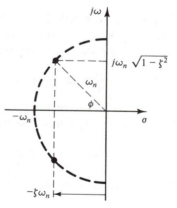

Figure 3.9 System poles quadratic lag.

From the geometry in Fig. 3.9, it is seen also that $\cos\phi = \zeta\omega_n/\omega_n = \zeta$. Hence

> The damping ratio $\zeta = \cos\phi$, where ϕ is the position angle of the poles with the negative real axis.

An angle of $\phi = 45°$ corresponds to $\zeta = 0.707$, angle $\phi = 60°$ to a damping ratio $\zeta = 0.5$.

For $\zeta > 1$, when the poles are real and distinct, the transient is a sum of two decaying exponentials, each with its own time constant.

Example 3.5.1

$$G(s) = \frac{2}{s^2 + 3s + 2} = \frac{2}{(s + 1)(s + 2)}$$

For a unit step,

$$C(s) = \frac{2}{s(s + 1)(s + 2)} = \frac{K_1}{s} + \frac{K_2}{s + 1} + \frac{K_3}{s + 2}$$

and it is found that

$$c(t) = 1 - 2e^{-t} + e^{-2t}$$

The nature of this result, that is, that the transient consists of exponentials with time constants $T_1 = 1$ and $T_2 = 0.5$, could have been predicted from inspection of the system pole–zero pattern in Fig. 3.10. The system is a series connection of two simple lags. The exponential corresponding to the pole closest to the origin has the largest time constant and takes longest to decay. This is called the *dominating pole*, and to increase the speed of response it would have to be moved to the left.

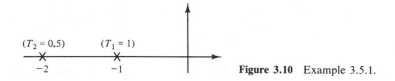

Figure 3.10 Example 3.5.1.

For $\zeta = 1$, a repeated root occurs at $-\omega_n$, and responses can be calculated as in Examples 3.3.1 and 3.3.2.

For $\zeta < 1$, the result in Table 1.6.1, verified by (3.15) and (3.16), applies:

$$c(t) = 1 - \frac{1}{\sqrt{1 - \zeta^2}} e^{-\zeta\omega_n t} \sin\left(\omega_n\sqrt{1 - \zeta^2}t + \tan^{-1}\frac{\sqrt{1 - \zeta^2}}{\zeta}\right) \qquad (3.30)$$

Figure 3.11 shows a normalized plot of this response for different values of the damping ratio ζ. The transient term is an oscillation of *damped natural frequency* $\omega_n\sqrt{1 - \zeta^2}$, of which the amplitude decays according to $e^{-\zeta\omega_n t}$. For an underdamped second-order system:

> The time constant T is the time in seconds for the amplitude of oscillation to decay to e^{-1} of its initial value: $e^{-\zeta\omega_n t} = e^{-1}$. Hence

$$T = 1/(\zeta\omega_n). \qquad (3.31)$$

Analogous to the simple lag, the amplitude decays to 2% of its initial value in $4T$ seconds.

*****

Important performance criteria of the response are identified in Fig. 3.12: *Settling time T_s* is the time required for the response to come permanently within a 5% or 2% band around the steady-state value.

$$T_s = 3T \ (5\%)$$
$$T_s = 4T \ (2\%) \qquad (3.32)$$

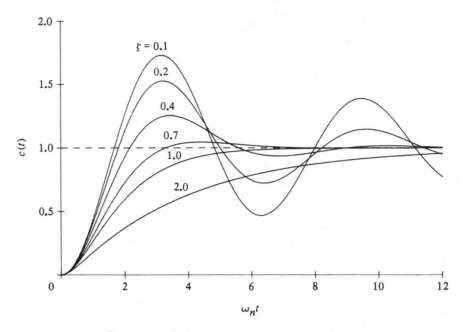

Figure 3.11 Unit step responses of a quadratic lag.

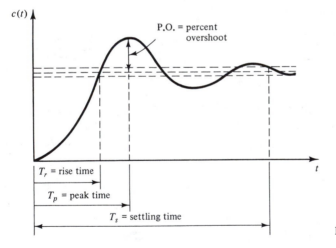

Figure 3.12 Performance criteria.

The maximum *percentage overshoot* (P.O.) over the steady-state response is a critical measure of performance. Equating the derivative of $c(t)$ in (3.30) to zero, to determine the extrema of the response, easily yields the equation

$$\tan \left(\omega_n \sqrt{1 - \zeta^2} t + \tan^{-1} \frac{\sqrt{1 - \zeta^2}}{\zeta} \right) = \frac{\sqrt{1 - \zeta^2}}{\zeta} \tag{3.33}$$

This implies that at the peaks

$$\omega_n \sqrt{1 - \zeta^2} t = i\pi \qquad i = 1, 3, \dots$$

since then left and right sides are equal. Hence the time at the maximum peak $(i = 1)$, the *peak time* T_p, is

$$T_p = \frac{\pi}{\omega_n \sqrt{1 - \zeta^2}} \qquad (3.34)$$

If the tan of the angle in (3.33) is $\sqrt{1 - \zeta^2}/\zeta$, its sin is $\pm \sqrt{1 - \zeta^2}$, and substituting (3.34) into (3.30) yields

$$\text{P.O.} = 100 \exp\left(\frac{-\pi\zeta}{\sqrt{1 - \zeta^2}}\right) \qquad (3.35)$$

The *rise time* T_r, identified in Fig. 3.12 as the time at which the response first reaches the steady-state level, is closely related to peak time T_p.

It is noted that while T_s, T_p, and T_r depend on both ω_n and ζ, P.O. depends only on the damping ratio ζ. Figure 3.13 shows a graph of P.O. versus damping ratio ζ. Permissible overshoot, and hence minimum acceptable ζ, depends on the application. For a machine tool slide, overshoot may cause the tool to gouge into the material being machined, so $\zeta \geq 1$ is required. But in most cases a limited overshoot is quite acceptable, and then $\zeta < 1$ is preferable, because it reduces peak time T_p in (3.34) and rise time T_r. For $\zeta = 0.7$ the overshoot is only 5% and Fig. 3.11 shows that the response approaches steady state much sooner.

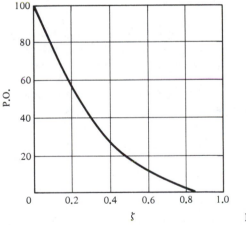

Figure 3.13 P.O. versus ζ.

If the damping ratio could be held constant while ω_n is increased, the poles would move radially outward and both settling time and rise time would decrease. An example is given to clarify how for complex poles both of these affect the system speed of response, and also to demonstrate what constraints may exist on how the poles of a closed-loop system can be adjusted.

Example 3.5.2   DC Motor Position Servo

Figure 3.14(a) shows the block diagram of the servo considered in Example 2.8.1. The closed-loop transfer function is

$$\frac{C(s)}{R(s)} = \frac{2K}{s^2 + 4s + 2K}$$

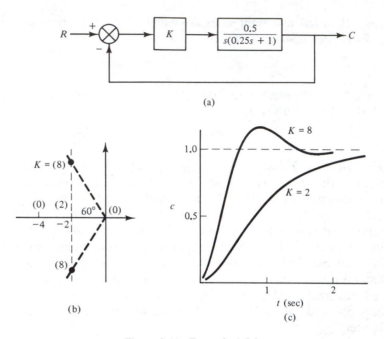

(a)

(b)

(c)

Figure 3.14   Example 3.5.2.

and the system poles are $s_{1,2} = -2 \pm \sqrt{4 - 2K}$. For $K < 2$ these poles lie along the negative real axis in the *s*-plane in Fig. 3.14(b). This corresponds to $\zeta > 1$, and the transient is a superposition of two decaying exponentials. (Figure 2.29 shows step responses for $K = 1$ and 1.9.)

For $K = 2$ both poles coincide at -2. By the technique used in Example 3.3.2, the unit step response, the inverse of $C(s) = 4/[s(s + 2)^2]$, is

$$c(t) = 1 - 2te^{-2t} - e^{-2t}$$

This is plotted in Fig. 3.14(c), where also the plot for $K = 8$ from Fig. 2.29 is repeated. Now, for $K = 8$ the poles are the roots of $s^2 + 4s + 16 = 0$, which corresponds to $\omega_n = 4$, $\zeta = 0.5$, and are located at $s_{1,2} = -2 \pm 3.464j$. The real part of the system poles is -2 for all $K \geq 2$, so that the time constant and settling time are the same for $K = 2$ and $K = 8$. Yet clearly the plot for $K = 8$ reflects a higher speed of response. The difference lies in the smaller peak time and rise time associated with the larger values of K and ω_n. The constraints on pole positions limit the poles to the real axis between 0 and -4 and the vertical at -2. This is, in fact, the root locus for the system, studied in Chapter 6.

As for the simple lag, it is again important to determine the correlations between dynamic behavior and the pole positions in the *s*-plane in Fig. 3.9:

1. *Absolute stability:* The real part of the pole positions is $-\zeta\omega_n$. If this is positive, then from (3.30) for $c(t)$, the transient will grow instead of decay due to $e^{-\zeta\omega_n t}$. Hence, for stability the poles must lie in the left half of the *s*-plane.

2. *Relative stability:* To avoid excessive overshoot and unduly oscillatory behavior, damping ratio ζ must be adequate. Since $\zeta = \cos\phi$, the angle ϕ may not be close to 90°.

3. *Settling time:* Time constant and settling time are reduced by increasing the real part of the pole positions.

4. *Frequency of transient oscillations* $\omega_n\sqrt{1 - \zeta^2}$: This frequency, also called the resonant frequency or damped natural frequency, equals the imaginary part of the pole positions.

5. *Undamped natural frequency* ω_n: This equals the distance of the poles to the origin. Moving the poles out radially (i.e., with ζ constant) reduces settling time, peak time, and rise time while the percentage overshoot remains constant.

6. *Speed of response:* For a constant real part, this is improved by increasing the imaginary part until ζ is reduced to a permissible level, thus reducing peak time and rise time.

3.6 PERFORMANCE AND STABILITY OF HIGHER-ORDER SYSTEMS

Higher-order systems arise all too easily. For example, in mechanical drives an electric or hydraulic motor may operate a rotating system that must be modeled by a number of inertias and interconnecting torsion springs, as in Example 2.2.3. Each added spring–inertia combination in effect adds two poles. An analogous situation exists for multimass translating, instead of rotating, systems along the lines of Example 2.2.2. Temperature control with multiple thermal capacitances as in Example 2.5.3 augmenting the dynamics of the heat source can also involve high-order transfer functions.

Although high-order dynamics are most often caused by the process, other elements in the loop also contribute. For example, a third-order transfer function was derived in Example 2.7.3 to model a valve-controlled hydraulic cylinder with allowance for the effect of oil compressibility. In this model the mechanical displacement of the spool valve was the input. However, in high-pressure hydraulic servos the forces required for positioning this valve are quite large and are generated instead by applying hydraulic pressures to the ends of the spool. These pressures are generated in a first stage of hydraulic power amplification, with the valve itself acting as the second stage. In two-stage electrohydraulic servos, used extensively in many areas of engineering for high-performance positioning of heavy loads, the first-stage amplifier often involves an electromagnetic torque motor, and the system input is an electrical signal. For high-performance design it may be necessary to allow for the effects of motor inertia and valve spool mass as well as oil compressibility, and the transfer function from input to valve spool position can be of sixth order, raising that from input to output from order 3 to 9.

Nature of the Transient Response and Dominating Poles

Whatever the order of the transfer function, it may be stated from the preceding sections and (3.19) that, since each real pole causes a decaying exponential transient and each complex pair a decaying oscillation:

> The *total transient response* is a superposition of exponential decays and decaying oscillations.

Repeated roots do not change this in an essential way. If all parameters are known, the response can be calculated, but its nature can also be judged without this by inspection of the pole positions. During design the parameters are not known and the aim is to use feedback which locates the poles in regions corresponding to satisfactory dynamics, that is, not too close to the imaginary axis and at a small enough angle to the negative real axis.

The dominating-poles concept is very important in this connection and simplifies design greatly. The response of many systems is dominated by one pair of complex poles relatively close to the imaginary axis. Design can therefore concentrate on locating this dominating pair satisfactorily. The fact that many systems behave approximately as second-order systems is also the reason the performance criteria for second-order systems discussed in Section 3.5 apply to higher-order systems as well.

Absolute and Relative Stability

The foregoing relative stability conditions on the locations of the dominating poles are much more stringent than those of absolute stability. *Absolute stability* requires only that all roots of the system characteristic equation

$$a_n s^n + a_{n-1} s^{n-1} + \cdots + a_1 s + a_0 = 0 \qquad (3.36)$$

(i.e., the poles of its transfer function) lie in the left-half *s*-plane. It is known that if any of the coefficients are zero or if not all coefficients have the same sign, there will be roots on or to the right of the imaginary axis. If all coefficients are present and have the same sign, which can be taken to be positive without loss of generality, the Routh–Hurwitz criterion discussed in the next section provides a quick method for determining absolute, but not relative, stability from the coefficients, without calculating the roots.

Computer-Aided Analysis and Design

For high-order systems and for routine work on those of lower order, computational aids are indispensable for analysis and design. For example, the techniques of Section 3.3 for transient response calculation clearly become laborious for high-order systems. Computer aids can range from batch-type programs for specific purposes, such as finding the roots of a characteristic equation or calculating the response for a given transfer function, to interactive analysis and design packages which include computer graphics.

Because computer aids are not essential to the development of concepts and techniques, it is desirable to allow for freedom of choice as to the degree to which such aids are exploited in the study of the subject. Computational aids, with examples of their use, have therefore been collected in Appendix B, and reference to particular programs will be made at appropriate points in the text.

In the context of the present chapter, Appendix B gives a program for transient response computation based on partial fraction expansion, assuming distinct poles. For closed-loop systems, the poles needed for this must usually be found from the system characteristic equation (3.36). Appendix B gives a routine for finding the roots of polynomials that can be used for this purpose.

3.7 ROUTH–HURWITZ STABILITY CRITERION

The Routh–Hurwitz stability criterion is a method for determining from the coefficients of the characteristic equation (3.36) how many system poles are in the right-half s-plane or on the imaginary axis.

Of the *Routh array* below, the first two rows are produced by arranging the coefficients a_n, . . ., a_0 in the order indicated by the arrows.

$$
\begin{array}{llcccc}
s^n: & a_n & a_{n-2} & a_{n-4} & \cdots & 0 \\[4pt]
s^{n-1}: & a_{n-1} & a_{n-3} & a_{n-5} & \cdots & 0 \\[2pt]
s^{n-2}: & b_1 & b_2 & b_3 & \cdots & 0 \\[2pt]
s^{n-3}: & c_1 & c_2 & c_3 & \cdots & 0 \\
& \vdots & & & & \\
s^1: & g_1 & 0 & & & \\
s^0: & h_1 & 0 & & &
\end{array}
\tag{3.37}
$$

Each of the remaining entries b_i, c_i, . . . is found from the two rows preceding it according to a pattern that can be recognized from the following equations:

$$
b_1 = \frac{-1}{a_{n-1}} \begin{vmatrix} a_n & a_{n-2} \\ a_{n-1} & a_{n-3} \end{vmatrix} = \frac{1}{a_{n-1}}(a_{n-1}a_{n-2} - a_n a_{n-3})
$$

$$
b_2 = \frac{-1}{a_{n-1}} \begin{vmatrix} a_n & a_{n-4} \\ a_{n-1} & a_{n-5} \end{vmatrix} = \frac{1}{a_{n-1}}(a_{n-1}a_{n-4} - a_n a_{n-5})
$$

$$
b_3 = \frac{-1}{a_{n-1}} \begin{vmatrix} a_n & a_{n-6} \\ a_{n-1} & a_{n-7} \end{vmatrix} = \frac{1}{a_{n-1}}(a_{n-1}a_{n-6} - a_n a_{n-7})
\tag{3.38}
$$

$$c_1 = \frac{-1}{b_1} \begin{vmatrix} a_{n-1} & a_{n-3} \\ b_1 & b_2 \end{vmatrix} = \frac{1}{b_1}(b_1 a_{n-3} - b_2 a_{n-1})$$

$$c_2 = \frac{-1}{b_1} \begin{vmatrix} a_{n-1} & a_{n-5} \\ b_1 & b_3 \end{vmatrix} = \frac{1}{b_1}(b_1 a_{n-5} - b_3 a_{n-1})$$

Calculations in each row are continued until only zero elements remain. In each of the last two rows the second and following elements are zero. It can be shown that the elements in any row can be multiplied by an arbitrary positive constant without affecting the results below. This can be useful to simplify the arithmetic. For large n, computer algorithms can be written, based on (3.38).

The *Routh–Hurwitz criterion* states:

1. A necessary and sufficient condition for stability is that there be no changes of sign in the elements of the first column of the array (3.37).
2. The number of these sign changes is equal to the number of roots in the right-half s-plane.
3. If the first element in a row is zero, it is replaced by a very small positive number ε, and the sign changes when $\varepsilon \to 0$ are counted after completing the array.
4. If all elements in a row are zero, the system has poles in the right-half plane or on the imaginary axis.

Example 3.7.1

$$s^5 + s^4 + 6s^3 + 5s^2 + 12s + 20 = 0$$

Using (3.37) and (3.38), the Routh array is

s^5:	1	6	12	0
s^4:	1	5	20	0
s^3:	1	-8	0	
s^2:	13	20	0	
s^1:	$\dfrac{-124}{13}$	0		
s^0:	20	0		

There are two sign changes in the first column, first from plus to minus and then from minus to plus. Hence this characteristic equation represents an unstable system, with two poles in the right-half s-plane.

Example 3.7.2

$$s^5 + s^4 + 5s^3 + 5s^2 + 12s + 10 = 0$$

The zero element in the first column for s^3 in the Routh array is replaced by a small positive number ε. Then

$$c_1 = \frac{5\varepsilon - 2}{\varepsilon} \approx \frac{-2}{\varepsilon} \qquad d_1 = \frac{2c_1 - 10\varepsilon}{c_1} \approx 2$$

$$
\begin{array}{llll}
s^5: & 1 & 5 & 12 & 0 \\
s^4: & 1 & 5 & 10 & 0 \\
s^3: & \varepsilon & 2 & 0 \\
s^2: & c_1 & 10 & 0 \\
s^1: & d_1 & 0 \\
s^0: & 10 \\
\end{array}
$$

Thus c_1 is a large negative number, implying two unstable poles, as in Example 3.7.1.

Example 3.7.3

For the electromechanical servo modeled in Fig. 3.15, find the limits on K for stability. The characteristic equation is $s^3 + 3s^2 + 2s + K = 0$ and the Routh array is

$$
\begin{array}{lll}
s^3: & 1 & 2 \quad 0 \\
s^2: & 3 & K \quad 0 \\
s^1: & \dfrac{6 - K}{3} & 0 \\
s^0: & K & 0 \\
\end{array}
$$

Evidently, $0 < K < 6$ is the range of K for a stable system.

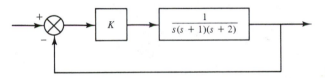

Figure 3.15   Example 3.7.3.

At both limits in Example 3.7.3, one row of the Routh array consists of only zero elements. It can be shown that this means that root pairs are present which are located symmetrically about the origin, usually on the imaginary axis. It can also be shown that such root pairs are the roots of an *auxiliary equation* formulated from the elements in the row of the array which precedes that with the zeros.

For Example 3.7.3, at the upper limit $K = 6$ the auxiliary equation is $3s^2 + 6 = 0$. The elements for row s^1 are zero, so it is formed from those for s^2. The highest power in the auxiliary equation is generally that of the row, and the powers of successive terms reduce by 2.

The roots of the auxiliary equation are $s = \pm j\sqrt{2} = \pm 1.414j$. So at the limit $K = 6$, system poles occur at these points on the imaginary axis. When several parameters vary, these techniques can be used to determine relations to be satisfied to ensure system stability.

3.8 EFFECT OF SYSTEM ZEROS

Much attention has been given to system poles and the correlations between their positions in the s-plane and the nature of the transient response. A natural question is therefore what correlations exist between the response and the positions of the system zeros. The unit step response of a system with transfer function

$$G(s) = \frac{\omega_n^2}{z_1} \frac{s + z_1}{s^2 + 2\zeta\omega_n s + \omega_n^2} \tag{3.39}$$

will be studied to examine the effect of adding a zero to an underdamped quadratic lag.

The pole–zero pattern of $C(s) = G(s)R(s)$ is shown in Fig. 3.16. The residues K_1 and $(\mathbf{K_2} \ \angle\phi_2)$ in the solution $c(t)$ given by (3.22) are, by the graphical rule,

$$K_1 = \frac{\omega_n^2}{z_1} \frac{z_1}{(\omega_n \ \angle -\phi)(\omega_n \ \angle \phi)} = 1$$

$$K_2 = \frac{\omega_n^2}{z_1} \frac{A \ \angle\alpha}{(\omega_n \ \angle \pi - \phi)(2\omega_n\sqrt{1 - \zeta^2} \ \angle \pi/2)} = \frac{A}{2z_1\sqrt{1 - \zeta^2}} \ \angle\alpha + \phi - 3\pi/2$$

and hence, by (3.22),

$$c(t) = 1 + \frac{A}{z_1\sqrt{1 - \zeta^2}} e^{-\zeta\omega_n t} \cos\left(\omega_n\sqrt{1 - \zeta^2}t + \phi + \alpha - \frac{3\pi}{2}\right) \tag{3.40}$$

$$= 1 - \frac{A}{z_1\sqrt{1 - \zeta^2}} e^{-\zeta\omega_n t} \sin(\omega_n\sqrt{1 - \zeta^2}t + \phi + \alpha)$$

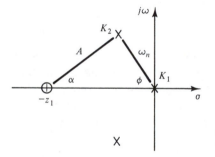

Figure 3.16   Pole–zero pattern of $C(s) = G(s)/s$ for (3.39).

The geometry of Fig. 3.16 shows that if z_1 is large compared to ω_n, then $A \approx z_1$ and $\alpha \approx 0$. Equation (3.40) then reduces to the response for a quadratic lag given by (3.30). Thus a zero far along the axis has little effect on the transient response of the quadratic lag. But if the zero is moved to the right, it will cause a gradually increasing percentage overshoot (P.O.) and the peak time T_p will reduce, as indicated in Fig. 3.17. The proof is analogous to that of (3.33) to (3.35) for the quadratic lag: With $\phi = \tan^{-1}(\sqrt{1 - \zeta^2}/\zeta)$, equating the derivative of $c(t)$ in (3.40) to zero yields

$$\tan[\omega_n\sqrt{1 - \zeta^2}t + \alpha + \tan^{-1}(\sqrt{1 - \zeta^2}/\zeta)] = \frac{\sqrt{1 - \zeta^2}}{\zeta} \tag{3.41}$$

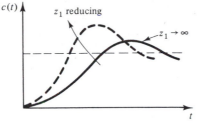

Figure 3.17 Effect of a zero.

At the extrema, $\omega_n\sqrt{1 - \zeta^2}t + \alpha = i\pi$, since then left and right sides are equal. The time T_p for the maximum peak ($i = 1$) is

$$T_p = \frac{\pi - \alpha}{\omega_n\sqrt{1 - \zeta^2}} \tag{3.42}$$

Since the tan of the angle in (3.41) is $\sqrt{1 - \zeta^2}/\zeta$, its sin is $\pm\sqrt{1 - \zeta^2}$, and using this and (3.42) in (3.40) yields the maximum overshoot:

$$\text{P.O.} = 100\frac{A}{z_1}\exp\left[-\frac{(\pi - \alpha)\zeta}{\sqrt{1 - \zeta^2}}\right] \tag{3.43}$$

Here A and α can be measured or calculated from Fig. 3.16. Examples:

1. If the zero is moved in from far left to the position where $\alpha = \pi/2$, (3.42) shows that the peak time is reduced by half.
2. With $\alpha = \pi/2$ and $\phi = 60$ ($\zeta = 0.5$), in Fig. 3.16 $A/z_1 = 1.732$, and (3.43) yields an overshoot of almost 70%, a large increase over the 16% when the zero is far away.

The effect of zeros can be quite significant even if the ratio A/z_1 does not exceed 1. This is demonstrated in the next example, in which a number of feedback systems of the same structure are designed to have the same (closed-loop) poles and different zeros.

Example 3.8.1

Figure 3.18(a) shows a control system with a simple lag process. The *PI controller* represents an extremely common form of control. The steady-state error for step inputs may be shown to be zero with this controller, by use of the final value theorem on $E(s)$ or $C(s)$. With a pure gain controller, this error would only approach zero for large values of gain. Except when the error E has become quite small, this would mean large process input signals (i.e., large flow rates of fluids or heat or current), implying a high cost of control.

Figure 3.18(a) could model a single-tank level control (Example 2.6.1), a single-capacitance temperature control (Example 2.5.2), a pneumatic pressure control (Example 2.6.3), or a speed control servo, among other possibilities. For a motor speed control, the system output transform is $s\theta(s)$, the transform of $\dot{\theta}(t)$, and the motor transfer function in (2.11) reduces to a simple lag. The feedback sensor

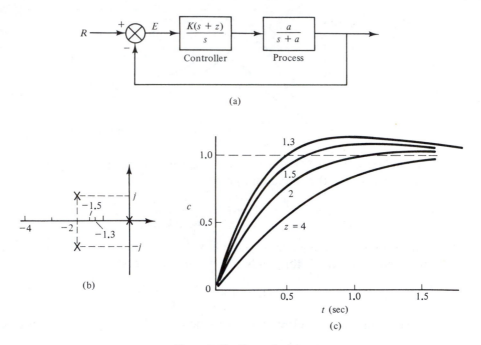

Figure 3.18   Example 3.8.1.

could be a small generator, giving a feedback voltage proportional to shaft speed $\dot{\theta}(t)$.

The closed-loop transfer function in Fig. 3.18(a) is

$$\frac{C(s)}{R(s)} = \frac{Ka(s + z)}{s^2 + a(1 + K)s + Kaz}$$

It is desired to make the closed-loop system poles equal to $-2 \pm j$; that is, the denominator of C/R should be $s^2 + 4s + 5$, or

$$a(1 + K) = 4 \qquad Kaz = 5$$

Systems for which $z = 4$, 2, 1.5, and 1.3, respectively, will then result for $Ka = 1.25$, 2.5, 3.3333, and 3.8462, with $a = 2.75$, 1.5, 0.6667, and 0.1538. For a unit step input $R(s) = 1/s$, the output transform is

$$C(s) = \frac{Ka(s + z)}{s(s^2 + 4s + 5)} = \frac{5(1 + s/z)}{s(s^2 + 4s + 5)}$$

The pole–zero pattern is shown in Fig. 3.18(b), and the graphical residue rule readily yields

$$c(t) = 1 + Ae^{-2t} \cos (t + \theta)$$

with the sets of values for (z, A, θ) equal to $(4, 1.25, -216.9°)$, $(2, 1.118, -153,4°)$, $(1.5, 1.667, -126.9°)$, and $(1.3, 2.1, -118.5°)$.

These responses are plotted in Fig. 3.18(c). The system poles at $(-2 \pm j)$ correspond to a damping ratio 0.894, for which the overshoot is negligible and the peak time, from (3.34), is 3.13 sec. Moving the zero in from the left is seen to reduce the peak time and increase the overshoot. For $z = 1.3$ the maximum overshoot is about 14% and occurs at about 1 sec. Thus care must be exercised in using the correlations between response and pole positions for a quadratic lag if zeros are present in relatively dominant locations.

The graphical technique for determining residues provides a particularly enlightening explanation for the effect of zeros.

Example 3.8.2

Find the unit step response of the system

$$G(s) = \frac{4}{a} \frac{s + a}{(s + 1)(s + 4)} \tag{3.44}$$

The pole–zero pattern of $C(s) = G(s)R(s)$ is shown in Fig. 3.19. The residues are

$$K_1 = \frac{(4/a)a}{(1)(4)} = 1$$

$$K_2 = \frac{(4/a)(a - 1)}{(-1)(3)} = \frac{-4(a - 1)}{3a}$$

$$K_3 = \frac{(4/a)(-4 + a)}{(-3)(-4)} = \frac{a - 4}{3a}$$

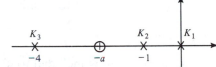

Figure 3.19   Example 3.8.2.

So the unit step response is

$$c(t) = 1 - \frac{4(a - 1)}{3a} e^{-t} + \frac{a - 4}{3a} e^{-4t} \tag{3.45}$$

It is seen that if $a = 1$, so if the zero coincides with, or cancels, the pole at -1, the transient corresponding to that pole is zero. Similarly, if $a = 4$, the zero cancels the pole at -4, making the corresponding transient zero.

The following remarks generalize the implications of this example for the significance of the zeros in the system pole–zero pattern:

1. A residue K_i at pole $-p_i$ corresponds to a transient term $K_i \exp(-p_i t)$, so the significance of the residue is that its magnitude is the initial size of the transient corresponding to the pole.
2. If a zero is close to a pole, the residue at the pole tends to be small, because of a short vector in the numerator, so that the corresponding transient is probably small.
3. If the zero coincides with the pole, it cancels it, and the transient term is zero.

4. Thus, the significance of zeros is that they affect the residues at the poles, and hence the sizes of the corresponding transients.
5. A "slow" pole (close to the imaginary axis, so with a large time constant) or a highly oscillatory pair (small damping ratio) may be acceptable if nearby zeros make the corresponding transients small.

Because pole–zero cancellation is so commonly used in design, it is important to observe that the response to initial conditions is not affected. The preceding results were based on the system transfer function, with the implied assumption of zero initial conditions. In Example 3.8.2, cross multiplication in $G(s) = C(s)/R(s)$ yields $(s^2 + 5s + 4)C = (4 + 4s/a)R$. For $r = 0$, the corresponding differential equation is $\ddot{c} + 5\dot{c} + 4c = 0$. Analogous to Example 1.7.1, transformation for nonzero initial conditions gives

$$C(s) = \frac{(s + 5)c(0) + \dot{c}(0)}{s^2 + 5s + 4}$$

This initial condition response is unaffected by the numerator of $G(s)$, so by any cancellations which may have been achieved in the input–output response.

*****

It is useful to clarify why the addition of a faraway zero does not greatly increase the sizes of residues and transients. By the final value theorem, the steady-state response of a system $G(s) = K/[(s + a)(s + b)]$ to a unit step is

$$c_{ss} = \lim_{s \to 0} sG(s)R(s) = \frac{K}{ab} \tag{3.46}$$

If a zero factor $(s + z)$ or pole factor $(s + p)$ is included to change the dynamic behavior, K must be multiplied by p/z if the steady state is not to be affected also; that is, $G(s)$ should be changed to

$$G_1(s) = K\frac{p}{z}\frac{s + z}{(s + p)(s + a)(s + b)} \tag{3.47}$$

A large vector $(s + z)$, say, now in effect becomes $(1 + s/z)$ and does not cause large residues.

This, as did the discussion below (3.40), again shows that the effect of faraway zeros on the transient response is small. The same is true for faraway poles, at which the residues will tend to be small. Such poles and zeros, except for their steady-state effects, are therefore often neglected on the many occasions when transfer function simplification is desirable for system analysis and design. It is fortunate that this is so, because for many systems the values of such faraway poles and zeros, and even their presence and the precise form of the model, constitute the most uncertain part of the transfer function.

3.9 CONCLUSION

In this chapter the dynamic response characteristics corresponding to given transfer functions have been studied. The features of the response have been interpreted in terms of the locations of system poles and zeros in the s-plane.

Note, however, that except in the examples little attention has been given to feedback and how it may be used to modify performance. Also, transient response calculations and stability determination, except by the Routh–Hurwitz criterion, of feedback systems with higher-order characteristic polynomials still requires discussion of techniques for finding their roots, the system poles. This is considered in Chapter 6, after study of the modeling, the performance, and the dynamic compensation of feedback systems.

PROBLEMS

3.1. Plot the pole–zero patterns of systems described by the following transfer functions.

(a) $\dfrac{K(2s + 1)}{s(4s + 1)(s + 3)}$

(b) $\dfrac{K(s^2 + 2s + 2)}{(s + 2)(s^2 + 2s + 10)}$

What are the root locus gains of these transfer functions?

3.2. The pole–zero pattern of a system with root locus gain 80 is shown in Fig. P3.2. What is its transfer function?

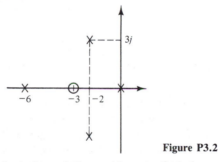

Figure P3.2

3.3. Use graphical calculation of the residues to find the unit step response of the following systems.

(a) $G(s) = \dfrac{4}{(2s + 1)(4s + 8)}$

(b) $G(s) = \dfrac{4(s + 1)}{(2s + 1)(4s + 8)}$

3.4. For the system

$$G(s) = \frac{2(s^2 + 9s + 19)}{s^2 + 6s + 8}$$

(a) What is the system characteristic polynomial?

(b) What is its characteristic equation?

(c) Find the system poles and zeros.

(d) Plot the system pole–zero pattern.

(e) Plot the pole–zero pattern of $C(s) = G(s)R(s)$ if the input $r(t)$ is a decaying exponential e^{-t}.

(f) Find $c(t)$ by graphical residue determination.

3.5. Find the unit ramp response of the system in Problem 3.3(a).

3.6. Find the unit ramp response for Problem 3.3(b).

3.7. Calculate the response of the system

$$G(s) = \frac{4}{2s^2 + 2s + 8}$$

to a unit ramp input. Note that the graphical residue rule can still be used at poles of $C(s) = G(s)R(s)$ other than the repeated pole, since moving the two poles slightly apart should not affect such results materially.

3.8. Calculate the unit step response of

$$G(s) = \frac{1}{(s + 2)^2(s + 1)}$$

3.9. Calculate the unit step response of the system

$$G(s) = \frac{54}{(2s + 6)(s^2 + 3s + 9)}$$

3.10. Calculate the unit step response for

$$G(s) = \frac{1.5(2s + 6)}{s^2 + 3s + 9}$$

3.11. A system is given by its transfer function

$$G(s) = \frac{5(1 - 0.4s)}{(s + 1)(0.2s + 1)}$$

(a) What are the time constants of the components of its transient response, and how long does it take for the transient to decay almost completely?

(b) The unit step response is readily found to be

$$c(t) = 5 - 8.75e^{-t} + 3.75e^{-5t}$$

Use the alternative definitions of a time constant to obtain reasonable sketches of the components of the response, and use these to sketch $c(t)$.

3.12. What should be the time constant of a simple lag system if it is specified that in 1 sec the transient must have reduced to half its initial value?

3.13. Calculate the unit step response of the system $G = 4/(s^2 + 5s + 4)$ and use the definitions of a time constant to sketch the transient components and to obtain a reasonable sketch of the overall response.

3.14. A system with transfer function

$$G(s) = \frac{10}{s^2 + s + 5}$$

is subjected to a step input.

(a) Plot the system pole–zero pattern and that of the system output.

(b) From the system pattern, find the undamped natural frequency of the transient, the damping ratio ζ, the frequency of oscillations, and the time constant of response.

(c) Determine the percentage peak overshoot and the time at which it occurs.

(d) Find the unit step response.

3.15. In Problem 3.14, what is the effect of halving the imaginary part of the system poles on:

(a) Settling time and time constant?

(b) ω_n and ζ?

(c) The number of oscillations during the decay (i.e., on how oscillatory the response is)?

(d) Percentage overshoot, peak time, and rise time?

(e) Discuss the differences with Problem 3.14.

3.16. From the results of Chapter 2, Fig. P3.16 could represent a simple motor position control system, with motor input being the amplified error between desired and actual shaft positions.

(a) Find C/R, E/R, and the characteristic equation.

(b) Use the final value theorem to express the steady-state shaft position error e_{ss} following unit step and unit ramp inputs. How does the value of K affect these errors?

(c) Find the value of K for which the system poles, the roots of the characteristic equation, will have a damping ratio $\zeta = 0.7$.

(d) How does the percentage overshoot for a step input change if K is raised above the value in part (c)?

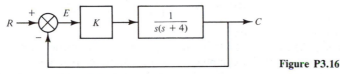

Figure P3.16

3.17. In Problem 3.16:

(a) Plot on one s-plane the system pole positions for $K = 0, 1, 4, 8,$ and 13.

(b) For the last four, find the percentage overshoot in response to a step input.

(c) Calculate the unit step responses for $K = 1, 4,$ and 8.

3.18. (a) Plot the required locations of the dominating pair of poles of a system if the damping ratio is to be about 0.5 and the time constant about 0.1.

(b) Show permissible pole locations if the system time constant may not exceed 1 sec.

(c) Similarly to part (b), if the damping ratio may not be less than about 0.7.

(d) Similarly to part (b), if conditions (b) and (c) must both be met.

3.19. Fig. P3.19 shows the pole–zero pattern of the system

$$G(s) = \frac{K}{(s + a)(s^2 + 2s + 2)}$$

for $a = 0.5, 1,$ and 2.

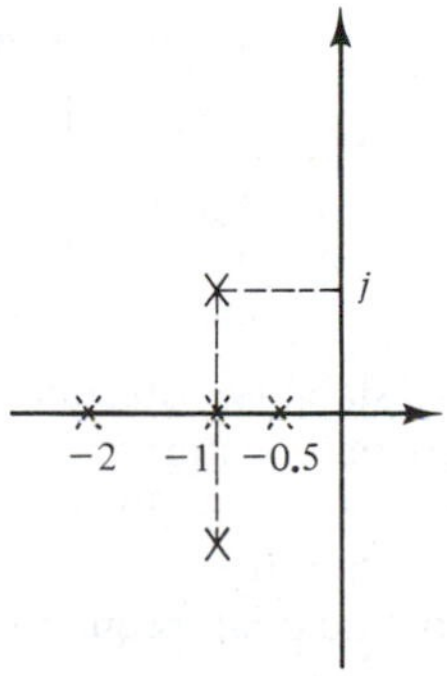

Figure P3.19

(a) Calculate the unit step responses for $K = 2a$.
(b) Compare these responses to gain an idea of the effect of the position of the real pole relative to the complex pair on the relative dominance of the transient terms, considering both size and speed of decay.

3.20. The pole–zero pattern of a system with root locus gain $3/a$ is shown in Fig. P3.20. The zero $-a$ lies between -1 and -3. Using graphical residue calculation:
(a) Calculate the unit step response.
(b) What happens if the zero is moved very close to either system pole? Which zero position is preferable, and why?

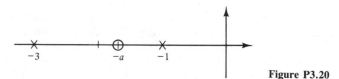

Figure P3.20

3.21. A system has the transfer function

$$G(s) = 4.8 \frac{s + 0.25}{(s + 0.3)(s^2 + 2s + 5)}.$$

(a) Plot its pole–zero pattern.
(b) What are the time constants or undamped natural frequency and damping ratio, as appropriate, of the components of the transient response to a step input? How long does the slowest component take to decay almost completely?
(c) Which component would you expect to dominate, and why?

3.22. For a system with root locus gain $\frac{2}{3}$ and pole–zero pattern as shown in Fig. P3.22:
(a) Express the transform of the system output for a unit step input.
(b) Express the residues for this input by the graphical rule and find their values.
(c) Express the unit step response.
(d) Which transient component dominates, and why?
(e) To verify part (d), calculate the value of time at which the exponential decay term and the amplitude of the oscillatory term are equal, and find this amplitude.

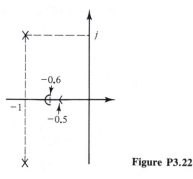

Figure P3.22

3.23. Use the Routh–Hurwitz stability criterion to determine the stability of systems with the following characteristic equations.

(a) $s^4 + 10s^3 + 33s^2 + 46s + 30 = 0$ (b) $s^4 + s^3 + 3s^2 + 2s + 5 = 0$
(c) $s^3 + 2s^2 + 3s + 6 = 0$ (d) $s^4 + 3s^3 + s^2 + 3s + 5 = 0$

For part (c), express the auxiliary equation and find one pair of roots.

3.24. For the system of Fig. P3.24 with $G(s) = 1/[s(s + 1)(s + 4)]$, use the Routh–Hurwitz criterion to find the limit on K for stability. At this limit, determine the position of the system poles on the imaginary axis of the s-plane.

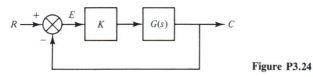

Figure P3.24

3.25. In Problem 3.24, determine the limit on the value of K for stability if the amplifier K is replaced by a dynamic compensator

$$\frac{K(0.5s + 1)}{0.1s + 1}$$

Where on the imaginary axis is a root pair located at this limit?

3.26. For the system in Fig. P3.24 with $G(s) = (s + 1)/(s + 2)$:
    (a) Find the value of K required for a system time constant $T = 0.667$ sec.
    (b) Plot the system pole–zero pattern for this K.
    (c) Calculate the corresponding unit step response.

3.27. In Fig. P3.24 with $G(s) = (s + 1)/[(s + 2)(s + 20)]$:
    (a) Find K so that the dominant system time constant will be $T = 0.667$ sec, and for this K also determine the second pole of the system.
    (b) Calculate the unit step response for K of part (a).
    (c) Compare this problem and the results with Problem 3.26, and discuss the observations.

3.28. The system in Fig. P3.24 with the simple lag plant $G(s) = 1/(s + 5)$ could represent, among others, a temperature, pressure, level, or speed control system.
    (a) Plot the root loci for varying gain K, that is, the loci showing how the closed-loop system pole changes with changing K.
    (b) Find the positions of the pole along the loci for $K = 10$ and $K = 45$.
    (c) From these positions, predict how the transient response is affected by the increase of K.
    (d) Compare this prediction with the step response results calculated in Problem 2.28.

3.29. Repeat Problem 3.28 for the same values of K if the plant model is refined to $G(s) = 20/[(s + 5)(s + 20)]$, for which step responses were calculated and plotted in Problem 2.29. For what value of K does the nature of the response change, and how does it change?

FEEDBACK SYSTEM MODELING
AND PERFORMANCE

4.1 INTRODUCTION

The first part of this chapter is concerned with the modeling of feedback systems, and the second with the motivations for the use of feedback and its effect on performance. The effect of changes of gain in the feedback loop on performance, and its limitations, are considered and will motivate the discussion of dynamic system compensation in Chapter 5.

As suggested earlier, the block diagram structure of a system may be more or less immediately evident from the system schematic diagram or the nature and even the existence of feedback may be rather difficult to see by inspection. In the latter case in particular, the derivation of a "good" block diagram, which clearly identifies the feedback, is an important aid in system analysis and design. Both types are considered through examples, beginning with the first.

4.2 FEEDBACK SYSTEM MODEL EXAMPLES

In this section examples are given where the system structure is rather evident from the schematic diagram.

Example 4.2.1   Water Level Control System

For a first example, it is appropriate to return to the level control system in Chapter 1. It would probably operate as a process control or regulator system; that is, the desired level is usually constant and the actual level must be held near it despite disturbances. The model should therefore allow for water supply pressure variations, probably the main disturbance.

A simplified schematic diagram is shown in Fig. 4.1. The level c is measured by means of a float and a lever is used as a summing junction to determine a measure e of the error with the desired level r. From this mechanical input, the controller and amplifier sets a pneumatic output pressure P_0 of sufficient power to operate the pneumatic actuator which adjusts the control valve opening x to control inflow q of the tank.

All subsystem transfer functions needed except that of the controller have been found in Chapter 2:

Mechanical lever (Fig. 2.1)

Pneumatic valve actuator (Fig. 2.23)

Control valve, with allowance for disturbances in supply pressure P_s (Fig. 2.25)

Tank and outflow valve [Equation (2.38)]

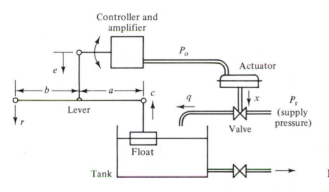

Figure 4.1   Water level control.

With these transfer functions and blocks, the translation of Fig. 4.1 to the block diagram in Fig. 4.2 is virtually immediate. The controller and amplifier is an off-the-shelf instrument, discussed in Chapter 5.

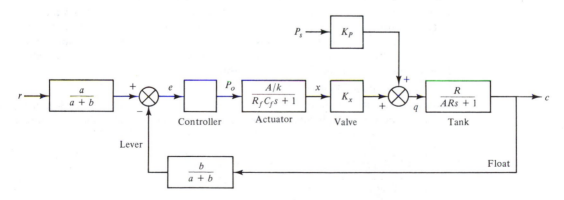

Figure 4.2   Water level control block diagram.

Example 4.2.2   Hydraulic Servo with Mechanical Feedback: Simple Model

In Fig. 4.3(a) the mechanical lever added to the hydraulic cylinder control in Fig. 2.26 acts as feedback because it causes piston motion c to affect valve position x. If, say, input r is moved to the right initially, the valve moves right, causing the piston to move left until the valve is again centered. Use of the simple valve–cylinder model in Fig. 2.27, which assumes small loads and neglects oil compressibility, now readily leads to the block diagram in Fig. 4.3(b).

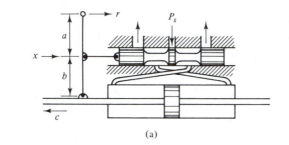

(a)

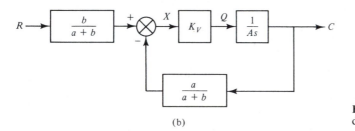

(b)

Figure 4.3 Hydraulic servo and block diagram.

Example 4.2.3   Simple Motor Position Servo

A simple motor position servo is shown in Fig. 4.4. The motor and its load are represented by the transfer function (2.11) for a field-controlled dc motor. The time constant $T_m = J/B$, where J is the inertia and B the damping constant. A potentiometer could be used to represent shaft position by a voltage, and an operational amplifier, discussed in Chapter 5, could serve as a summing junction to determine the error voltage E with the voltage R representing desired position. Alternatively, electrical bridge circuits could be used. The controller G_c may be a simple amplifier and generates a low-power output voltage V_c. Its power is raised in a power amplifier of which the output is applied to the motor.

Example 4.2.4   Servo with Velocity Feedback

In speed or position control servomechanisms such as Example 4.2.3, design for satisfactory performance is often complicated by a lack of adequate inherent damping

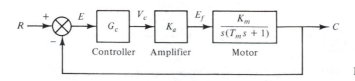

Figure 4.4 Motor position servo.

in motor and load. The difficulty of positioning a large inertia J rapidly without severe overshoot in response to a step input in Fig. 4.4 if the damping constant B is small can be appreciated intuitively. One possible solution is to install a mechanical damper on the motor shaft. However, a better and more elegant solution is possible by the use of feedback. A damping torque is a torque proportional to shaft speed $\dot{c}$ and in the opposite direction. Such a torque can also be generated by mounting a small tachometer–generator on the motor shaft to obtain a signal proportional to speed:

$$b = K_g \dot{c} \qquad B(s) = K_g s C(s) \tag{4.1}$$

and feeding this back negatively to the power amplifier input.

In the block diagram, this can be represented by a minor loop feedback path. In this case the minor loop is a velocity feedback, as shown in Fig. 4.5. Equation (1.35) gives the reduction of this type of diagram. Its importance may be judged from the availability of motors with integrally mounted tachometers on the shaft.

Example 4.2.5   Motor Position Servo with Load Disturbance Torques

To improve the model in Fig. 4.4 for a field-controlled dc motor position servo by making allowance for load disturbance torques T_l acting on the motor shaft, it is necessary to return to the motor equations in Example 2.4.1. Field voltage E_f and motor developed torque T are related to field current I_f by

$$E_f = (R_f + L_f s)I_f \qquad T = K_t I_f \tag{4.2}$$

where R_f and L_f are field resistance and inductance and K_t is the motor torque constant. If T_l is taken as positive in a direction opposite that of T, a net torque $(T - T_l)$ is available to accelerate motor and load inertia J and overcome their damping B:

$$T - T_l = J\ddot{c} + B\dot{c} \qquad T(s) - T_l(s) = s(Js + B)C(s)$$

Hence

$$\frac{C}{T - T_l} = \frac{1}{s(Js + B)} = \frac{1/B}{s(T_m s + 1)} \qquad T_m = \frac{J}{B} \tag{4.3}$$

Figure 4.4 is now modified to the diagram in Fig. 4.6. As was noted, often the field time constant $T_f = L_f/R_f \ll T_m$ and can be neglected. The factor $1/R_f$ can then be considered to be incorporated into K_a and $E_f = I_f$ assumed in the diagram for purposes of analysis. Figure 4.6 shows T_l as a second, disturbance, input to the block diagram. The reduction of such a diagram was discussed in Example

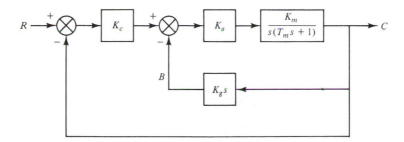

Figure 4.5   Servo with velocity feedback.

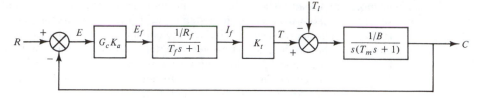

Figure 4.6   Position servo with disturbance torque.

1.8.4. Severe disturbance torques can arise, for example, due to wind in antenna positioning systems, or in steel rolling mill speed controls when slabs enter and leave the rolls.

4.3 DIRECT BLOCK DIAGRAM MODELING OF FEEDBACK SYSTEMS

In Section 4.2 feedback was realized by separate physical elements and the structure of the block diagram was easy to perceive. However, on many occasions the feedback is generated by the use of signals or physical elements which are an intrinsic part of the system. The precise nature of the feedback, or even its presence, may then be far from obvious. In such cases block diagrams can be derived directly from the system equations, and serve an important function in clarifying system behavior.

Example 4.3.1   Pneumatic Pressure Regulator

Figure 4.7 shows a schematic diagram of this very common device. Its purpose is to keep the pressure P_l to the load serviced by the controller constant, equal to a value set by manual adjustment, despite variations of the flow W_l required by the load. Physically, the action is that a reduction of P_l reduces the pressure against the bottom of the diaphragm. This permits the spring force to push it downward to increase valve opening x, and hence increase valve flow from a supply with constant pressure P_s. This increase serves to raise P_l back toward the set value.

Frequently, system equations can be represented by a variety of possible block diagrams, which are all mathematically correct but not all equally useful.

A *good block diagram* is one that clearly identifies the components and parameters in the feedback loop.

To obtain such a diagram, the system equations are written first. This is done as

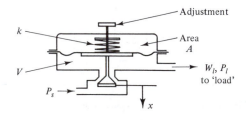
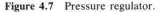

Figure 4.7   Pressure regulator.

in Chapter 2, for the parts of the system, taken in rather arbitrary order. Only after all have been written is consideration given to their combination into a good block diagram model on the basis of the physical operation of the system.

A linearized model is used, implying that the variables are deviations from operating point values. Hence the constant supply pressure P_s will not appear. For the weight flow W_v through the valve the linearized model (2.51) is used:

$$W_v = K_x x - K_p P_l \qquad W_v(s) = K_x X(s) - K_p P_l(s) \qquad (4.4)$$

The net flow entering the volume below the diaphragm, equal to V at the operating point, is $(W_v - W_l)$. This flow must equal

$$W_v - W_l = C_g \dot{P}_l - \rho A \dot{x} \qquad W_v(s) - W_l(s) = C_g s P_l(s) - \rho A s X(s) \qquad (4.5)$$

Here $C_g = V/(nRT)$ is the capacitance of V according to Fig. 2.17 and $C_g \dot{P}_l$ the compressibility flow. The term $\rho A \dot{x}$ is the flow rate corresponding to the change of V, equivalent to, say, (2.57) (which is in terms of volume flow rates). The minus sign arises because Fig. 4.7 defines x as positive in the direction of decreasing volume below the diaphragm.

Force equilibrium on the moving parts, modeled as a spring–mass–damper system k, m, b, gives the equation

$$m\ddot{x} + b\dot{x} + kx = -AP_l \qquad (ms^2 + bs + k)X(s) = -AP_l(s) \qquad (4.6)$$

This model is approximate because it neglects the flow forces on the poppet. The minus sign is needed because the pressure force AP_l on the bottom of the diaphragm acts in the direction of negative x.

To combine (4.4) to (4.6) into a block diagram, it is noted first that the output is the controlled variable P_l, and the input is the disturbance W_l, the unknown flow to the load. Thus, following convention, it is desirable to show W_l at the left and P_l at the right in the diagram. The feedback should then show how changes of P_l are used to make valve flow W_v "follow" W_l. Pressure P_l determines x via (4.6), but x and P_l together determine W_v, and for model simplicity it appears preferable to eliminate W_v between (4.4) and (4.5):

$$K_x X - K_p P_l - W_l = C_g s P_l - \rho A s X \qquad \text{or}$$

$$(C_g s + K_p)P_l = -W_l + (\rho A s + K_x)X \qquad (4.7)$$

Figure 4.8 now shows the block diagram, obtained directly from (4.6) and (4.7), in classical form. Valve flow W_v could be shown at the expense of more loops and a less simple appearance.

Example 4.3.2   Motor with IR-Drop Compensation

Figure 4.9 shows the armature-controlled dc motor of Fig. 2.9(b) (Example 2.4.2) with a resistance R added in the armature loop, the voltage across which is fed

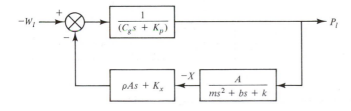

Figure 4.8 Block diagram for Fig. 4.7.

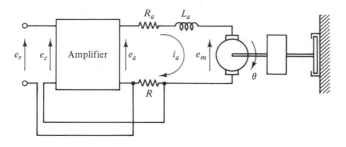

Figure 4.9 Motor with IR-drop compensation.

back as shown to the input of the power amplifier. A block diagram should clarify the nature of this feedback and allow its effect to be studied. The transfer function θ/E_a given by (2.12), with R_a replaced by $R_a + R$, does not show I_a, so that it is necessary to return to the equations from which it was derived. These are repeated here, with R_a replaced by R:

$$E_a = (R_a + R + L_a s)I_a + E_m \qquad E_m = K_e s\theta \qquad T = K_T I_a = s(Js + B)\theta \qquad (4.8)$$

The additional equations, in Fig. 4.9, are

$$E_c = E_r - I_a R \qquad E_a = K_a E_c \qquad (4.9)$$

The block diagram in Fig. 4.10(a) is readily obtained from these equations, and allows the effect of the IR-drop compensation to be studied. In Fig. 4.10(b) this diagram has been partially reduced, by replacing feedback R by a feedback $K_a R$ to the second summing junction and replacing this feedback loop around

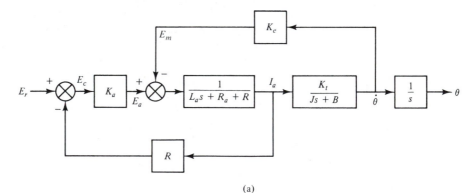

(a)

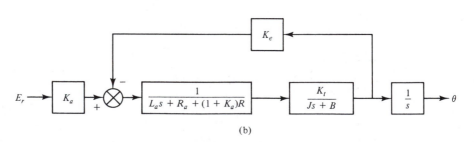

(b)

Figure 4.10 Original (a) and reduced (b) diagram for Fig. 4.9.

$1/(L_a s + R_a + R)$ by its closed-loop transfer function $1/[L_a s + R_a + (1 + K_a)R]$. Compared to the case where $R = 0$, it is seen that the armature time constant has been reduced from L_a/R_a to $L_a/[R_a + (1 + K_a)R]$.

The dynamic controllers in Chapter 5 provide additional examples of direct modeling.

4.4 EFFECT OF FEEDBACK ON PARAMETER SENSITIVITY AND DISTURBANCE RESPONSE

The motivations for the use of feedback were listed in Section 1.4:

1. Reducing the effects of parameter variations
2. Reducing the effects of disturbance inputs
3. Improving transient response
4. Reducing steady-state errors

The first two are the prime reasons why feedback is needed, and are discussed first.

Sensitivity to Parameter Variations

Consider the standard feedback loop in Fig. 4.11. G is the transfer function of the plant or process to be controlled, G_c is that of a controller which may be just a gain or dynamic as discussed in Chapter 5, and H may represent the feedback sensor. The plant model G is usually an approximation to the actual dynamic behavior, and even then the parameter values in the model are often not precisely known and may also vary widely with operating conditions. An aircraft at low level responds differently to control surface deflections than at high level. A power plant model linearized about the 30% of full power operating point has different parameter values than that linearized about the 75% point. For very wide parameter variations, adaptive control schemes, which adjust the controller parameters, may be necessary, but a prime advantage of feedback is that it can provide a strong reduction of the sensitivity without such changes of G_c.

These properties in Fig. 4.11 are determined primarily by the *loop gain* $G_c GH$.

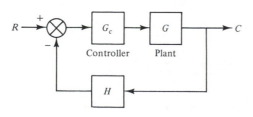

Figure 4.11 Standard loop.

$G_c GH$ = loop gain function (i.e., the product of the transfer functions around the loop)

The closed-loop transfer function is

$$T = \frac{C}{R} = \frac{G_c G}{1 + G_c GH} \tag{4.10}$$

and shows that if $G_c GH \gg 1$,

$$T = \frac{C}{R} \approx \frac{G_c G}{G_c GH} = \frac{1}{H} \tag{4.11}$$

Hence

If the loop gain $G_c GH \gg 1$, C/R depends almost entirely on the feedback H alone, and is virtually independent of the plant and other elements in the forward path and of the variations of their parameters.

This is a major reason for the use of feedback. The feedback H must be chosen for small parameter variations, but unlike G, it is usually under the control of the designer.

Formally, the sensitivity properties can be studied by use of the *sensitivity function S*. For example, the sensitivity of the closed-loop transfer function T to changes in the forward path transfer function $G_f = G_c G$ is the percentage change $\partial T/T$ divided by the percentage change $\partial G_f/G_f$ of G_f which causes it:

$$S = \frac{\partial T/T}{\partial G_f/G_f} = \frac{G_f}{T} \frac{\partial T}{\partial G_f} = \frac{G_f}{T} \frac{\partial}{\partial G_f} \frac{G_f}{1 + G_f H}$$

$$= \frac{G_f}{T} \frac{1}{(1 + G_f H)^2} = \frac{1}{1 + G_f H} \tag{4.12}$$

The *static sensitivity* is the value of S for $s \to 0$. *Dynamic sensitivities* are usually calculated by replacing s by $j\omega$ and plotting S as a function of frequency ω. Such results indicate, as later work will show, how sensitivity changes with the frequency of a sinusoidal input R.

Example 4.4.1

In Fig. 4.12(a) the closed-loop transfer function is

$$T = \frac{KA}{\tau s + 1 + KAh}$$

The sensitivities S_a, S_h, and S_τ for small changes of A, h, and τ are as follows:

$$S_a = \frac{\partial T/T}{\partial A/A} = \frac{A}{T} \frac{\partial T}{\partial A} = \frac{A}{T} \frac{K(\tau s + 1)}{(\tau s + 1 + KAh)^2} = \frac{\tau s + 1}{\tau s + 1 + KAh} \tag{4.13a}$$

$$S_h = \frac{\partial T/T}{\partial h/h} = \frac{h}{T} \frac{\partial T}{\partial h} = \frac{h}{T} \frac{-(KA)^2}{(\tau s + 1 + KAh)^2} = \frac{-KAh}{\tau s + 1 + KAh} \tag{4.13b}$$

$$S_\tau = \frac{\partial T/T}{\partial \tau/\tau} = \frac{\tau}{T} \frac{\partial T}{\partial \tau} = \frac{\tau}{T} \frac{-KAs}{(\tau s + 1 + KAh)^2} = \frac{-\tau s}{\tau s + 1 + KAh} \tag{4.13c}$$

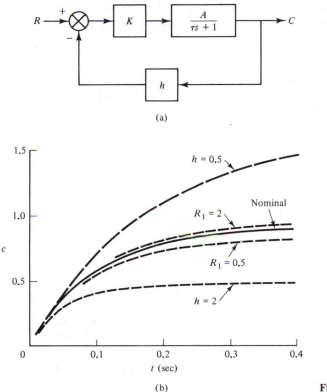

(a)

(b)

Figure 4.12 Example 4.4.1.

The static sensitivity S_{as} is seen to reduce as loop gain KAh increases, but the magnitude of S_{hs} will approach 1. It is seen also that the static sensitivity $S_{\tau s}$ is zero.

Frequently, the variations of A and τ will be interrelated. From Example 2.6.1, if Fig. 4.12(a) models a single-tank level control, $A = K_v R_1$, $\tau = A_1 R_1$, where K_v is the control valve gain and A_1 and R_1 are the tank area and outlet valve resistance. R_1 could vary due to obstructions or inadvertent adjustment, and level sensor gain h could change due to a malfunction. To illustrate the effects, unit step responses when these parameters are halved and doubled from their nominal values will be compared, assuming that $K = 9$, $K_v = 1$, $A_1 = 1$, R_1(nom.) = 1, h(nom.) = 1. Leaving R_1 and h as free parameters, the closed-loop transfer function yields the unit step response

$$c(t) = \frac{9}{9h + 1/R_1}\left\{1 - \exp\left[-\left(9h + \frac{1}{R_1}\right)t\right]\right\}$$

This is plotted in Fig. 4.12(b) for the nominal parameters and when one of R_1 and h is 2 or 0.5 times the nominal value. As expected, the response is far more sensitive to h than to R_1. In view of the large parameter changes, the differences between the curves for $R_1 = 1$, 2 and 0.5 appear small. If K were reduced, reducing loop gain, these differences would be larger. This trend is evident in the curves in that loop gain is smaller for $h = 0.5$ and $R_1 = 0.5$. Accordingly, these responses differ more from the nominal plot than those for $h = 2$ and $R_1 = 2$, respectively.

Effect of External Disturbances

Three of the examples earlier in this chapter included the modeling of disturbance inputs. These fit the model shown in Fig. 4.13, which is equivalent to Fig. 1.12 in Section 1.8, on block diagram reduction. The effect of disturbance D on output C is given by the transfer function

$$\frac{C}{D} = \frac{G_2 L}{1 + G_c G_1 G_2 H} = -\frac{E}{D} \qquad \text{for } H = 1 \qquad (4.14)$$

This shows the following:

If loop gain $G_c G_1 G_2 H \gg 1$, then

$$\frac{C}{D} \approx \frac{L}{G_c G_1 H} \qquad \left(\approx \frac{-E}{D} \quad \text{for } H = 1 \right) \qquad (4.15)$$

so feedback strongly reduces the effect of a disturbance D on C if the loop gain is much greater than 1 due to high gain in the feedback path between C and D.

This also shows why a system may respond well to an input R while the response to another input D is small. For $H = 1$, when $C/R = 1$ is desired, loop gains $G_c G_1 G_2 \gg 1$ will ensure that $C/R = G_c G_1 G_2/(1 + G_c G_1 G_2)$ is near 1. By locating the high gains in $G_c G_1$, between the points where R and D enter the loop, both requirements can be met.

Disturbances may also enter the feedback path, due to the sensor measuring C. The importance of measures to avoid this is clear, since normally there will not be a high gain in the feedback between C and such disturbances to attenuate their effect.

The steady-state response, for $t \to \infty$, to a disturbance is clearly an important

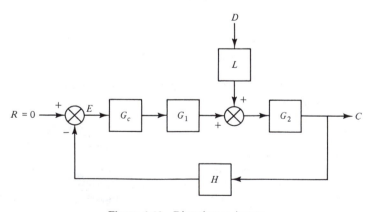

Figure 4.13 Disturbance inputs.

measure of system quality, and is readily found using the final value theorem. From (4.14), for a unit step input $D = 1/s$,

$$\lim_{t\to\infty} c(t) = \lim_{s\to 0} sC(s) = \lim_{s\to 0} \frac{G_2L}{1 + G_cG_1G_2H} \qquad (4.16)$$

Example 4.4.2   Motor Position Control (Fig. 4.6)

For this system, from Fig. 4.6, if G_c is an amplifier with gain K_c, $G_c = K_c$, then

$$G_cG_1 = \frac{K_cK_aK_t/R_f}{T_fs + 1} \qquad L = H = 1 \qquad G_2 = \frac{1/B}{s(T_ms + 1)}$$

Hence, with the negative sign at the load disturbance torque $T_l = 1/s$,

$$\lim_{t\to\infty} c(t) = \lim_{s\to 0} \frac{-(T_fs + 1)/B}{s(T_ms + 1)(T_fs + 1) + K_cK_aK_t/(R_fB)} \qquad (4.17)$$

$$= -\frac{R_f}{K_cK_aK_t}$$

High controller gain, therefore, not only makes the system less sensitive to parameter variations but also improves its stiffness against load disturbances. Note, incidentally, that the feedback to the summing junction for D in Fig. 4.13 is not positive, as the plus sign might suggest, since the minus sign at R can be moved to D when evaluating C/D.

It is useful for the sake of insight to verify the result (4.17) directly from Fig. 4.6. In the steady state, the net torque $T - T_l$ must be zero, so if $T_l = 1$, T must be 1 as well. From the diagram, this requires that $e = R_f/(K_cK_aK_t)$. Since input r is zero while considering the response to T_l, $e = r - c = -c$, so indeed $c = -R_f/(K_cK_aK_t)$.

4.5 STEADY-STATE ERRORS IN FEEDBACK SYSTEMS

High loop gains were shown to be advantageous to reduce sensitivity of performance to both parameter variations and disturbance inputs. They will now prove equally desirable from the point of view of the reduction of steady-state errors in feedback systems. Intuitive reasoning in Section 1.4 and several examples already suggested this, and it will now be verified by consideration of the *unity feedback system* in Fig. 4.14. $E = R/(1 + G)$, and the steady-state error e_{ss} can be found directly, without the need for inverse transformation, by the final value theorem:

$$e_{ss} = \lim_{t\to\infty} e(t) = \lim_{s\to 0} sE(s) = \lim_{s\to 0} \frac{sR(s)}{1 + G(s)} \qquad (4.18)$$

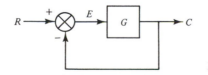

Figure 4.14 Unity feedback system.

For $G(s)$, the following general form is assumed:

$$G(s) = \frac{K}{s^n} \frac{a_k s^k + \cdots + a_1 s + 1}{b_l s^l + \cdots + b_1 s + 1} \tag{4.19}$$

In this equation:

1. K as given, with the constant terms in numerator and denominator polynomials made unity, is formally the *gain* of the transfer function G. It should be distinguished from the *root locus gain*, defined earlier as that for which the highest power coefficients are unity, and equal to $K a_k / b_l$ in (4.19).

2. The *type number* of G is the value of the integer n. As discussed in Example 2.4.1, a factor s in the denominator represents an integration, so the type number is the number of integrators in G.

3. Gain $K = \lim_{s \to 0} s^n G(s)$, and a common practice associates the following names and notations with K, depending on n:

$$
\begin{aligned}
n = 0: \quad & K_p = \text{position error constant} \\
n = 1: \quad & K_v = \text{velocity error constant} \\
n = 2: \quad & K_a = \text{acceleration error constant}
\end{aligned} \tag{4.20}
$$

Equation (4.19) shows that $\lim_{s \to 0} G(s) = \lim_{s \to 0} (K/s^n)$, so that (4.18) can be written as follows:

$$e_{ss} = \lim_{s \to 0} \frac{sR(s)}{1 + (K/s^n)} \tag{4.21}$$

This readily yields Table 4.5.1 for the steady-state errors corresponding to different type numbers and inputs, of which the transforms are given in Table 1.6.1. For example, for a type 2 system with a unit ramp input,

$$e_{ss} = \lim_{s \to 0} \frac{s(1/s^2)}{1 + (K/s^2)} = \lim_{s \to 0} \frac{s}{s^2 + K} = 0$$

$$*****$$

TABLE 4.5.1   STEADY-STATE ERRORS

	Type number: $n = 0$	$n = 1$	$n = 2$
Step $u(t)$; $R = 1/s$	$\dfrac{1}{1 + K_p}$	0	0
Ramp t; $R = 1/s^2$	∞	$\dfrac{1}{K_v}$	0
Acceleration $t^2/2$; $R = 1/s^3$	∞	∞	$\dfrac{1}{K_a}$

For insight into these results it is useful to consider Fig. 4.15, which shows type 0 and type 1 systems for $s \to 0$. For $n = 0$, $C = K_p E$, so there cannot be a nonzero output without a proportional, nonzero error. If the output must increase along a ramp, the error must increase according to a ramp as well. So a type 0 system has a steady-state error for a step input and cannot follow a ramp.

For $n = 1$,

$$C = \frac{K_v}{s}E \qquad \text{so } c(t) = K_v \int e(t)\, dt$$

This means that the output cannot level off to a constant value unless the error levels off at zero. A steady-state with nonzero steady-state error cannot exist for a step input because the integrator would cause the output to change.

From a design point of view, this means that if, as is often the case, the performance specifications require zero steady-state error after a step input, the designer must ensure that the system is at least of type 1. A type 2 system would be necessary if zero steady-state errors following both steps and ramps are specified.

The hydraulic servo and motor control examples in Section 4.2 are type 1 systems, but the level control is type 0, unless it is made into type 1 via the controller.

The nonzero finite errors in Table 4.5.1 decrease as gain is increased, as do parameter sensitivity and disturbance response. As suggested in Section 1.4, for larger K a smaller error can achieve the same effect on the output.

Example 4.5.1

In Fig. 4.14:

(a) $G(s) = \dfrac{K_1(as + b)}{(cs^2 + ds + e)(fs + g)}$:   type 0   gain $K_p = \dfrac{K_1 b}{eg}$

Unit step: $e_{ss} = \dfrac{1}{1 + K_p}$ unit ramp: $e_{ss} \longrightarrow \infty$

(b) $G(s) = \dfrac{K_1(as + b)}{s(cs^2 + ds + e)(fs + g)}$:   type 1   gain $K_v = \dfrac{K_1 b}{eg}$

Unit step: $e_{ss} = 0$ unit ramp: $e_{ss} = \dfrac{1}{K_v}$

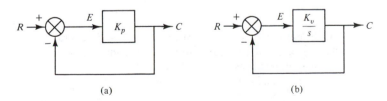

(a)   (b)

Figure 4.15 Types 0 (a) and 1 (b) systems.

Example 4.5.2   Steady-State Errors for Unit Steps in R and D (Fig. 4.16)

The system is type 1, so $e_{ss} = 0$ for a step input R. For input D, the steady-state value m_{ss} of M is also zero (since otherwise c_{ss} could not be constant), but the feedback from C to D is not unity but K. For a unit value of D the condition for $m_{ss} = 0$ is $1 + Ke_{ss} = 0$, so $e_{ss} = -1/K$. Direct application of the final value theorem to $C(s)$ will verify these results.

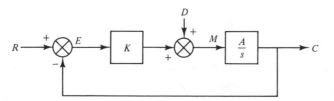

Figure 4.16   Example 4.5.2.

4.6 TRANSIENT RESPONSE VERSUS STEADY-STATE ERRORS

Preceding sections have shown that parameter sensitivity, disturbance response, and steady-state errors are all improved by increased gain. The fourth motivation for the use of feedback is to improve dynamic response.

Section 1.4 has suggested, and examples such as 2.8.1 and 3.5.2 supported this, that dynamic response considerations usually limit the permissible gain, and that feedback design is generally concerned with achieving a satisfactory compromise between relative stability and accuracy (i.e., small errors).

Two examples will be given to show how transient response and accuracy are related. The first is unusual in that both improve with increasing gain. It is useful however in its striking demonstration of the effects of feedback.

Example 4.6.1   Proportional Control of a Simple Lag

Figure 4.17(a) shows a simple lag plant with a pure gain controller. This is called *proportional control,* or *P control.* The closed-loop transfer function is

$$\frac{C}{R} = \frac{KG}{1 + KG} = \frac{K}{Ts + 1 + K} = \frac{K}{1 + K} \frac{1}{[T/(1 + K)]s + 1} \tag{4.22}$$

Hence the closed-loop system is also a simple lag, but with time constant $T/(1 + K)$ instead of T. So the speed of response increases with K. Steady-state errors for a unit step input are $1/(1 + K)$, since the system is type 0 with gain K, so also reduce with increasing K. The unit step response is

$$c(t) = \frac{K}{1 + K}(1 - e^{-(1+K)t/T}) \tag{4.23}$$

and is plotted in Fig. 4.17(b) for $T = 1$ and several values of K. It is apparent that for larger K the response not only comes closer to the desired value of unity, but also approaches the steady-state sooner.

The steady-state value of the output,

$$c_{ss} = r - e_{ss} = 1 - \frac{1}{1 + K} = \frac{K}{1 + K}$$

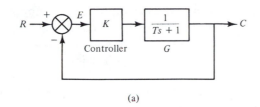

(a)

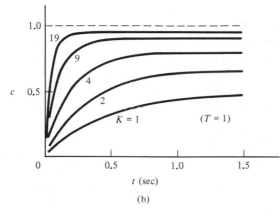

(b)

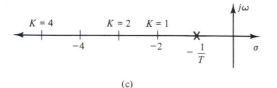

(c)

Figure 4.17 Example 4.6.1: P control.

can also be verified from (4.22), which gives

$$c_{ss} = \lim_{s \to 0} sC(s) = \lim_{s \to 0} \frac{K}{1 + K} \frac{s(1/s)}{[T/(1 + K)]s + 1} = \frac{K}{1 + K}$$

Evidently, increasing gain K does in this example not reduce relative stability. To allow correlation of the change of transient response with change of system pole position in the s-plane, Fig. 4.17(c) shows the root locus of the system. It indicates, for $T = 1$, how the system pole $-(1 + K)/T$ moves as K is increased from zero. This plot is in fact enough to show that there is no relative stability problem and that, because of increasing distance of the pole to the imaginary axis, the system responds faster with increasing K.

Apparently, improving the transient response can be a major motivation for the use of feedback. In this example a possibly very slow process could be made to respond at any desired speed!

However, a practical restriction should be emphasized. Changing a water level or temperature in a tank fast requires large-capacity and costly water or heat supply systems. The results for the example imply the assumption of infinite capacity supplies. In practice, the use of excessive gain will cause the supplies to saturate at their maximum output levels. The system then operates in a nonlinear regime and its response changes from the predictions based on linear analysis.

Often the maximum value of K permitted by these considerations is not large enough to meet specifications on the steady-state error $1/(1 + K)$. Indeed, in many cases this error is required to be zero. This motivates the use of dynamic compensation, introduced in Chapter 5, since a constant-gain controller is then no longer adequate. The results in Section 4.5 indicate that a type 1 system is required. Since the plant in Example 4.6.1 is type 0, this means that a separate factor s must be introduced into the denominator of the controller. By far the most common solution is, in fact, the PI controller $K(s + z)/s$ already introduced in Example 3.8.1 and Fig. 3.18 for the control of a simple lag plant. Figure 3.18 also presents step response curves showing the effect of the choice of the zero $-z$.

Example 4.6.2 P Control with Two Simple Lags

In Fig. 4.18(a) a second simple lag has been included in the system of Fig. 4.17(a). Among other possibilities, it could represent a two-tank level control as in Fig. 2.19, a temperature control with two heat capacitances in series, or a motor speed

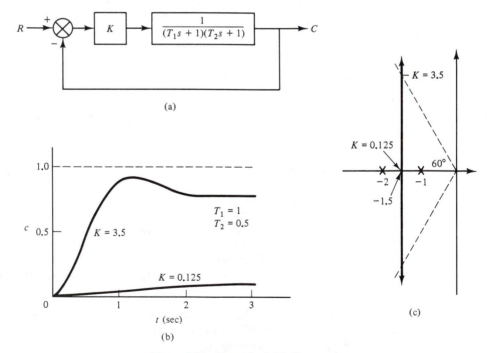

Figure 4.18   Example 4.6.2: P control.

control. As in Example 4.6.1, the system is type 0 with gain K, so the steady-state error after a unit step is $1/(1 + K)$. Again dynamic compensation, probably PI control, is needed if this error is required to be zero.

Consider now stability, using $T_1 = 1$, $T_2 = 0.5$ for a numerical example. The closed-loop transfer function and the system characteristic equation are, respectively,

$$\frac{C}{R} = \frac{2K}{s^2 + 3s + 2 + 2K} \tag{4.24}$$

$$s^2 + 3s + 2 + 2K = 0$$

The closed-loop poles are

$$s_{1,2} = -1.5 \pm \sqrt{0.25 - 2K} \tag{4.25}$$

and a root locus showing how these pole positions in the s-plane change with K can easily be plotted in this case. The plot is shown in Fig. 4.18(c). For $K = 0$ the poles coincide with those of the loop gain function, -1 and -2. As K is increased, the poles initially are still real, and move together until for $K = 0.125$ both are located at -1.5. For larger K the real part is constant at -1.5 and the imaginary part increases with K. The cosine of the angle of the poles with the negative real axis is the damping ratio ζ, and decreases with increasing K. Since the real part is constant, the time constant and settling time are the same for all $K \geq 0.125$. But as discussed in Example 3.5.2, speed of response is improved by reducing rise time, and this suggests that ζ be reduced to the lowest value for which the step response overshoot is acceptable. If this is $\zeta = 0.5$, equating the characteristic equation with $s^2 + 2\zeta\omega_n s + \omega_n^2$ yields $\omega_n = 3$, $K = 3.5$.

Thus the increase of K which is desirable to increase the speed of response and to reduce steady-state errors, parameter sensitivity, and disturbance response is limited by considerations of relative stability.

The unit step response can be calculated from (4.24) and is shown in Fig. 4.18(b) for $K = 0.125$ and $K = 3.5$. The improvement in accuracy and speed of response for the higher gain is evident, but the steady-state error is still large and PI control, as mentioned earlier, would probably be necessary, to change the system to type 1 and achieve zero steady-state error.

Example 4.6.2 is more typical than Example 4.6.1, in that relative stability considerations usually limit the permissible gain. It is still not representative, however, of most practical systems since at least absolute stability is maintained for all K. Usually, systems actually become unstable beyond a certain value of K. As discussion of the root locus technique in Chapter 6 will show, if the vertical branches of the root locus in Fig. 4.18(c) would instead bend off to the right, then beyond a certain value of K the closed-loop poles would be located in the right-half s-plane.

The examples also motivate the use of dynamic compensation, introduced in Chapter 5. PI control $K(s + z)/s$ was mentioned as a possible solution to replace a pure gain or P control if at the maximum gain permitted by relative stability the steady-state errors are still too large.

4.7 CONCLUSION

In this chapter the modeling of systems with feedback in block diagram form was considered first. The examples include those where the structure of the system and the nature of the feedback are clear from the schematic diagram, as well as cases where even the existence of feedback may not be obvious. In particular in the latter case, a good block diagram is an important aid in system analysis and design.

The motivations for the use of feedback were examined next, and it was found that considerations of relative stability usually limit the gain increase desirable to increase the speed of response and to reduce sensitivity to plant parameter variations, response to disturbances, and errors.

Examples were given that motivate the use of dynamic compensation, introduced in Chapter 5, to overcome this limitation. The emphasis in Chapter 5 will be on the use of proportional plus integral plus derivative control (PID). This form is extremely common in practice, and its basic control actions are fundamental to dynamic compensation generally.

The physical realization of such controllers will also provide additional examples of the derivation of block diagrams for systems with feedback. In this context the operational amplifier mentioned in Chapter 2, and some of its applications in controller realization and system simulation, will be discussed as well.

The concept of a root locus was introduced in the examples of Section 4.6. The limitation to second-order systems, for which the roots of the characteristic equation can easily be calculated, suggests the need for more advanced techniques in practical applications. The root locus technique in Chapter 6 and the frequency response methods in Chapters 7 and 8 will fulfill this need. However, dynamic compensation is considered first because, to enhance insight, it is important to introduce this subject in a simple mathematical framework.

PROBLEMS

4.1. A water level control of the type of Fig. 4.1 with a proportional pneumatic controller (P control) is indicated in Fig. P4.1. Air and water supply pressures are constant. Variations p_o of nozzle back pressure are proportional to variations x_f of the flapper-to-nozzle distance. Motion x_b is proportional to changes p_o of pressure in the bellows. For the pneumatically actuated water control valve, the simple lag transfer function $K/(Ts + 1)$ can be assumed. Introducing parameters as needed, write the necessary equations and represent the system by a block diagram. The desired level is set by screw adjustment on the float post.

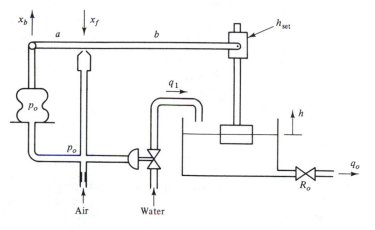

Figure P4.1

4.2. Another form of water level control is suggested in Fig. P4.2. Here the valve–cylinder modeled in Fig. 2.27 is used to control the water valve opening. The oil and water supply pressures are constant. Write the necessary equations, introducing parameters as needed, and obtain a block diagram model. What is the essential difference with the control of Fig. P4.1, and how does it affect steady-state errors?

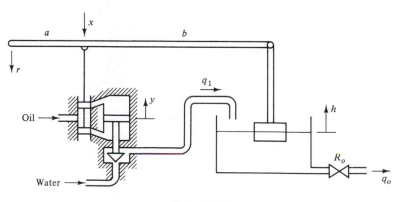

Figure P4.2

4.3. Write the linearized equations and derive a block diagram for the hydraulic servo in Fig. P4.3. Introduce parameters as needed and assume the load on the cylinder to be small.

4.4. Refine the hydraulic servo model for Fig. P4.3 derived in Problem 4.3 by assuming the mass m to be so large that the loading on the cylinder can no longer be neglected and oil compressibility must be included. Introduce parameters as needed, write all linearized equations and obtain a block diagram representation.

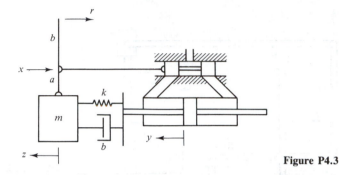

Figure P4.3

4.5. An armature-controlled dc motor is used in a speed control system shown in Fig. P4.5. The voltage r representing desired speed is obtained by the setting of a potentiometer to which a constant voltage V is applied. Actual speed is represented by the voltage from a tachometer/generator on the motor shaft. The difference of these voltages, the speed error signal, is applied to a voltage amplifier of gain K_a, of which the output is raised in power in the power amplifier K_p. Using (2.12) to represent the motor and load, obtain a block diagram and express the steady-state speed error for step demands. Why is this not a type 1 system even though a motor is present?

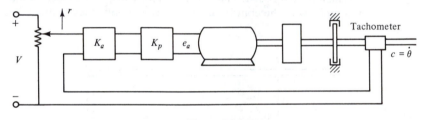

Figure P4.5

4.6. A field-controlled dc motor is used in the position control system indicated in Fig. P4.6. Using the results of Fig. 4.6 for Example 4.2.5 where appropriate, write the modified equations necessary and obtain a block diagram representation. Shaft position θ_o is measured by means of a potentiometer. The torsion spring on the output shaft has spring constant k.

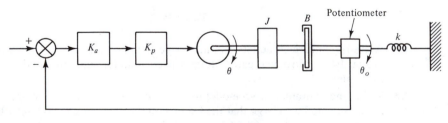

Figure P4.6

4.7. Figure P4.7 shows a schematic diagram of a temperature control system. The fluid is incompressible, with specific heat c, the mass in the tank is W, and the mass

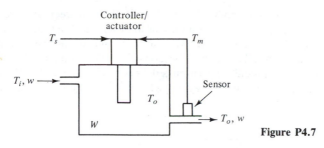

Figure P4.7

flow rate through the tank is w. The temperature sensor has a significant time constant, τ_m. The desired temperature T_s is the input to a controller with transfer function G_c, and heat flow rate q from the heater is proportional to controller output.

Write the equations and obtain a block diagram from which the effect on T_o of each of T_s, T_i, and w could be studied. Remember that for small variations x and y about x_0 and y_0, the expression $(x_0 + x)(y_0 + y)$ is linearized to $(x_0 y + y_0 x)$, considering only the changes from operating-point values.

4.8. Obtain a block diagram for study of the dynamic behavior of a ship stabilizer from the following description. The differential equation for the roll angle θ due to a disturbance torque T_d on the ship in rough seas is $J\ddot{\theta} + b\dot{\theta} + k\theta = T_d$, where J is the inertia, b a generally small damping constant, and k a spring constant representing the self-righting effect due to ship geometry. To minimize roll, many ships are equipped with stabilizer fins on the sides, of which the angles can be adjusted in opposite directions to produce a restoring torque T_r, much like the flaps on the wings of an aircraft. T_r can be taken to be proportional and opposite to roll angle θ, but with a simple lag delay with time constant τ_f due to the fin actuator. Roll angle is measured by a vertical gyro roll sensor, and a roll rate sensor is included to measure the rate of change of θ. Explain why the rate sensor is needed and how it is used in the system, and include this in the block diagram.

4.9. The pneumatic pressure regulator in Fig. P4.9 is a modification of that in Fig. 4.7 discussed in Example 4.3.1. A seal with an orifice of resistance R through it has been incorporated between the output pressure p_l and the space under the diaphragm. The capacitance of this space, where the pressure is p, is C_d, and that of the volume under pressure p_l is C_g. Introducing additional parameters as needed, write the equations and derive a block diagram. What should be the key difference with that in Fig. 4.8?

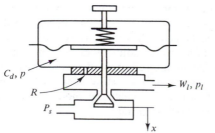

Figure P4.9

4.10. The hydraulic pressure controller in Fig. P4.10 must keep the pressure p_l to the load constant, regardless of variations of load flow q_l. Spring, damping, and mass constants for the valve spool are k, b, and m, and its surface area is A. The flow q_p from a pump may be taken to be constant. The volume of oil under pressure p_l is V and its bulk modulus β. Discuss the operation physically. Write all linearized equations and set up a good diagram which clearly identifies the feedback.

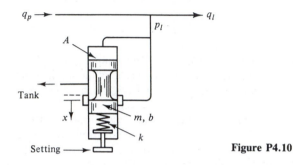

Figure P4.10

4.11. Improve the model obtained in Problem 4.10 by making allowance for a drop in supply flow q_p with increasing pressure due to leakage, taking it to be proportional to pressure.

4.12. The pressure-reducing valve in Fig. P4.12 must keep p_l constant, at a value lower than the constant available supply pressure P_s, regardless of variations of the flow q_l to the load. Spring, damping, and mass constants for spool motion are k, b, and m, and the volume of oil, with bulk modulus β, under pressure p_l is V. The compressibility effect in the small end chamber of the spool may be neglected. Discuss the operation physically, write all linearized equations and obtain a good block diagram.

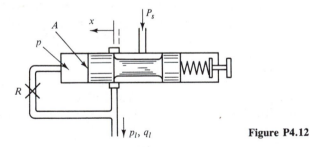

Figure P4.12

4.13. A pressure-compensated pump system is indicated in Fig. P4.13. It uses a variable-displacement pump, of which the stroke, and hence the delivery per revolution, can be varied by a mechanical adjustment x. Variations q_p of pump delivery can be taken to be proportional to the adjustments x. The equivalent spring, damping, and mass load on the piston are represented by the constants k, b, and m, and the volume of oil under pressure p_l is V. How does the system operate? Introducing parameters as needed, write the linearized model equations, with allowance for leakage, and derive the block diagram.

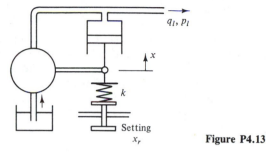

Figure P4.13

4.14. A slightly generalized version of the water level control block diagram of Fig. 4.2 is shown in Fig. P4.14(i). A simplified model, in which the controller is a gain and supply pressure disturbances P_s' and the time constant τ_f of the actuator are assumed to be negligibly small, is shown in Fig. P4.14(ii). In this model the valve flow gain K_x, incorporated into K, and the tank outflow valve resistance R are linearized gains that change with operating point and are generally not precisely known. K may also change due to gain variations of the valve actuator, and the level sensor gain H may vary from its nominal value of 1. In this model:

(a) Express the sensitivity functions to changes in K, R, and H.

(b) Determine the static sensitivities and hence comment on the effect of K, R, and H under static conditions.

(c) Compare the sensitivity to R with that for open-loop control and hence discuss the effect of feedback.

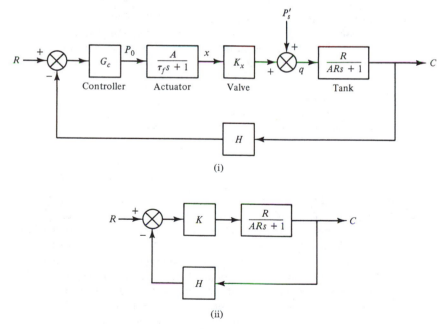

Figure P4.14

4.15. In Fig. P4.15 the actuator time constant τ_f in Fig. P4.14(i) has not been neglected as it was in Fig. P4.14(ii).

(a) Determine the sensitivity function of the closed-loop transfer function C/R with respect to the parameter τ_f, and find the static value of this function.

(b) In frequency response methods $s = j\omega$, and the magnitude of the sensitivity function as a function of radian frequency ω is of interest. If $AR = 1$ and $\tau_f = 0.2$, find the magnitude of the sensitivity function at $\omega = 1$ if KRH is such that the system damping ratio is 0.5. What is the value of S_τ at very high frequencies?

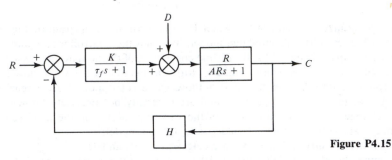

Figure P4.15

4.16. In Fig. P4.15 determine the sensitivities of the transfer function C/D for the disturbance input D to variations of the parameters R and K. What are the static values of these sensitivities, and what is the significance of these results for large loop gains KRH?

4.17. In the motor position servo of Fig. P4.17, let $G(s) = 1/[s(s + 1)]$ represent the motor and load and $G_c = K$ the controller/amplifier. Calculate the steady-state errors of the system for unit step and unit ramp signals applied in turn to reference input R and disturbance input D. Explain the results obtained physically.

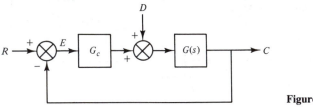

Figure P4.17

4.18. In Fig. P4.17, let $G(s)$ be a simple lag plant $G(s) = K_p/(Ts + 1)$ and G_c an integral controller $G_c = K/s$. Calculate the steady-state errors for unit step and unit ramp inputs applied in turn to R and D. Compare the results and discuss the desired gain distribution over K and K_p. Explain the differences with Problem 4.17.

4.19. In Fig. P4.17, let $G_c = K$ and let $G(s)$ model a thermal or hydraulic process which can be approximated by two simple lags: $G(s) = K_p/[(s + 1)(0.2s + 1)]$. Find the steady-state errors for unit step inputs applied in turn to R and D, and compare the results. Discuss the desired gain distribution over K and K_p.

4.20. In Fig. P4.20, with $G(s) = A/(Ts + 1)$ the parameters A and T are nominally 1, but each may vary by a factor of 2 in either direction with operating conditions.

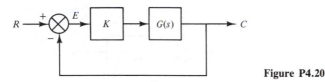

Figure P4.20

Find K so that despite these variations the steady-state errors for step inputs will not exceed 10% and the system time constant will stay below 0.2 sec.

4.21. For the system in Fig. P4.20 with $G(s) = (s + 1)(s + 3)/[s(s + 2)(s + 4)]$:
  (a) What is the system type number?
  (b) What is the gain of the loop gain function?
  (c) What are the steady-state errors following unit step and unit ramp inputs?

4.22. In Fig. P4.20 with $G(s) = 1/[(s + 1)(s + 7)]$:
  (a) How long would transients take to decay almost completely if there were no feedback?
  (b) What is the characteristic equation of the closed-loop system, and where must the dominating system pole be located if the time in part (a) is to be halved?
  (c) What value of K will achieve this?
  (d) What is the corresponding steady-state error for a unit step input?

4.23. Let in Problem 4.22 an increase of gain be desirable to reduce the steady-state error.
  (a) Express the system poles as functions of K.
  (b) Find the lowest value of K that will minimize the settling time.
  (c) Find K and the corresponding steady-state error for a unit step to obtain a system damping ratio of about 0.7.
  (d) Compare the settling times of parts (b) and (c). Which is the best to minimize rise time, and why?

4.24. For the system in Fig. P4.20 with $G(s) = 1/[(s + 1)(s + 4)]$:
  (a) What is the dominating time constant of the plant?
  (b) For what value of K will feedback make the dominating system time constant half of that in part (a)?
  (c) Calculate the unit step response for K in part (b) and find the steady-state error.
  (d) What limits the increase of K desirable to reduce the steady-state error in part (c)? Find K and the corresponding steady-state error for a system damping ratio of about 0.7.

4.25. The open-loop system in Fig. P4.25(a) responds the same to reference inputs R and disturbance inputs D. In Fig. P4.25(b), calculate K so that the steady-state response to R is the same as in part (a), and compare parts (a) and (b) on the basis of dynamic response to R and steady-state and dynamic performance in response to D.

4.26. In Fig. P4.20 with $G(s) = 1/[(s + 2)(s + 10)]$, to examine the effect of feedback on performance:
  (a) Calculate the unit step responses for $K = 7$ and $K = 20$.
  (b) Verify the steady-state error values of these responses directly.
  (c) Compare the responses on the basis of settling time and nature of the response.

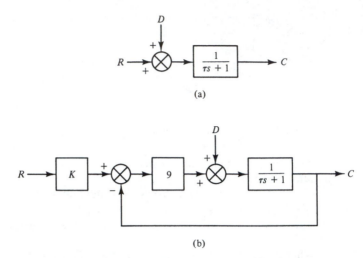

Figure P4.25

4.27. In Fig. P4.20 with $G(s) = (s + 1)/[s(s + 3)]$:
(a) Find K so that the dominating system time constant will be 2 sec, and find for this K the second system pole.
(b) What are the resulting steady-state errors following unit step and unit ramp inputs?
(c) Calculate the unit step response from the results in part (a).

4.28. A plant $G(s)$ has the transfer function $1/(s - 2)$.
(a) Determine stability from its pole–zero pattern.
(b) Show that the dynamic behavior may be changed by adding a feedback loop with an amplifier K around G.
(c) Find K to obtain a stable system with a time constant $T = 0.1$ sec.
(d) What is the corresponding steady-state error for a unit step input?

4.29. The system in Fig. P4.20 with $G(s) = (s + 2)/[(s - 2)(s + 4)]$ would be unstable without feedback. Why? Show that it can be stabilized by feedback as shown, and find K so that the dominating system time constant is 1 sec.

5

DYNAMIC COMPENSATION
OF FEEDBACK SYSTEMS

5.1 INTRODUCTION

In proportional control, or P control, as discussed in Section 4.6, only gain adjustment is available to improve performance. Its limitations for achieving both satisfactory accuracy and acceptable relative stability were pointed out. If performance specifications cannot be met by P control, it must be replaced by a dynamic controller or dynamic compensator to provide more flexibility.

One common system configuration is minor loop feedback compensation, encountered in Fig. 1.12 and Fig. 4.5 and shown in Fig. 5.1(b). The most common configuration, however, is *series compensation,* indicated in Fig. 5.1(a). The controllers $G_c(s)$ and $H_c(s)$ are usually very simple transfer functions. The tools of the root locus and frequency response methods, still to be provided, are needed for most practical applications, but are not required to gain insight into the essential concepts of how such controllers can change system behavior.

After a discussion of feedback compensation, the emphasis in this chapter will be on series compensation using proportional plus integral plus derivative (PID) controllers as dynamic compensators. PID control is the most commonly used standard form of dynamic compensation in practice. Electronic, pneumatic, and digital PID controllers are available off the shelf in a wide variety of makes and types. Furthermore, its basic proportional, integral, and derivative actions are also fundamental to other dynamic compensators, such as the phase-lag and phase-lead compensators discussed at length later. A good understanding of the P, I, and D actions, both in analytical and physical terms, is therefore important for an appreciation of how dynamic compensation can change system behavior.

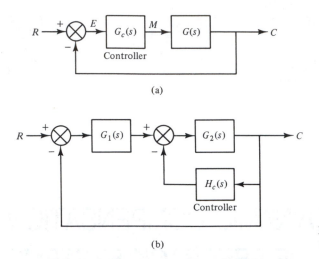

Figure 5.1 Series (a) and feedback (b) compensation.

The physical realization of PID control by pneumatic means or by the use of operational amplifiers is also discussed, adding to the examples in Chapter 4 on the modeling of systems with feedback by block diagrams.

The operational amplifier is a very important general-purpose device, used extensively in many applications. It is the basic element of general-purpose analog computers, which may contain hundreds of such amplifiers. These computers can simulate the blocks or the differential equations of systems with or without feedback, and show the dynamics of system variables in terms of analog voltages. Operational amplifiers can serve as summing junctions and to realize dynamic compensators, and have many other applications. In this chapter the use in dynamic compensators is emphasized, but analog computers are also discussed briefly.

5.2 FEEDBACK COMPENSATION: VELOCITY FEEDBACK

The most common example of feedback compensation is velocity feedback. Figure 5.2 repeats the block diagram in Fig. 4.5 which was derived in Example 4.2.4 for a motor position servo with velocity feedback. T_m has been replaced by its definition J/B in terms of inertia J and damping constant B, and K_m is equivalent to $K_m B$ in Fig. 4.5. The physical motivation for velocity feedback was discussed in Example 4.2.4. If damping B is small, velocity feedback can provide the equivalent of a damping torque, proportional to velocity and in the opposite direction. To verify the improved servo damping, the minor loop in Fig. 5.2 is reduced to obtain the transfer function

$$\frac{C}{E} = \frac{K_c K_a K_m}{s(Js + B + K_a K_m K_g)} \tag{5.1}$$

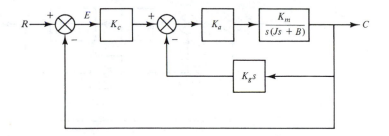

Figure 5.2 Velocity feedback.

This is the loop gain function in the unity feedback major loop, and yields the system characteristic equation

$$s^2 + \frac{B + K_a K_m K_g}{J} s + \frac{K_c K_a K_m}{J} = 0 \tag{5.2}$$

on dividing through by J. Comparison with the normalized form $s^2 + 2\zeta\omega_n s + \omega_n^2 = 0$ now shows immediately that the term with velocity feedback gain K_g has indeed increased the effective damping constant of the system and improved the damping ratio.

In systems where the inherent damping B is quite small, the characteristic equation without velocity feedback would be

$$s^2 + \frac{K_c K_a K_m}{J} = 0$$

so that the system poles would lie on the imaginary axis, where the system damping ratio is zero.

Example 5.2.1   Velocity Feedback

The loop gain function in Fig. 5.3(a) is

$$\frac{C}{E} = \frac{25}{s(s + 2 + 25K_g)}$$

and the characteristic equation

$$s^2 + (2 + 25K_g)s + 25 = 0$$

Let a system damping ratio $\zeta = 0.7$ be desired. Comparison with $s^2 + 2\zeta\omega_n s + \omega_n^2 = 0$ shows that $\omega_n = 5$ and $2\zeta\omega_n = 10\zeta = 2 + 25K_g$. Thus the required velocity feedback gain $K_g = 0.2$. The unit step response is shown in Fig. 5.3(b), determined from the closed-loop transfer function

$$\frac{C}{R} = \frac{25}{s^2 + (2 + 25K_g)s + 25}$$

The overshoot is less than 5%, as expected from Fig. 3.13 for a damping ratio of 0.7.

Since the characteristic equation is of second order, a root locus showing how the system poles and damping ratio change with K_g can easily be plotted. It

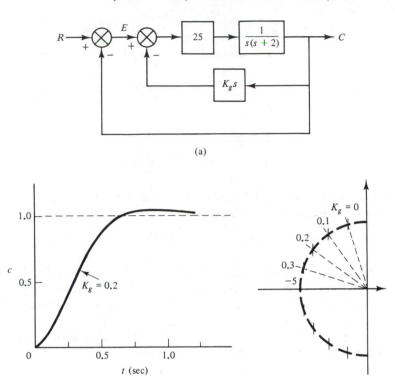

Figure 5.3 Example 5.2.1: velocity feedback.

is shown in Fig. 5.3(c). Except for large enough K_g, the poles lie on a circle of radius $\omega_n = 5$. The angle ϕ of the poles with the negative real axis depends on K_g according to $\zeta = \cos \phi = 0.2 + 2.5K_g$. The improvement due to velocity feedback is evident. The angle ϕ for $K_g = 0$ is not far from 90° and corresponds to a damping ratio of only 0.2.

However, the effect of this compensation on steady-state errors must also be considered. The loop gain function C/E shows that the system is still of type 1, so steady-state errors after step inputs remain zero. But the gain is $25/(2 + 25K_g)$, so the steady-state error following a unit ramp is $(0.08 + K_g)$ and increases with increasing velocity feedback gain K_g. Between $K_g = 0$ and $K_g = 0.2$ it changes from 0.08 to 0.28. Thus K_g should not be made larger than necessary to raise ζ to a satisfactory level.

5.3 SERIES COMPENSATION USING PID CONTROLLERS

If the series compensator $G_c(s)$ in Fig. 5.1(a) is a proportional plus integral plus derivative controller, its output m and input e are related by the equation

$$m = K_c e + K_i \int e \, dt + K_d \dot{e} \tag{5.3}$$

From Table 1.6.1, this implies the transfer function

$$G_c(s) = K_c + \frac{K_i}{s} + K_d s \qquad (5.4)$$

PID controllers are available off the shelf in a wide variety of pneumatic and electronic implementations. Alternatively, their action may be programmed in the form of a *control algorithm* on digital computers or microprocessors.

Not all terms need be present. P control, where $G_c(s) = K_c$, was used in Chapter 4, where its limitations were also examined.

PI control is extremely common in *process control,* a very large and important area of application of control systems. Process control is usually concerned with regulator systems, to maintain controlled variables such as temperatures, pressures, and levels at constant values despite disturbances and parameter variations. The water level control system in Fig. 4.1 is an example where the controller is most likely to be of the PI type.

The need for PI control to reduce steady-state errors was raised in both Examples 4.6.1 and 4.6.2, for plants consisting of one and two simple lags. With one simple lag, the high gains needed for small errors imply large actuating signals to the plant, so a high cost of control. With two simple lags, Fig. 4.18 shows excessive steady-state error even when a system damping ratio of 0.5 is allowed. To explain the basic control actions, the P control in these examples will be replaced with PI control and PD control.

Example 5.3.1   PI Control of Simple Lag Plant

In Fig. 5.1(a) let

$$G(s) = \frac{1}{s + 1} \qquad G_c(s) = \frac{K_c(s + z)}{s} \qquad (5.5)$$

The integral of the error as well as the error itself is used for control. The loop gain function

$$G_c G = \frac{K_c(s + z)}{s(s + 1)}$$

shows immediately the essential improvement due to the addition of I control. The system has unity feedback and Table 4.5.1, on steady-state errors, applies. Thus:

> By adding I control, the system has been changed from type 0 to type 1, and hence now has zero steady-state error following a step input.

This improvement is due to the factor s in the denominator of $G_c(s)$. As explained in Section 4.5, for a step input a steady state cannot exist unless the input to the integrator, the error e, is zero. A direct physical reason will be given as well, when the pneumatic PI controller is discussed.

The closed-loop transfer function is

$$\frac{C}{R} = \frac{K_c(s + z)}{s^2 + (1 + K_c)s + K_c z}$$

and design involves the choice of K_c and z. In Example 3.8.1 the effect of zero location was considered by designing a number of simple lag plants with PI control which all have the same closed-loop poles but different zeros. In the present case the effect of different choices of z will be compared for designs of which the closed-loop poles all have real part -2. So all have time constant 0.5 and identical settling times. If the imaginary part is a, the system characteristic polynomial is then

$$(s + 2 - ja)(s + 2 + ja) = s^2 + 4s + 4 + a^2$$

and equating this with the denominator of C/R yields

$$K_c = 3 \qquad a^2 = 3z - 4$$

The pole–zero patterns of C/R corresponding to the choices $z = 2$, 3, and 4 are shown in Fig. 5.4(b). Also shown is that for the special case $z = 1$, when the zero cancels the plant pole at -1 and the closed-loop transfer function with the closed-loop pole at -2 is $C/R = 2/(s + 2)$.

The unit step response for $z = 1$ is $c(t) = 1 - e^{-2t}$, and is shown in Fig. 5.4(a) together with those for $z = 2$, 3, and 4. The latter may be found using graphical determination of the residues or by means of the computer programs in Appendix B. The responses are of the form $c(t) = 1 + Ae^{-2t} \cos(at + \theta)$, where for $z = 2$, 3, and 4 the values of (A, a, θ) are, respectively, $(1.22, 1.414, -144.7°)$, $(1.096, 2.236, -155.9°)$, and $(1.060, 2.828, -160.6°)$.

Comparison of the responses with the pole–zero patterns again demonstrates that due to the zeros in relatively dominant positions, overshoot cannot be predicted on the basis of the damping ratios associated with the pole positions. These ratios are indicated on Fig. 5.4(b) and would for $z = 2$, 3, and 4 predict overshoots much smaller than the 7.5%, 13.5%, and 18.5% actually present. Noting that all designs have the same settling time, the smaller rise time and greater speed of response associated with larger imaginary part of the poles is also again in evidence.

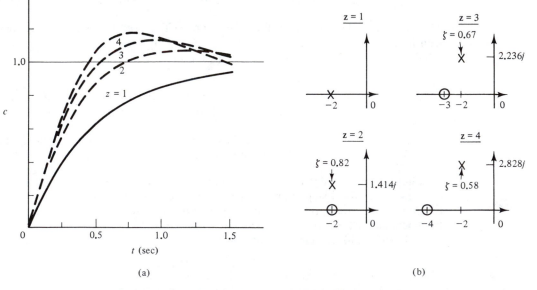

Figure 5.4 Example 5.3.1: PI control.

Example 5.3.2   PI Control of Plant with Two Simple Lags

With the same plant as in Example 4.6.2, in Fig. 5.1(a) let

$$G(s) = \frac{1}{(s + 1)(0.5s + 1)} \tag{5.6}$$

$$G_c(s) = K_c + \frac{K_i}{s} = K_c \frac{s + z}{s} \tag{5.7}$$

where $z = K_i/K_c$. The loop gain function

$$G_cG = K_i \frac{1 + s/z}{s(s + 1)(0.5s + 1)} = 2K_c \frac{s + z}{s(s + 1)(s + 2)}$$

shows that the system has again been made to be of type 1, with zero steady-state error following step inputs as discussed in Example 5.3.1. The gain in the sense of Section 5.4, with all constant terms made unity, is K_i, so that the steady-state error following a unit ramp input is $1/K_i$ and reduces as the integral gain factor $K_i = K_c z$ is increased. This applies also in Example 5.3.1, where, since $K_c = 3$, this error is smallest for the design with $z = 4$. Figure 5.4(a) shows, as one would expect, that the improvement is limited by considerations of relative stability.

The closed-loop transfer function

$$\frac{C}{R} = \frac{K_c(s + z)}{0.5s^3 + 1.5s^2 + (1 + K_c)s + K_c z}$$

shows a third-order characteristic equation. Design to determine suitable values of K_c and z can be carried out using the root locus and frequency response methods discussed in the following chapters. This will show that careful design is necessary to avoid oscillatory or slow system dynamic behavior. For given values of K_c and z, of course, the programs in Appendix B can be used to compute transient responses. Alternatively, the program for finding the roots of a polynomial may be used to determine the closed-loop poles, and the graphical residue method to calculate the transient response.

However, it is important to point out that the use of controllers in practice does not depend on finding values of K_c and K_i by analytical means. Instead, these are set by following established procedures for tuning a controller after installation. This is why experienced personnel with knowledge of the process can apply PID controllers without any formal knowledge of mathematical modeling or control theory. This important reason for the success of PID controllers should be recognized as a major advantage, even in a book on these theories.

Example 5.3.3   PD Control of Plant with Two Simple Lags

Here both the error and its derivative are used for control. In Fig. 5.1(a), with the same plant as in Example 5.3.2:

$$G(s) = \frac{1}{(s + 1)(0.5s + 1)} \qquad G_c(s) = K_c + K_d s \tag{5.8}$$

The loop gain function is, with $z = K_c/K_d$,

$$G_cG = K_c \frac{1 + s/z}{(s + 1)(0.5s + 1)} = 2K_d \frac{s + z}{(s + 1)(s + 2)} \tag{5.9}$$

This is a type 0 system with gain K_c, so if K_c is the same as K for P control in Fig. 4.18, the addition of D control will have no effect on steady-state errors. The physical explanation of this result is straightforward:

> The controller output is $m = K_c e + K_d \dot{e}$. Under steady-state conditions for a step input, $e = $ constant, so $\dot{e} = 0$, and hence the derivative control component has no effect.

However, the steady-state error can be reduced indirectly. This is because the addition of D control improves relative stability and therefore allows K_c to be larger than with P control alone.

Let a damping ratio 0.5 be required. The closed-loop transfer function is

$$\frac{C}{R} = \frac{2K_d(s + z)}{s^2 + (3 + 2K_d)s + 2 + 2K_d z} = \frac{2K_d(s + K_c/K_d)}{s^2 + (3 + 2K_d)s + 2 + 2K_c}$$

For $K_d = 0$, the characteristic equation is $s^2 + 3s + 2 + 2K_c = 0$, and to achieve $\zeta = 0.5$ requires $\omega_n^2 = 9 = 2 + 2K_c$, so $K_c = 3.5$. The steady-state error for a unit step input is $1/(1 + K_c)$, so equals 22%, and the system poles are located at A in Fig. 5.5. If the specifications permit an error of 10%, $K_c = 9$ is required, and for $K_d = 0$ the poles move to B, at the same distance as A to the imaginary axis and at a distance $\omega_n = \sqrt{2 + 2K_c} = \sqrt{20} = 4.472$ to the origin. The damping ratio associated with B is an inadequate 0.335 ($= 1.5/4.472$). To see the effect of D control, it is observed by equating the denominator of C/R with $(s^2 + 2\zeta\omega_n s + \omega_n^2)$ that the distance ω_n of the poles to the origin depends only on K_c, and that to the imaginary axis only on K_d:

$$\omega_n^2 = 2 + 2K_c \qquad 2\zeta\omega_n = 3 + 2K_d \tag{5.10}$$

Thus, varying K_c for constant K_d moves the poles vertically, and varying K_d for constant K_c moves them along a circle. The pole location C in Fig. 5.5 will realize both the desired damping and the desired steady-state accuracy, and is obtained, with $K_c = 9$, if $\zeta\omega_n = 1.5 + K_d = 2.236$, so if $K_d = 0.736$. The zero of the closed-

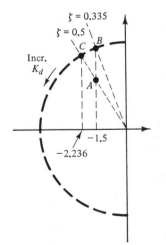

Figure 5.5   Example 5.3.3: PD control.

loop transfer function is $-K_c/K_d = -9/0.736 = -12.23$. This is about 5.5 times as far from the imaginary axis as the poles at C, so that the unit step responses should closely resemble those for a quadratic lag in Fig. 3.11. It is apparent that the addition of D control improves the damping for a given value of K_c. From a different point of view, PD control may also be used to improve accuracy, because it permits an increase of K_c without loss of system damping. The reduction of settling time from the systems A or B to C should be noted as well.

A physical appreciation of why derivative control improves damping may be gained from the following:

> The derivative control component responds to the rate of change of error, and hence gives a stronger control signal if the error changes faster. Thus it anticipates large errors, and attempts corrective action before they occur.

5.4 PNEUMATIC PID CONTROLLERS

Pneumatic controllers that provide approximations to the idealized PID control of Section 5.3 are available in a wide variety of designs and makes. The operation of a schematic design is discussed in this section, and the behavior modeled by means of block diagrams.

The design may be seen as a possible implementation of the controller block identified in Fig. 4.1 for the water-level control system. This block, shown in Fig. 5.6(a), has a mechanical position input e, which is a measure of the error in the control system, and a pneumatic pressure output P_o, which, after a power amplification stage which is not indicated, operates the actuator to adjust the valve.

The simplest possible configuration to control P_o from e is indicated in Fig. 5.6(b), and is not PID control, but *on-off control*. The operation is based on the flapper–nozzle device. Air is supplied through a constant restriction from

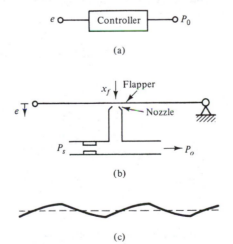

(a)

(b)

(c)   **Figure 5.6** On-off control.

a source with constant pressure P_s. When the flapper is moved closer to the nozzle, the resistance to airflow out of the nozzle is increased, and the back pressure P_o, which is also the output pressure, rises. Linearized models will be used, assuming small variations about an operating point. P_o and x_f in Fig. 5.6(b) represent changes from operating-point values of output pressure and flapper–nozzle distance, with x_f positive in the direction shown. Hence, if x_f increases, P_o increases, and a logical linearized model is

$$P_o = K_f x_f \qquad (5.11)$$

The problem is, however, that the gain K_f of the flapper–nozzle amplifier is extremely large. This means that only a very small change of e will already cause P_o to change from its minimum to its maximum value, or vice versa. Hence the control valve in the level control in Fig. 4.1 switches back and forth between fully open and fully closed for only very small variations of e. Hence the name "on-off control," and as a result the level will fluctuate about the desired value as suggested by Fig. 5.6(c). P control is the first step toward improved performance.

Proportional Control

A P controller is shown in Fig. 5.7(a). The right end of the flapper is controlled by the pressure P_o via a bellows that expands proportional to pressure:

$$x_b = K_b P_o \qquad (5.12)$$

The mechanical lever is modeled as in Fig. 4.3 for the hydraulic servo:

$$x_f = \frac{a}{a + b} e - \frac{b}{a + b} x_b \qquad (5.13)$$

With input e at the left and output P_o at the right, these equations immediately yield the block diagram in Fig. 5.7(b). It identifies the lever as the summing junction, and the bellows and lever as feedback elements. Because K_f is very large, the closed-loop transfer function is

$$\frac{P_o}{e} = \frac{a}{a + b} \frac{K_f}{1 + K_f K_b b/(a + b)} = \frac{a K_f}{a + b + K_f K_b b} \approx \frac{a}{b K_b} \qquad (5.14)$$

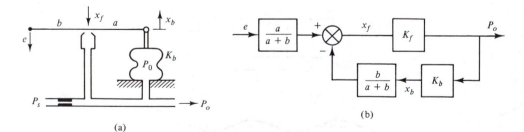

(a)   (b)

Figure 5.7 P controller.

This proves that the P control gain $K_c = a/(bK_b)$, and can be set to a desired value by adjusting the lever ratio.

The steady-state error for step inputs, predicted for this type 0 system by Section 4.5, can be explained from Fig. 5.7(a). As noted earlier, the motion x_f is extremely small and the lever essentially pivots around a fixed point at the nozzle. A larger valve flow requires larger P_o, so raises the right end of the lever, and hence depresses the left end, which represents the error in the level control system. Zero steady-state error would require the possibility for P_o to have any value without a change of the steady-state position of the left end of the lever. This can be realized by expanding the P controller to PI control.

Proportional Plus Integral (or Reset) Control

A PI controller is shown in Fig. 5.8(a). A bellows connected to P_o via a severe resistance R_i is added to oppose the proportional bellows. This resistance is so severe that P_i rises only very slowly after a step increase of e, causing the system to operate initially much like a P controller and causing x_b to rise. When this transient operation is largely complete, P_i is still increasing, pushing x_b down. Finally, in the steady state P_i equals P_o, balancing the two bellows and ensuring that $x_b = 0$. Since, as discussed earlier, the changes of x_f needed to produce

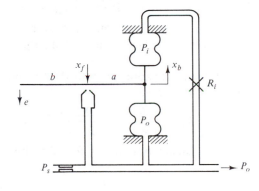

(a)

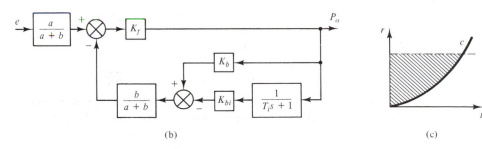

(b)   (c)

Figure 5.8   PI controller.

the range of output pressures P_o is extremely small as well, this means that in the steady state the left end of the lever is in virtually unchanged position, $e = 0$.

This physical explanation of the zero steady-state errors can be confirmed mathematically. Similarly to P control:

$$P_o = K_f x_f \qquad x_f = \frac{a}{a + b} e - \frac{b}{a + b} x_b \tag{5.15}$$

$$x_b = K_b P_o - K_{bi} P_i \tag{5.16}$$

The model for the pressure P_i is analogous to that for the pneumatic tank in Fig. 2.21. The airflow rate through resistance R_i is

$$q = \frac{P_o - P_i}{R_i} \tag{5.17}$$

Assuming small bellows motion, the capacitance C_i is about constant, so that q raises pressure P_i according to

$$q = C_i \dot{P}_i \tag{5.18}$$

Eliminating q now yields the model

$$R_i C_i \dot{P}_i + P_i = P_o \qquad \frac{P_i(s)}{P_o(s)} = \frac{1}{T_i s + 1} \qquad T_i = R_i C_i \tag{5.19}$$

Equations (5.15), (5.16), and (5.19) immediately provide the block diagram in Fig. 5.8(b), which shows that both bellows and the lever operate as feedback elements.

To prove that this system approximates PI control, the two parallel feedback loops are combined first, to give

$$\frac{X_b}{P_o} = K_b - \frac{K_{bi}}{T_i s + 1} = \frac{K_b T_i s}{T_i s + 1} \qquad \text{if } K_b = K_{bi} \tag{5.20}$$

Then the closed-loop transfer function is found to be

$$\frac{P_o(s)}{E(s)} = \frac{a K_f}{a + b} \Bigg/ \left(1 + \frac{b}{a + b} \frac{K_b T_i s}{T_i s + 1} K_f \right)$$

$$= \frac{a K_f (T_i s + 1)}{(a + b)(T_i s + 1) + b K_b T_i s K_f}$$

Since K_f is quite large, this approximates

$$\frac{P_o(s)}{E(s)} \approx \frac{a}{b K_b} \frac{T_i s + 1}{T_i s} = \frac{a}{b K_b} + \frac{a}{b K_b T_i} \frac{1}{s} \tag{5.21}$$

This is indeed PI control, and it should be noted that the proportional and integral gain constants K_c and K_i in (5.4) can only be adjusted independently via $T_i(R_i)$:

$$K_c = \frac{a}{b K_b} \qquad K_i = \frac{a}{b K_b T_i} \tag{5.22}$$

Reset windup or *integral windup* is a problem that can arise in all physical implementations of integral control, including those by digital computers and microprocessors. For large changes of input or large disturbances it could lead to severe transient oscillations. Referring to Fig. 5.8(c), if a large step input r is applied, the error $e = r - c$ will be of the initial sign for a considerable period, until the output c passes the level of r. During this period the integrator accumulates a large output. It winds up to a much larger value than that which makes the actuator go to its physical limit, such as the fully open position of a control valve. After c passes r, it takes long for the integrator to wind down to a value where the controller output is back inside the range where the valve even begins to close. This delay in closing can cause a very large overshoot and severe oscillations. To avoid this, provisions are usually made to clamp the integrator output and prevent values outside selected high and low limits.

Proportional Plus Derivative Control

A PD controller is shown in Fig. 5.9(a). It is seen to be identical to the P controller in Fig. 5.7(a) except that a resistance R_d is added in the line to the bellows. In P control, if a step increase of e is applied, x_b rises immediately to counteract its effect on x_f. In PD control, the presence of R_d delays the rise of P_d, and hence of x_b. If e changes faster, P_d lags more behind P_o, and the flapper is correspondingly closer to the nozzle. This represents PD control because P_o contains a component proportional to the rate of change $\dot{e}$.

The block diagram in Fig. 5.9(b) can be verified in a manner similar to that for PI control, and its approximate transfer function can be shown to be

$$\frac{P_o(s)}{E(s)} \approx K_c + K_d s \qquad K_c = \frac{a}{bK_b} \qquad K_d = \frac{aT_d}{bK_b} \qquad (5.23)$$

PID Control

The PI controller in Fig. 5.8(a) becomes a PID controller when a resistance is added in the line to the proportional bellows. The I and D actions do not interfere

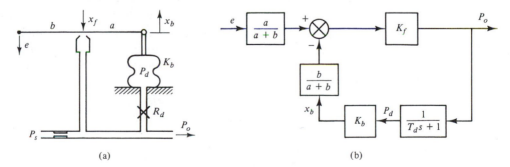

(a)   (b)

Figure 5.9 PD controller.

with each other because the time constant T_i is much larger than T_d; that is, the integral action is made to occur much more slowly by making resistance R_i much larger than R_d.

5.5 DYNAMIC COMPENSATION AND SYSTEM SIMULATION USING OPERATIONAL AMPLIFIERS

The operational amplifier, discussed briefly in Section 5.1, is a dc amplifier with very high gain, of the order of 10^5 to 10^8, and very high input impedance. It is shown symbolically in Fig. 5.10(a), where the ground connections have been ignored. The amplifier changes the sign, so that amplifier output e_o and input e_g are related by the equation

$$e_o = -Ae_g \qquad A \approx 10^5 \text{ to } 10^8 \tag{5.24}$$

Because A is so large, and e_o is limited to a maximum of 100 volts, 10 volts, or less, depending on the amplifier, the input voltage e_g is virtually zero. The amplifier input current i_g is also virtually zero, because the input impedance is quite large. Hence, in the application of the amplifier, the following can be assumed:

$$e_g = 0 \qquad i_g = 0 \tag{5.25}$$

Figure 5.10(b) shows the basic building block in which the operational amplifier is used. An input impedance Z_i is connected to the input, and a feedback impedance Z_f between output and input.

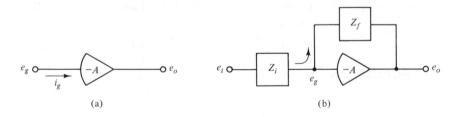

(a)   (b)

Figure 5.10 (a) Operational amplifier; (b) system building block.

From the equations for the elements of electrical systems in Fig. 2.7(a), voltage v and current i are related by $v = iR$ for a resistance R and by $i = C\dot{v}$ for a capacitance C. The impedance Z can be expressed as the ratio of the Laplace transforms $V(s)$ and $I(s)$:

$$Z = \frac{V}{I} \qquad \begin{array}{l} \text{resistance: } V = IR; \quad Z = R \\ \text{capacitance: } I = CsV; \quad Z = \dfrac{1}{Cs} \end{array} \tag{5.26}$$

Since $e_g = 0$, the current through Z_i in Fig. 5.10(b) is E_i/Z_i and that through Z_f is $-E_o/Z_f$, where E_i and E_o are the transforms of e_i and e_o. Also, since $i_g = 0$, these two currents must be the same. Hence

$$\frac{E_i}{Z_i} = -\frac{E_o}{Z_f} \qquad \frac{E_o}{E_i} = -\frac{Z_f}{Z_i} \tag{5.27}$$

*****

By choosing Z_i and Z_f, operational amplifiers can be used to realize active dynamic compensation transfer functions for use in control systems, for example to replace the passive RC networks in Fig. 2.8. They can also serve as summing junctions or as P controllers.

The most common cases are listed below.

1. *Constant gain (P controller):*

$$Z_i = R_i \qquad Z_f = R_f$$

$$\frac{e_o}{e_i} = -\frac{R_f}{R_i}$$

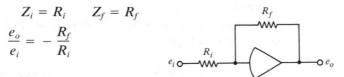

For a gain $a \le 1$, a potentiometer can be used:

$$a = \frac{R_p}{R_t} \qquad e_o = ae_i$$

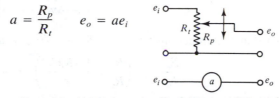

R_f/R_i is only adjustable in steps, usually between 0.1 and 10, limited by available resistor values. To realize a gain of, say, 3.45, a series connection of an amplifier with $R_f/R_i = 5$ and a potentiometer with $a = 0.69$ can be used.

2. *Summer (summing junction):* Analogous to (5.27), the current $-e_o/R_f$ through R_f must equal the sum of the currents e_j/R_j through the parallel input resistors. Hence

$$e_o = -\frac{R_f}{R_1}e_1 - \frac{R_f}{R_2}e_2 - \frac{R_f}{R_3}e_3$$

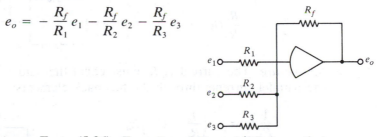

3. *Integrator:* From (5.26), $Z_i = R_i$ and $Z_f = 1/(C_f s)$, so that

$$\frac{E_o}{E_i} = \frac{-1}{R_i C_f s} \qquad e_o = \frac{-1}{R_i C_f}\int_0^t e_i\, dt + e_o(0)$$

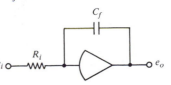

An initial condition e_o (0) can be applied as an initial voltage across the capacitor.

4. *Differentiator:*

$$Z_i = \frac{1}{C_i s} \qquad Z_f = R_f$$

$$\frac{E_o}{E_i} = -R_f C_i s \qquad e_o = -R_f C_i \dot{e}_i$$

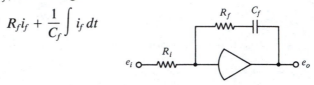

This element is very sensitive to noise which may be superimposed on the input signal, because the derivative represents the slopes of the irregularities in the output.

5. *Proportional plus integral (PI) controller:* $Z_i = R_i$. For the feedback, if the current is i_f, the voltage is

$$R_f i_f + \frac{1}{C_f} \int i_f \, dt$$

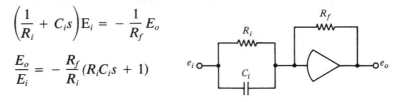

so that the impedance is $Z_f = R_f + 1/(C_f s)$, and

$$\frac{E_o}{E_i} = -\left(\frac{R_f}{R_i} + \frac{1/(R_i C_f)}{s}\right) = -\frac{K(\tau s + 1)}{s}$$

with $K = 1/(R_i C_f)$ and $\tau = R_f C_f$

6. *Proportional plus derivative (PD) controller:* The sum $[(e_i/R_i) + C_i \dot{e}_i]$ of the parallel currents through R_i and C_i must equal the current $-e_o/R_f$ through R_f:

$$\left(\frac{1}{R_i} + C_i s\right) E_i = -\frac{1}{R_f} E_o$$

$$\frac{E_o}{E_i} = -\frac{R_f}{R_i}(R_i C_i s + 1)$$

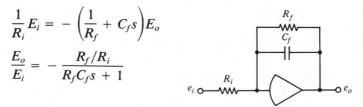

7. *Simple lag:* The current e_i/R_i must equal the sum $-[(e_o/R_f) + C_f \dot{e}_o]$ of the parallel currents through the feedback elements:

$$\frac{1}{R_i} E_i = -\left(\frac{1}{R_f} + C_f s\right) E_o$$

$$\frac{E_o}{E_i} = -\frac{R_f/R_i}{R_f C_f s + 1}$$

8. *Phase-lead or phase-lag compensators:* Analogous to the above:

$$\left(\frac{1}{R_i} + C_i s\right) E_i = -\left(\frac{1}{R_f} + C_f s\right) E_o$$

$$\frac{E_o}{E_i} = -\frac{R_f R_i C_i s + 1}{R_i R_f C_f s + 1}$$

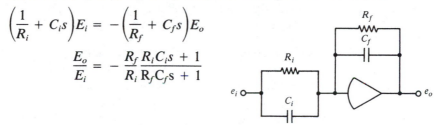

9. *Lead compensator (high-pass filter):* Virtually by inspection, after the preceding examples,

$$\frac{E_o}{E_i} = -\frac{R_f C_i s}{R_f C_f s + 1}$$

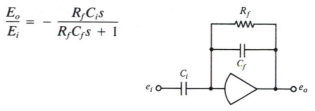

Analog Computer Simulation

Block diagram simulation.   In this approach for simulating a feedback system on an analog computer, the individual blocks are modeled, using the preceding examples, as shown in Fig. 5.11 for the case of PI control of a process with two simple lags discussed in Example 5.3.2. The numbers in the simulation refer to one of the cases listed previously, of which the equations would be used to determine the resistor and capacitor values required.

It is likely that at least one potentiometer will be needed to match the loop

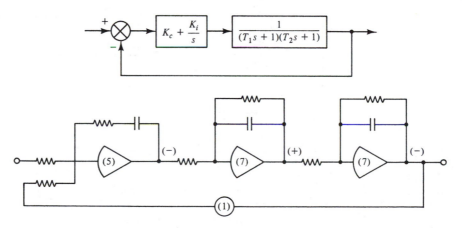

Figure 5.11   Block simulation using operational amplifiers.

gain. The sign inversions are indicated in parentheses. If the last one had been positive, a constant-gain element $R_f/R_i = 1$ would have to be added in the feedback to ensure that it will be negative.

General-purpose simulation.   The preceding approach is natural and convenient for control system simulation. General-purpose analog simulation, however, is usually concerned with one or more differential equations, and uses only the elements 1, 2, and 3 of those listed previously.

To explain the idea, Fig. 5.12 shows a simulation of the differential equation

$$\ddot{x} + 6\dot{x} + 22x = r \tag{5.28}$$

In Fig. 5.12, the values shown with resistors are the resistances in megaohms, and those with capacitors the capacitances in microfarads. If both are 1 for an integrator, its multiplication factor is

$$-\frac{1}{R_i C_f} = -\frac{1}{10^6 \times 10^{-6}} = -1$$

Equation (5.28) is first solved for $\ddot{x}$:

$$\ddot{x} = -6\dot{x} - 22x + r \tag{5.29}$$

The idea of the simulation is that $\ddot{x}$ is integrated once to obtain $\dot{x}$, and then again to generate x. These variables are then fed back to generate $\ddot{x}$ in a summer amplifier according to (5.29). Because of the sign inversion in the summer amplifier, it is necessary to add an "inverting" amplifier in the feedback from $\dot{x}$. The reason for generating $-5\dot{x}$ instead of $-\dot{x}$ (another choice could have been made) is to obtain a reasonable distribution of gains and avoid, for example, the need for amplifiers in the feedback from x.

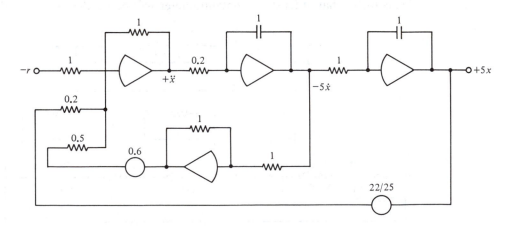

Figure 5.12   Simulation of (5.28).

This is only a brief introduction into the subject of analog computers. Discussion of the problems of amplitude scaling, to ensure that output voltages do not exceed permissible levels, and of time scaling, to simulate slow processes in less than real time, has been omitted. Analog computers can be used as well to simulate and study the behavior of multivariable systems, which have more than one input and output and are described by sets of differential equations. Function generators are available to include the effect of system nonlinearities in such simulations. The proliferation of digital computers and the availability of techniques for digital simulation, however, have had a great deal of impact on this area.

5.6 CONCLUSION

Chapters 4 and 5 have been concerned with the modeling and the performance of systems with feedback. The key questions raised in Section 1.4, that is, how systems behave and how this behavior may be changed, were both considered. Indeed, Chapter 4 showed that profound changes of performance result by changing only the gain of the loop gain function. However, the discussion also indicated the limitations of this P control in achieving both adequate relative stability and satisfactory accuracy.

In Chapter 5, therefore, more flexibility was provided by replacing the P controller by a dynamic compensator. The physical realization of such compensators by pneumatic means and active circuits based on operational amplifiers was treated. The latter are an alternative to the passive electrical networks in Fig. 2.8. Digital computer realization in the form of computer control algorithms will be discussed later.

The essential effects of the basic integral and derivative control actions were emphasized in the sections on PID control. Phase-lag and phase-lead compensators, also very widely used and discussed later, rely on approximations to these basic actions.

However, in these chapters the inadequacy of the tools made available so far for system analysis and design has become apparent. For example, how can the effect of gain or compensator parameters on relative stability be determined if the system characteristic equation is of order 3 or higher? Further, while the final value theorem easily yields steady-state errors, what about accuracy under dynamic conditions? For example, if the frequency spectrum of the system input is known to be in a certain range, what is the largest error that may occur, or how does one design systems for which this error satisfies given specifications?

The root locus method and frequency response techniques which are discussed next are the classical tools for both system analysis and design. Frequency response considerations will also provide new and important insights into the behavior of feedback systems.

PROBLEMS

5.1. (a) For the motor position servo in Fig. P5.1 without the rate or velocity feedback $(K_g = 0)$, find K for a system damping ratio 0.5 and the corresponding steady-state error following unit ramp inputs.

(b) Still with $K_g = 0$, what value of K will give a steady-state unit ramp following error of 0.1, and what is corresponding damping ratio?

(c) With K as in part (b), what value of K_g will give a system damping ratio 0.5? How does the steady-state error compare with that in part (b)?

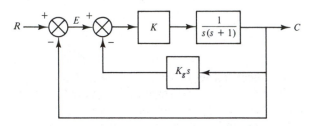

Figure P5.1

5.2. In Fig. P5.2, the system in Fig. P5.1 for $K = 10$ has been extended by including the amplifier K_a. Determine whether it is now possible to achieve both the steady-state error 0.1 in Problem 5.1(b) and the damping ratio 0.5 in Problem 5.1(c). If so, find the values of K_a and K_g required.

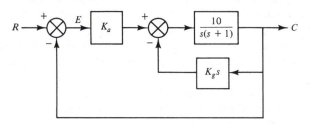

Figure P5.2

5.3. The transient velocity feedback in the motor position servo in Fig. P5.3 is a modification of the velocity feedback in Fig. P5.2. It may be realized by following the tachogenerator by a simple lag RC circuit, and may be used to reduce the effect of high-frequency noise, as a later chapter will show.

(a) Determine the steady-state errors for unit step and unit ramp inputs, and compare with those for pure velocity feedback for the same values of K_c and K_g.

(b) Use the Routh–Hurwitz criterion to investigate stability for all values of K_c.

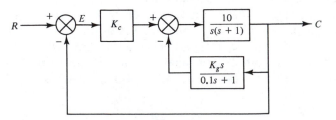

Figure P5.3

5.4. The block diagram of a roll stabilizer for a ship such as considered in Problem 4.8 is shown in Fig. P5.4. Minor loop rate feedback is included because of the low damping associated with the ship dynamics.

   (a) Express the transfer function for the effect of wave disturbance torque T_d on ship roll angle C.

   (b) Find the equations that must be satisfied by K_a, K_1, and K_g to ensure both a steady-state value of no more than 0.1 for C in response to a unit step T_d and a system damping ratio 0.5.

   (c) Which of K_1 and K_a must be adjustable to enable both specifications in part (b) to be met?

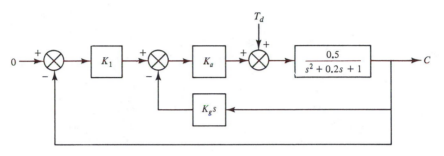

Figure P5.4

5.5. In Fig. P5.5 the plant is a simple lag $G = 1/(s + 1)$. To demonstrate why stability considerations usually dictate the choice of PI control $G_c = K_c + K_i/s$ over pure I control $G_c = K_i/s$ if zero steady-state errors for constant inputs are desired, try to design these controllers for a system damping ratio 0.5 and either of the following conditions:

   1. A steady-state error of 0.25 following unit ramp inputs.
   2. A settling time of about 4 sec.

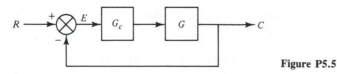

Figure P5.5

5.6. In Problem 5.5, compare P control and PI control on the basis of steady-state errors and nature of the response if both are designed for a time constant $T = 0.5$ sec. Note the design freedom still left with PI control, and use it to minimize rise time subject to a constraint of about 0.7 on damping ratio, a desirable solution.

5.7. In Problem 5.5:

   (a) For I control $G_c = K_i/s$, find K_i for a steady-state error of 0.25 for unit ramp inputs, and the corresponding damping ratio.

   (b) Suppose that D control is added in part (a) to improve damping; that is, $G_c = (K_i/s) + K_d s$ is used. Investigate the stability and accuracy properties with this ID control, and conclude whether this is a desirable addition.

(c) Evaluate whether D control by itself, $G_c = K_d s$, realizes desirable properties for steady-state error and settling time. Explain the steady-state error results physically.

5.8. Compare the effect of the location of the integrator in the loop gain functions in Fig. P5.8 on stability and steady-state errors for step and ramp inputs of both reference input R and disturbance input D. Express and compare all steady-state errors:

(a) How do the errors for R and D compare in part (i)?

(b) How do these errors compare for part (ii)?

(c) Explain these differences in behavior.

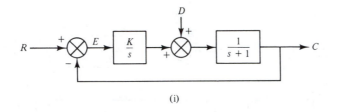

(i)

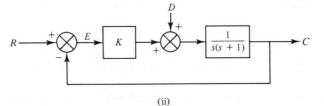

(ii)   **Figure P5.8**

5.9. In Fig. P5.5, Let $G(s)$ be the open-loop unstable plant $G(s) = 1/(s - 1)$. Design the simplest possible controller $G_c(s)$ that will satisfy all of the following specifications:

1. The steady-state error for constant inputs must be zero.
2. The system settling time must be about 4 sec.
3. The system damping ratio should be 0.5.

5.10. Let Fig. P5.5 model a temperature control system with plant transfer function $G(s) = 1/[(s + 1)(s + 5)]$.

(a) With P control $G_c = K_c$, what is the system type number, and what is the gain?

(b) For $G_c = K_c$, find K_c for a damping ratio 0.5 and the corresponding steady-state error for a unit step input.

(c) Choose the form of controller that will make this steady-state error zero, and write the characteristic equation to note why this choice complicates stability analysis.

(d) Use the Routh–Hurwitz criterion to determine for what relations among the parameters, if any, either of these systems may be unstable.

5.11. (a) In Problem 5.10, compare P control $G_c = K_c$ and PD control $G_c = K_c + K_d s$. A system damping ratio 0.5 is required and the steady-state error for step inputs should not exceed 5%.

(b) If a 15% steady-state error is acceptable, compare the solutions on the basis of settling time and rise time.

5.12. Similar to the motor position servo with a load disturbance torque T_l in Fig. 4.6, Fig. P5.12 has been extended to include velocity feedback.

(a) If $G_c = K_c$, a gain, find K_g and K_c to obtain a system damping ratio 0.5 and 5% steady-state error for step inputs T_l.

(b) Does K_g affect this steady-state error directly? If not, why not?

(c) How does the velocity feedback affect steady-state errors?

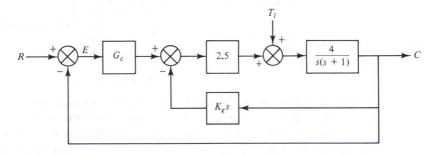

Figure P5.12

5.13. In Fig. P5.12:

(a) Why is there a difference in steady-state error behavior to step inputs of R and of T_l for this type 1 system?

(b) PI control for G_c, $G_c = K_c + K_i/s$, would be a natural choice if the steady-state error for step changes of T_l must be zero. Verify that this is indeed achieved.

(c) Write the system characteristic equation for the choices of K_c and K_g made in Problem 5.12(a), and use the Routh–Hurwitz criterion to determine the limiting value of K_i for stability.

5.14. A diagram of a PID controller is shown in Fig. P5.14.

(a) Write the equations and derive the block diagram representation.

(b) Show that for large flapper–nozzle gains the transfer function P_o/E approaches idealized PID control if the bellows spring constants are equal.

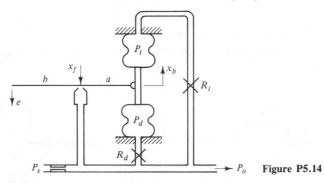

Figure P5.14

5.15. The water level control system of Fig. P4.1 has been extended in Fig. P5.15 by an additional lever and bellows.

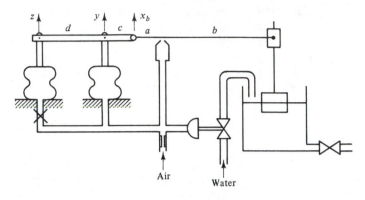

Figure P5.15

(a) Amend the block diagram for Fig. P4.1 to represent this system.

(b) Assuming as before that K_f is very large, derive the condition on the parameters for which the controller has the most common form for this type of system.

5.16. Realize the following dynamic compensators by means of operational amplifiers, giving values for resistors and capacitors in terms of megaohms ($= 10^6$ ohm) and microfarads ($= 10^{-6}$ farad):

1. PI controller: $-(10 + 20/s)$.

2. Phase-lead compensator: $-2(0.5s + 1)/(0.1s + 1)$.

Note that capacitors and resistors are not available in odd sizes.

5.17. Velocity or rate feedback is also used to improve damping in systems of which the plant does not include a motor and the rate of change of the output cannot be measured by a tachometer–generator. The derivative of a measured output must then be found, and a simple lag is often added to reduce the effect of high-frequency noise in the measured output. The desired compensator is then a transient rate or transient velocity minor loop feedback transfer function of the form of the following example:

$$-\frac{5s}{0.1s + 1}$$

Realize this compensator using operational amplifiers.

5.18. Derive the transfer functions corresponding to the operational amplifier circuits in Fig. P5.18.

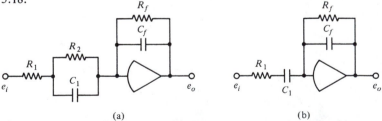

Figure P5.18

5.19. Show the general form of an operational amplifier block diagram type simulation of the motor position control system in Fig. P5.19. The position sensor has a significant time constant T_s, and a phase-lag or phase-lead compensator is used.

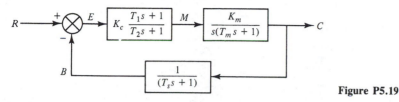

Figure P5.19

5.20. Derive a block diagram type of analog computer simulation for the temperature control system of Problem 5.10 if G_c is a PI control $G_c = 13(1 + 2/s)$. Give numerical values for resistors, capacitors, and potentiometers present, but ignore the problem of amplitude scaling which is normally part of analog computer simulation to ensure that amplifier output voltages do not exceed permissible levels, inside the linear range.

5.21. Derive an analog computer simulation for the differential equation

$$\ddot{x} + 3.5\dot{x} + 9.25x = r$$

(a) Of the type of Fig. 5.11.

(b) An alternative using one less operational amplifier, by using the integrator that generates $\dot{x}$ also as a summing amplifier.

As in Problem 5.20, ignore amplitude scaling and also time scaling, which can be used to simulate slow processes in less than real time.

5.22. A system for the control of water level in the steam drum of a power station boiler is shown schematically in Fig. P5.22(i), and Fig. P5.22(ii) gives its block diagram model. W_f and W_s are the mass flow rates of feedwater to the drum and steam from the drum. G_c is the controller transfer function, and the time constant of the feedwater control valve is neglected. The transfer function of the drum, of which the net inflow is $W_f - W_s$ in the model, is $1/(As)$, as for a hydraulic cylinder. Variations of steam flow W_s due to changes of steam turbine control valve opening are the main disturbances affecting the system. If $K_v = 10$, $A = 5$, examine stability and determine the steady-state errors for step changes of L_r and W_s for P control $G_c = K_c$ and PI control $G_c = K_c + K_i/s$. For PI control, find the relation between K_c and K_i for a system damping ratio of about 0.7. Ignore the dashed links in Fig. P5.22.

5.23. In the drum-level control of Fig. P5.22, the drum model $1/(As)$ is often inadequate, because it does not reflect the swell and shrink of water level that can occur in dynamic operation. *Swell* occurs when an increase of W_s causes a temporary drop of drum pressure, which in turn causes steam bubbles in the water to grow. *Shrink* causes the level to fall temporarily when relatively cool feedwater, to balance the increased steam flow, enters and shrinks the bubbles. These level changes can introduce severe transients and often make PI control inadequate.

Feedforward control is very important in practice to reduce the effect of measurable disturbances. In the present case it is indicated by the dashed link K_s, and in effect introduces a second, parallel, path from W_s to L.

(a) Ignoring link K_f, find the gain K_s in terms of the parameters of a P or PD

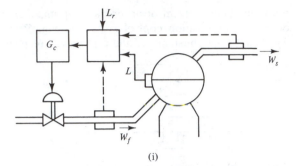

(i)

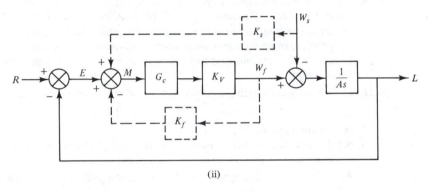

(ii)

Figure P5.22

controller G_c such that the steady-state effect of step changes of W_s on L will be zero.

(b) What would happen in the steady state if in part (a) G_c would include integral control?

5.24. In Fig. P5.22, to allow for disturbances in the feedwater system, the feedwater flow W_f is often also measured and fed back via K_f. The system can be viewed as a flow control system, to make W_f equal to W_s, with an outer correction loop to ensure that W_f and W_s will balance at the desired level.

(a) Find the conditions on the parameters needed to ensure zero steady-state error for step changes of W_s, both with P or PD control and with PI control.

(b) In view of parameter uncertainties, are zero errors actually likely in either case? Where would you locate the PI controller?

5.25. In Fig. P5.5, let $G(s) = 1/[(s + 1)(s + 4)]$ and let the compensator G_c have the form $G_c = K(T_1s + 1)/(T_2s + 1)$. This is called phase-lead compensation if $T_1 > T_2$ and phase-lag compensation if $T_1 < T_2$. Both are very common and their design will be discussed at length. One approach, considered here, is to choose the zero of G_c to cancel one of the plant poles. If the system is to be designed for a damping ratio 0.5 using

1. Proportional control $G_c = K$
2. Phase-lag compensation $K(s + 1)/(5s + 1)$
3. Phase-lead compensation $K(0.25s + 1)/(0.05s + 1)$

then:

(a) Determine the values of K required.

(b) Find and compare the steady-state errors following unit step inputs.

(c) Determine the closed-loop system time constants and compare the speeds of response.

5.26. In Problem 5.25, evaluate the effect of phase-lag compensation by calculating and plotting, on the same graph, the unit step responses for compensators 1 and 2.

5.27. Compare the phase-lead compensation 3 in Problem 5.25 by adding the unit step response curve for this compensator to the graph in Problem 5.26.

6

THE ROOT LOCUS METHOD

6.1 INTRODUCTION

Figure 6.1 shows a system with loop gain function G_cGH. The closed-loop transfer function is $C/R = G_cG/(1 + G_cGH)$, and the closed-loop poles are the roots of the characteristic equation $1 + G_cGH = 0$. The root loci show how these poles move in the s-plane when a parameter of G_cGH is varied. Calculation is easy when the characteristic equation is of first or second order, and loci for such cases were already constructed and used in the examples in preceding chapters. Figures 4.17, 4.18, 5.3, and 5.5 show, respectively, loci for the following loop gain functions:

$$\frac{K}{s + 1} \qquad \frac{K}{(s + 1)(0.5s + 1)} \qquad \frac{25(K_gs + 1)}{s(s + 2)} \qquad \frac{K_c + K_ds}{(s + 1)(0.5s + 1)}$$

For the first two, and this is the most common case, the root loci show how the closed-loop poles change when gain K is changed. For the third, they show the effect of velocity feedback gain K_g on pole position, and for the fourth both loci for varying K_c with K_d constant and for varying K_d with K_c constant were constructed.

These examples already illustrate the power of the root locus method in analysis and design, because the loci give a graphic picture of the effect of selected parameters on the system poles, and suggest what values should be chosen to meet specifications on time constant and damping ratio and how the speed of response can be improved. The closed-loop poles are also needed to determine system stability and to calculate transient responses by the partial fraction expansion technique.

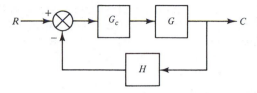

Figure 6.1 System configuration.

While in the examples mentioned the systems are of first or second order, most practical systems are unfortunately of at least third order. In Example 5.3.2 on PI control of a plant consisting of two simple lags, for instance, $H = 1$, $G_c = K(s + z)/s$, and $G = 1/[(s + p_1)(s + p_2)]$, so that the closed-loop transfer function is

$$\frac{C}{R} = \frac{K(s + z)}{s(s + p_1)(s + p_2) + K(s + z)} \tag{6.1}$$

and the closed-loop poles are the roots of a cubic characteristic equation.

The root locus method originated as a graphical technique for determining how the system poles move when a parameter, say K in (6.1), is changed, and to find these poles for particular values of the parameter.

This method and the graphical construction rules are developed in this chapter and applied to analysis and design. The graphical construction technique provides the insight needed to enable the general shape of root loci to be sketched rapidly. The ability to do this remains quite important for analysis and design even though computer methods are much faster and provide accurate plots, and are commonly used for root locus plotting in practice.

An interactive computer program using graphics is given in Appendix B, with an example. It is noted that given a program to calculate the roots of polynomials, these roots can be generated for a range of values of the varying parameter and the results plotted.

*****

In the development, the loop gain function in Fig. 6.1 will be assumed to be of the general form

$$G_c(s)G(s)H(s) = K \frac{(s - a_1)(s - a_2) \cdots (s - a_m)}{(s - b_1)(s - b_2) \cdots (s - b_n)} \tag{6.2}$$

First the following general note:

Any transfer function, to be physically realizable, must have at least as many poles as zeros (i.e., $n \geqslant m$).

For very large s the constant terms in (6.2) are negligible relative to s, so $G_cGH \approx Ks^{m-n}$. Later work will show that to determine the response to sinusoidal inputs of frequency ω, s can be replaced by $j\omega$ and that $|G_cGH| \approx |K(j\omega)^{m-n}| = K\omega^{m-n}$ then gives the amplification which the function applies to a sinusoidal input of a high frequency ω. If $m > n$ this would mean infinite amplification of an input of infinite frequency. This is not possible for physical systems. If the frequency of the input is increased, sooner or later the system will in effect give up and its output amplitude will stop growing. PD control ($K_c + K_d s$), for example, is actually realized as ($K_c + K_d s$)/($Ts + 1$), but T is so small that it would only cause appreciable deviations from idealized PD control at frequencies beyond those of interest in operation.

Next, some definitions regarding (6.2):

1. $a_1, \ldots, a_m$ are the open-loop zeros of $G_c GH$.
2. $b_1, \ldots, b_n$ are the open-loop poles of $G_c GH$.
3. The open-loop pole–zero pattern is the s-plane plot of open-loop poles and zeros.
4. K is the root locus gain, defined in Section 3.2 as the gain factor that results if the coefficients of the highest powers of s in the numerator and denominator polynomials of $G_c GH$ are made unity, as in (6.2).

With this form, as discussed in Section 3.2, ($s - a_i$) is a vector from a_i to s, and ($s - b_k$) from b_k to s. As indicated in Fig. 6.2, these vectors may be expressed alternatively as follows:

$$s - a_i = A_i e^{j\alpha_i}$$
$$s - b_k = B_k e^{j\beta_k}$$
(6.3)

Here A_i and B_k are vector lengths and α_i and β_k vector angles, measured positive counterclockwise from the direction of the positive real axis.

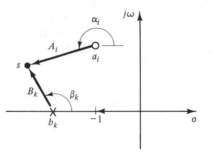

O : open-loop zeros

X : open-loop poles   **Figure 6.2**   Vectors in the s-plane.

6.2 ROOT LOCI

The closed-loop poles are the roots of the system characteristic equation $G_c GH + 1 = 0$, so of the equation

$$G_c GH = -1$$
(6.4)

Interpreting both sides as vectors in the s-plane, the vector -1 indicated on Fig. 6.2 is a vector from the origin to the point -1, on the negative real axis. This vector has a length, or magnitude, of unity, and a phase angle that is an odd multiple of $\pm 180°$, or $\pm(2n + 1)180°$, where n is any integer. Therefore, the closed-loop poles are the values of s for which the vector G_cGH has a length of unity and a phase angle of $\pm(2n + 1)180°$. Substituting (6.3) into (6.2) gives, assuming that K is positive:

$$\text{magnitude } (G_cGH) = K\frac{A_1A_2 \cdots A_m}{B_1B_2 \cdots B_n} \tag{6.5a}$$

$$\text{phase } (G_cGH) = \alpha_1 + \cdots + \alpha_m - \beta_1 \cdots - \beta_n \tag{6.5b}$$

Hence

The closed-loop poles are the values of s that satisfy both of the following conditions:

1. *Angle condition*:

$$\text{phase } (G_cGH) = \alpha_1 + \cdots + \alpha_m - \beta_1 - \cdots - \beta_n = \pm(2n + 1)180° \tag{6.6a}$$

2. *Magnitude condition*:

$$\text{magnitude } (G_cGH) = K\frac{A_1 \cdots A_m}{B_i \cdots B_n} = 1 \quad \text{or} \quad K = \frac{B_1 \cdots B_n}{A_1 \cdots A_m} \tag{6.6b}$$

Equations (6.6) lead to a two-stage process for construction of the loci of the closed-loop pole positions in the s-plane for varying root locus gain K:

1. The root loci are constructed from the angle condition alone, as the loci of all points s for which the sum of the vector angles α_i from all open-loop zeros to s minus the sum of the vector angles β_i from all open-loop poles to s equals an odd multiple of $\pm 180°$.

2. After the loci have been constructed, the magnitude condition shows that the value of K for which a closed-loop pole will be located at a given point s along a locus equals the product of the vector lengths B_i from all open-loop poles to s divided by the product of the vector lengths A_i from all open-loop zeros to s.

As this discussion suggests, root locus work starts with the construction of the open-loop pole–zero pattern. The simple example below is given to clarify the ideas.

Example 6.2.1   A Design Problem

For a system with loop gain function $G_cGH = K/(s + a)$, use root loci to find K for which the (closed-loop) system time constant will be T seconds. The open-loop pole–zero pattern, plotted first, consists of just a pole at $-a$, shown in Fig. 6.3. Following the procedure, a trial point s is chosen, and the vector is drawn from the open-loop pole to s. In this example, the sum of vector angles from the open-loop zeros minus that from the open-loop poles is $-\beta$. Here it is clear that only trial points on the real axis to the left of $-a$ satisfy the angle condition that the net sum be an odd multiple of $\pm 180°$. All such points satisfy the angle condition. Hence, and this completes the first stage, the root locus is the real axis to the left of the open loop pole at $-a$.

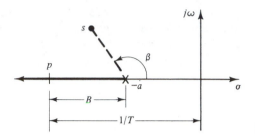

Figure 6.3 Loci for $K/(s + a)$.

To solve the design problem, for a time constant T, the system pole is required to be at p, at a distance $1/T$ from the imaginary axis. The magnitude condition immediately gives the value of K needed for the pole to be at this point along the locus:

$$K = \frac{B_1 B_2 \cdots}{A_1 A_2 \cdots} = B$$

Here B is the distance from $-a$ to p in Fig. 6.3. (Without zeros, the A_i factors are not present; that is, the denominator is in effect unity.)

Although construction of the loci, and solution of the design problem, were easy for this example, it is clear that something better than arbitrarily choosing trial points is required to make the technique feasible for less simple systems.

The *180° locus* which is emphasized in this chapter is based on the assumption that the varying gain of the loop gain function for which the loci are plotted is positive. When the effect of a parameter other than the root locus gain is studied, allowance must also be made for negative gains. This will be encountered in Section 6.7 and on occasion elsewhere. In such cases the *0° locus* is desired, because a trial point s will lie on the locus if the specified sum of vector angles is an even multiple of $\pm 180°$.

6.3 RULES FOR ROOT LOCUS PLOTTING

The following guides are provided to facilitate the plotting of root loci:

1. For $K = 0$ the closed-loop poles coincide with the open-loop poles, since then $B_i = 0$.
2. For $K \rightarrow \infty$ closed-loop poles approach the open-loop zeros, since then $A_i \rightarrow 0$.
3. There are as many locus branches as there are open-loop poles. A branch starts, for $K = 0$, at each open-loop pole. As K is increased, the closed-loop pole positions trace out loci, which end, for $K \rightarrow \infty$, at the open-loop zeros.
4. If there are fewer open-loop zeros than poles ($m < n$), those branches for which there are no open-loop zeros left to go to tend to infinity along

asymptotes. The number of asymptotes is equal to the number of open-loop poles minus the number of open-loop zeros, $n - m$.

5. The directions of the asymptotes are found from the angle condition. In Fig. 6.4, choose a trial point s at infinity at angle α. The vectors from all m open-loop zeros and n open-loop poles to s then have angle α, so the net sum of the vector angles is $(m - n)\alpha$. For s to lie on the locus, this must equal an odd multiple of $\pm 180°$. Hence the asymptote angles α must satisfy

$$\alpha = \frac{\pm(2i + 1)180}{n - m} \qquad i = \text{any integer} \tag{6.7}$$

If $n - m = 1$, α is $180°$; if $n - m = 2$, α is $+90°$ and $-90°$; if $n - m = 3$, α is $+60°$, $-60°$, and $180°$; and so on. The angles are uniformly distributed over $360°$.

6. All asymptotes intersect the real axis at a single point, at a distance ρ_0 to the origin:

$$\rho_0 = \frac{(\text{sum of o.l. poles}) - (\text{sum of o.l. zeros})}{(\text{number } n \text{ of o.l. poles}) - (\text{number } m \text{ of o.l. zeros})} \tag{6.8}$$

The proof will be omitted. If ρ_0 is positive, the intersection occurs on the positive real axis. Note that, say, the sum of a complex conjugate pair of open loop (o.l.) poles is real:

$$(a + bj) + (a - bj) = 2a$$

7. Loci are symmetrical about the real axis since complex open-loop poles and zeros occur in conjugate pairs.

8. Sections of the real axis to the left of an odd total number of open-loop poles and zeros on this axis form part of the loci. This is because any trial point on such sections satisfies the angle condition. For example, at a trial point s in Fig. 6.5(a), the vectors from the poles left of point s contribute zero angles, and the contributions from the complex poles cancel each other. The pole and zero on the axis to the right of s contribute ($\pm 180 \mp 180$), which is $0°$ or $\pm 360°$, not an odd multiple of $\pm 180°$, so s is not part of the loci.

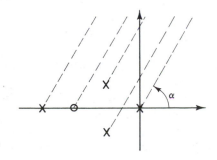

Figure 6.4 Asymptote angles.

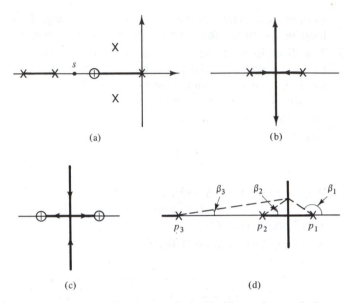

Figure 6.5 (a) Root locus rule 8; (b)–(d) root locus rule 9.

9. Points of breakaway from, or arrival at, the real axis may also exist. If, as indicated in Fig. 6.5(b) and (c), the part of the real axis between two o.l. poles (o.l. zeros) belongs to the loci, there must be a point between them where the loci break away from (arrive at) the axis.

   The loci that start at each open-loop pole as K is increased from zero cannot disappear into thin air, nor can the locus branches that must approach the zeros as $K \to \infty$ appear out of it. Algebraic rules for determining the locations of these points are available in numerous books. However, usually, as will be discussed later, these locations are not of much interest and a rough approximation is satisfactory.

   If no other poles and zeros are close by, the breakaway point will be halfway. In Fig. 6.5(d), if pole p_3 is not present, only points on a vertical halfway between p_1 and p_2 satisfy the angle condition $\beta_1 + \beta_2 = 180°$. From the geometry in Fig. 6.5(d), adding the pole p_3 pushes the breakaway point away. But even if the distance $p_2 p_3 = p_1 p_2$, the point only moves from $0.5(p_1 p_2)$ to $0.42(p_1 p_2)$ distance to p_1. A zero at the position of p_3 would similarly attract the breakaway point. If better accuracy is desired, sections of the breakaway branches constructed elsewhere in the s-plane can be extrapolated by applying the angle condition to trial points at decreasing distance to the real axis.

10. The angle of departure of loci from complex o.l. poles (or of arrival at complex o.l. zeros) is a final significant feature. Apply the angle condition to a trial point very close to p_1 in Fig. 6.6. Then the vector angles from the other poles and the zero are the same as those to p_1 shown in the plot.

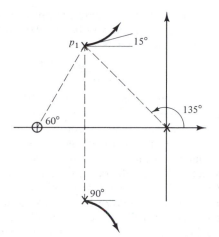

Figure 6.6 Angle of departure.

If the point is offset from p_1 in the direction of $\gamma°$, it will lie on the locus if $60° - 90° - 135° - \gamma° = \pm(2i + 1)180°$, so if $\gamma = 15°$. Thus the loci will depart from pole p_1 at an angle of $15°$.

6.4 ROOT LOCUS EXAMPLES: PLOTTING AND SKETCHING

The Spirule is available to facilitate the addition of vector angles for a trial point in the graphical technique. It consists of a protractor with a rotating arm pinned at its center, which also helps in vector-length measurements. However, when its use is warranted, the computer-aided construction method in Appendix B is more appropriate for the task.

Example 6.4.1 Simple Motor Position Servo (Fig. 6.7)
(a) Find the loci of the (closed-loop) system poles for varying K.

(b) *Design:* Find the value of K to obtain a system damping ratio $\zeta \approx 0.7$.

First the open-loop pole–zero pattern is plotted, consisting of poles at the origin and $-a$, as shown in Fig. 6.7(b). There are two poles, so two locus branches,

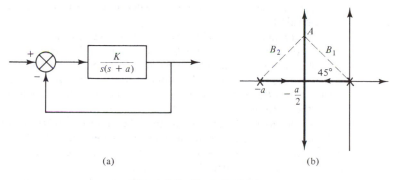

(a)   (b)

Figure 6.7 Example 6.4.1.

starting at 0 and $-a$ for $K = 0$. Since there are no open-loop zeros, there must be two asymptotes. From rule 5, these will be at $+90°$ and $-90°$, and from (6.8) of rule 6, they will intersect the real axis at

$$\rho_0 = \frac{0 - a - 0}{2 - 0} = -\frac{a}{2}$$

By rule 8, the real axis between 0 and $-a$ is part of the locus, because it lies to the left of one (an odd number) pole on this axis. Since this part is between two poles, there must be a breakaway point. According to rule 9, with no other poles and zeros present, breakaway will occur halfway, at $-a/2$. Indeed, in this example asymptotes and loci coincide because any point on this vertical satisfies the angle condition. This completes part (a), the construction of the loci.

For part (b), Chapter 3 has shown that to achieve $\zeta \approx 0.7$, the (closed-loop) system poles must be at an angle ϕ given by $\zeta = \cos \phi$ to the negative real axis, that is, at $45°$. So the poles must lie where lines at $45°$ intersect the loci, at point A and its complex conjugate position. The value of K needed to locate the poles at these points is found immediately from the magnitude condition (6.6b):

$$K = \frac{B_1 B_2 \cdots}{A_1 A_2 \cdots} = B_1 B_2 = (\tfrac{1}{2}a\sqrt{2})(\tfrac{1}{2}a\sqrt{2}) = \frac{a^2}{2}$$

It may be recalled that Fig. 3.14 shows loci for a numerical example of this type, and also compares step responses for $K = a^2/4$ and the value of K corresponding to a damping ratio 0.5. Both have the same settling time and time constant, but the last has greater speed of response due to a smaller rise time.

Example 6.4.2

For the system in Fig. 6.8(a):

(a) Find the loci of the system poles for varying K.

(b) *Design:* Find K to realize a damping ratio $\zeta \approx 0.7$ for the dominating pair of closed-loop poles, as defined in Section 3.6.

(c) Find the closed-loop poles for K of part (b).

(d) Determine the limiting value of K for stability.

The open-loop pole–zero pattern consists of poles at 0, -1, and -2. There are three o.l. poles ($n = 3$) and no o.l. zeros ($m = 0$), so there are three asymptotes. Rule 5 gives their directions as $+60°$, $-60°$, and $180°$, and rule 6 their intersection with the real axis at $\rho_0 = (0 - 1 - 2 - 0)/(3 - 0) = -1$. Figure 6.8(b) shows these asymptotes. The real axis between 0 and -1 and left of -2 belongs to the loci, by rule 8.

There must be a breakaway point between 0 and -1. But for the pole at -2, it would be halfway, at -0.5. This third pole pushes it away to the right, so it is taken to be somewhat to the right of -0.5. (In rule 9, the true value for this case was given as -0.42.) It is logical that the breakaway branches should move toward the $+60°$ and $-60°$ asymptotes, and that the branch from the pole at -2 forms the third asymptote.

A root locus sketch such as that shown in Fig. 6.8(b) can now be completed, without any use of trial points and angle measurements. It is very useful to remember that the loci need only be constructed with satisfactory accuracy where they are needed to solve the particular design problem. In the present case, the dominating poles, on the complex branches, are required to have a damping ratio of 0.7. This

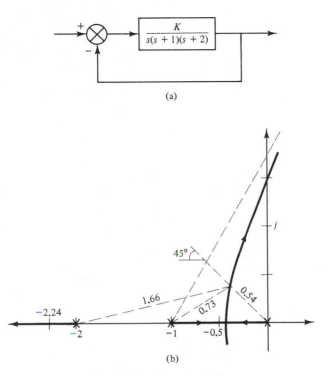

(a)

(b)

Figure 6.8 Example 6.4.2.

means, analogous to Example 6.4.1, that it is only necessary to know with reasonable accuracy where the locus intersects the line at 45° to the negative real axis. Using the sketch for a first guess:

> This can be done by applying the angle condition to some trial points along the 45° line and interpolating.

With this point found, the value of K needed so that the closed-loop poles will be at this location along the locus can be obtained from the magnitude condition, using the measured vector lengths indicated on Fig. 6.8(b):

$$K = \frac{0.54 \times 0.73 \times 1.66}{1} = 0.65$$

It is observed that in the present case, where the system characteristic equation does not exceed fourth order in s, both the intersection and K can readily be found analytically. The characteristic equation, the denominator of C/R, is

$$s(s + 1)(s + 2) + K = s^3 + 3s^2 + 2s + K = 0 \qquad (6.9)$$

The line at 45° to the negative real axis can be described by the equation $s = \omega(-1 + j)$. Then $s^2 = \omega^2(-1 + j)^2 = -2j\omega^2$ and $s^3 = -2j\omega^3(-1 + j) = 2(1 + j)\omega^3$. Substituting these into (6.9), since the s-value of interest must lie on the line, and separating the real and imaginary parts, yields

$$(2\omega^3 - 2\omega + K) + 2j\omega(\omega^2 - 3\omega + 1) = 0 \qquad (6.10)$$

Both the real and imaginary parts must be zero, and solving the quadratic for ω gives $\omega = 0.382$. So the locus must intersect the 45° line at $0.382j$ distance to the real axis. For this ω, the real part in (6.10) gives $K = 0.65$, as before.

The position of the third pole for this value of K must still be found. One way is to find K for a number of points along the third locus branch, using the magnitude condition, and interpolating. A simpler way is available for the example. If the system poles are λ_1, λ_2, and λ_3, the characteristic equation is

$$(s - \lambda_1)(s - \lambda_2)(s - \lambda_3) = s^3 - (\lambda_1 + \lambda_2 + \lambda_3)s^2 + \cdots = 0 \qquad (6.11)$$

This illustrates a general result:

> The sum of the system poles is equal to the negative of the coefficient of the next-to-highest power term of the characteristic equation.

Comparing (6.9) and (6.11), $\lambda_1 + \lambda_2 + \lambda_3 = -3$. But the sum of two of these, say $\lambda_1 + \lambda_2$, has already been found in Fig. 6.8(b): $\lambda_1 + \lambda_2 = -2 \times 0.54 \cos 45°$ $= 0.76$. Hence $\lambda_3 = -3 + 0.76 = -2.24$ is the position of the third pole.

With the closed-loop poles now known, the closed-loop transfer function is

$$\frac{C}{R} = \frac{0.65}{(s + 2.24)(s + 0.38 + 0.38j)(s + 0.38 - 0.38j)} \qquad (6.12)$$

and the transient response for given inputs could be calculated by the methods of Chapter 3.

By applying the magnitude condition to other points, the loci can also be used to determine pole sensitivity to gain changes.

The value of K at which the loci cross the imaginary axis is the limit for stability because above it two system poles are inside the right half of the s-plane. It can be found graphically by applying the angle condition to trial points along the imaginary axis and then using the magnitude condition. Alternatively, the Routh–Hurwitz stability criterion can be used, as was done for this case in Example 3.7.3. A third technique is analogous to that which led to (6.10). Points along the imaginary axis satisfy $s = j\omega$. Substituting this into the characteristic equation (6.9) and separating real and imaginary parts leads to the equation

$$K - 3\omega^2 + j\omega(2 - \omega^2) = 0 \qquad (6.13)$$

Both real and imaginary parts must be zero, so that $\omega = \sqrt{2}$, $K = 6$. Hence the limiting value for stability is $K = 6$, and the loci cross the imaginary axis at $\pm 1.414j$, as was found in Example 3.7.3.

These techniques can also be used if the design problem would be to find K corresponding to a specified time constant T for the dominating poles. In this case the poles are required to be on a vertical at $-1/T$ on the real axis. Trial points along this vertical can provide the crossing point, for which the magnitude condition then gives the gain K.

The ability to sketch the general shape of the loci, without regard for accuracy, is very useful in analysis and design. In analysis, it can often provide a quick explanation of why, say, the step response changes in a certain direction with changes of gain, even if the numerical values of the coefficients in the transfer function are not known. In design, considered in the next section, it may show quickly whether a form of compensation being contemplated could be successful. Some sketch examples follow, with the loop gain functions and the corresponding loci shown in Fig. 6.9.

Example 6.4.3

In Fig. 6.9(a), three asymptotes, at $+60°$, $-60°$, and $180°$, intersect the real axis at the average real part of the poles. The real axis left of $-a$ is part of the loci. Branches from the complex poles approach the other asymptotes. An approximate equation for the angle of departure α is $-90 - 110 - \alpha = -180$, so $\alpha \approx -20°$.

Example 6.4.4

Consider Fig. 6.9(b). Since $n = 3$ and $m = 1$, there are two asymptotes, at $+90°$ and $-90°$, intersecting the real axis at $0.5(-b - c + a)$. Breakaway occurs somewhat to the left of halfway between 0 and $-b$, because the zero at $-a$ pulls it more than the pole at $-c$ pushes it. The breakaway branches approach the asymptotes.

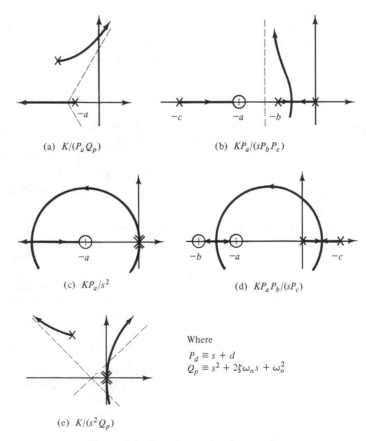

(a) $K/(P_a Q_p)$ (b) $KP_a/(sP_b P_c)$

(c) KP_a/s^2 (d) $KP_a P_b/(sP_c)$

(e) $K/(s^2 Q_p)$

Where

$$P_d \equiv s + d$$
$$Q_p \equiv s^2 + 2\zeta\omega_n s + \omega_n^2$$

Figure 6.9 Root locus sketch examples.

Example 6.4.5

In Fig. 6.9(c) there is a double pole at the origin. Verify that only trial points close to the origin above and below the double pole satisfy the angle condition, so the loci must depart from the origin in vertical direction. Since $n = 2$, $m = 1$, only the negative real axis is an asymptote. The real axis left of $-a$ is part of the locus since it lies left of three poles and zeros on this axis. But there must be an arrival point of the loci here, because as $K \to \infty$ one branch must approach $-a$ and the other tend to infinity along the asymptote. For this example it is easily verified, by applying the angle condition to trial points slightly off the axis, that arrival must occur at $-2a$. The locus can also be shown to be a circle.

Example 6.4.6

Figure 6.9(d) illustrates that nothing changes if open-loop poles, or zeros, occur in the right-half s-plane. A system with an open-loop pole in the right-half plane is *open-loop unstable;* that is, it is unstable unless a suitably designed feedback loop is closed around it. A tall rocket is an example that requires feedback control for stability.

Example 6.4.7

In Fig. 6.9(e) there are four poles and no zeros, so four asymptotes, at $+45°$, $-45°$, $+135°$, and $-135°$, intersect the real axis at the average real part. Departure from the double pole at the origin is in the vertical direction.

6.5 ROOT LOCI AND SYSTEM DESIGN

In the preceding sections, root loci were used for design to the extent of choosing the gain to obtain a specified damping ratio or time constant. Such P control design does not change the shape of the loci. But if dynamic compensation is used, such as a series compensator $G_c(s)$ in Fig. 5.1, then G_c will add poles and zeros to the open-loop pole–zero pattern, in order to change the shape of the loci in a desirable direction.

Example 6.5.1   Effect of Adding a Pole or Zero

Figure 6.10(a) shows loci that could represent P control of a process consisting of two simple lags, as in Fig. 4.18 for Example 4.6.2. These loci are equivalent to those for a simple motor position servo in Fig. 6.7. The loci in Fig. 6.10(b) are equivalent to those in Fig. 6.8 and show the effect of adding a pole. The loci in Fig. 6.10(c) are much like those in Fig. 6.9(c) and (d), and show the effect of adding a zero. The following general effect is evident:

Adding a pole pushes the loci away from that pole, and adding a zero pulls the loci toward that zero.

These effects increase in strength with decreasing distance. A zero can improve relative stability because it can pull the loci, or parts thereof, away from the imaginary axis, deeper into the left-half plane.

Example 6.5.2   PD Controller

$$G_c(s) = K_c + K_d s$$

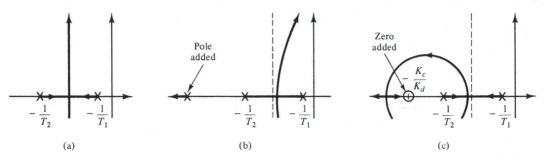

Figure 6.10   Effect of adding a pole or a zero.

The discussion above explains on a more general basis the stabilizing effect of PD control on the dual simple lag plant discussed in Example 5.3.3 of Section 5.3. For this example $T_1 = 1$, $T_2 = 0.5$, and the loop gain function G_cG in (5.9) is

$$G_cG = 2K_d \frac{s + K_c/K_d}{(s + 1)(s + 2)} \qquad (6.14)$$

This shows that PD control has added the zero indicated in Fig. 6.10(c) to the open-loop pole–zero pattern of Fig. 6.10(a). Equation (6.14) also serves to show that the root locus gain, of which the magnitude condition gives numerical values, is $2K_d$.

The differences of the loci for the same system in Fig. 5.5 and Fig. 6.10(c) illustrate the flexibility of the method. Those in Fig. 5.5 show the effect if each one of K_c and K_d is changed in turn, with the other held constant. In Fig. 6.10(c) the root locus gain $2K_d$ is the varying parameter along the loci. But if K_d changes, the zero $-K_c/K_d$ would change as well. Thus Fig. 6.10(c) implies the assumption that K_c and K_d change simultaneously and proportionally.

Example 6.5.3   PI Controller

$$G_c(s) = K_c + \frac{K_i}{s}$$

In Example 5.3.2, PI control $G_c = K_c + K_i/s$ was applied to the plant $G = 2/[(s + 1)(s + 2)]$, and the advantage for steady-state accuracy was discussed. But the characteristic equation is third order, and suitable tools to consider the effect on stability were not yet available.

The loop gain function is

$$G_cG = 2K_c \frac{s + z}{s(s + 1)(s + 2)} \qquad \left(z = \frac{K_i}{K_c}\right) \qquad (6.15)$$

The controller has added a pole at the origin, which makes the system into type 1, and a zero at $-z$ to the open-loop pole–zero pattern. It may be noted that the loop gain function is that in Fig. 6.8 for Example 6.4.2 with a zero added.

The shape of the loci is determined by the choice of z, and the positions of the closed-loop poles along the loci by the root locus gain $2K_c$.

System step responses for three different choices of z will be compared: 4, 0.5, and 1.2. In each case the system is designed for a damping ratio 0.5 of the

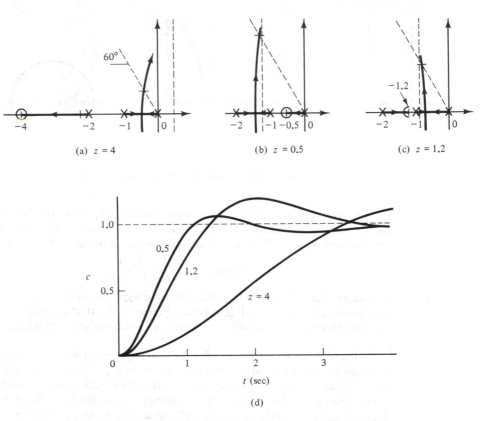

(a) $z = 4$ (b) $z = 0.5$ (c) $z = 1.2$

(d)

Figure 6.11   Example 6.5.3: choice of zero in PI control.

complex pair of closed-loop poles. Figure 6.11(a) to (c) shows the upper halves of
the root locus sketches. For the design, only the intersections with the 60° lines
are needed. These may be found graphically, using trial points along the line,
assuming that the programs in Appendix B are not used to produce loci which are
accurate everywhere, or by the following calculation. Substituting $s = -a +$
$1.732ja$ ($s^2 = -2a^2 - 3.464ja^2$; $s^3 = 8a^3$), for points along the 60° line, into the
characteristic equation

$$s^3 + 3s^2 + (2 + 2K_c)s + 2K_i = 0$$

and equating the real and imaginary parts to zero yields

$$K_c = 3a - 1 \qquad K_i = 6a^2 - 4a^3$$

The sum of all poles is -3, from the second coefficient in the characteristic equation.
Since the sum of the complex pair is $-2a$, the third pole is $(-3 + 2a)$.

If the time constant $1/a$ were specified instead of z, the poles and $z = K_i/K_c$
could be calculated directly. With z specified, some trial values of a, guided by
the sketches, and interpolation yield the following results for K_c and K_i, the poles

$p_{1,2}$ and p_3, and $\omega_n = |p_{1,2}|$:

z	K_c	K_i	$p_{1,2}$	p_3	ω_n
4	0.17	0.676	$-0.39 \pm 0.676j$	-2.22	0.78
0.5	2.852	1.425	$-1.284 \pm 2.224j$	-0.432	2.568
1.2	1.577	1.892	$-0.859 \pm 1.488j$	-1.282	1.718

The output transform for a unit step input is

$$C(s) = \frac{2K_c(s + z)}{s(s - p_3)(s^2 + \omega_n s + \omega_n^2)}$$

and responses may be calculated using the program in Appendix B or the graphical residue technique. The latter yields

$$c(t) = 1 + A_1 e^{p_3 t} + A_2 e^{-at} \cos(bt + \theta)$$

where $a = \text{Re}(p_{1,2})$, $b = \text{Im}(p_{1,2})$, and corresponding sets of values (z, A_1, A_2, $\theta°$) are (4, -0.072, 1.215, $-220°$), (0.5, -0.158, 0.989, $-212°$), and (1.2, $+0.084$, 1.218, $-207°$).

These responses are plotted in Fig. 6.11(d), and their nature agrees with what inspection of the root locus sketches, even without the pole locations at 60° found, would suggest:

Figure 6.11(a): For $z = 4$ the zero is too far left. The asymptotes are in the right-half plane ($\rho_0 > 0$), and the branches to these asymptotes cross into the right-half plane at a relatively low gain. These branches leave the real axis to the right of -0.5, suggesting the relatively slow response confirmed by Fig. 6.11(d).

Figure 6.11(b): For $z = 0.5$, the complex branches are advantageously far into the left-half plane, but there is now a complete root locus branch between the origin and the zero at -0.5. Somewhere on this branch, depending on the root locus gain, must be a closed-loop pole. Its time constant is relatively large, and the overall transient may be visualized as consisting of an oscillatory component due to the complex poles being superimposed on a slowly decaying exponential due to this real pole. For this reason, the zero is usually not chosen to the right of the pole at -1. However, if the gain is high enough that the closed-loop pole is close to the zero at -0.5, which is also a zero of C/R, the residue and hence the size of the transient corresponding to the pole will be small. The plot in Fig. 6.11(d) approaches this. The slow decay above about 2.5 sec is evident but is rather small. Moreover, the overshoot is about 5% instead of the 16% corresponding to $\zeta = 0.5$ because at the time of the peak the exponential term reduces the total response appreciably.

Figure 6.11(c): The zero for $z = 1.2$ is somewhat to the left of the pole at -1. The value of ρ_o is to the right of that for part (b), suggesting the somewhat slower response verified in Fig. 6.11(d), which also shows an overshoot of about 18%, somewhat higher than that for $\zeta = 0.5$. This is due to the zero, but is small in both parts (b) and (c) because a nearby pole nearly cancels its effect.

Example 6.5.4   Use of Pole–Zero Cancellations

A common practice if G has poles in undesirable locations which cannot easily be changed by feedback alone is to choose zeros of G_c at the same or nearby locations.

A root locus branch, of ideally zero length, is associated with such pole–zero pairs. The zero of G_c is also a closed-loop zero and will be close to the closed-loop pole along this branch. By Section 3.8, this implies a small, ideally zero, residue at the pole and therefore a small transient.

In Example 6.5.3, the choice $z = 1$ would cancel the pole at -1 and the loop gain function would be $K/[s(s + 2)]$. Design for $\zeta = 0.5$ would yield the step response for an underdamped quadratic in Fig. 3.11. To allow for variations of or uncertainty about the actual parameter value, the zero is usually chosen to the left of the nominal pole position.

Two points should be emphasized. The first, already raised in Section 3.8 above (3.46), is that the system response to initial conditions is not affected by any cancellations which may have been achieved in the input–output response. The second is that open-loop unstable poles of G, that is, poles of G in the right-half s-plane, may never be canceled in this way by zeros of G_c. However short the locus branch between such a pole–zero pair, there would be a closed-loop pole along it, in the right-half plane. Here feedback must be used to pull the pole into the left-half plane, as in Fig. 6.9(d).

Example 6.5.5 Feedback Compensation

Figure 6.12 shows a more complex motor position servo with velocity feedback than that of Example 5.2.1, for which the step response and root loci for varying K_g are shown in Fig. 5.3. G is the transfer function (2.10) of the motor and its load if the load damping is assumed to be negligible. If error analysis is required, the loop gain function C/E must be used to keep E in evidence, but for stability analysis the two feedback loops can be combined into $(K_g s + 1)$. By inspection, the loop gain function is then

$$\frac{KK_m(K_g s + 1)}{s^2(T_f s + 1)} = \frac{KK_m K_g}{T_f}\,\frac{s + 1/K_g}{s^2(s + 1/T_f)} \tag{6.16}$$

The second form gives the root locus gain, of which the magnitude condition provides numerical values at points along the loci.

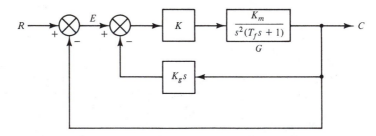

Figure 6.12 Example 6.5.5: velocity feedback.

Figure 6.13 shows root locus sketches for a range of choices of the zero $-1/K_g$. For $K_g = 0$ the system is apparently unstable for any value of gain K. With the zero to the left of $-1/T_f$, in (b), the system is still unstable for all K, and for $K_g = T_f$, in (c), it is marginally stable. The loci (d) and (e) cover a suitable range of values of K_g. The change of the nature of the loci from (d) to (e) as the zero moves close enough to the origin illustrates a situation where sketching is no

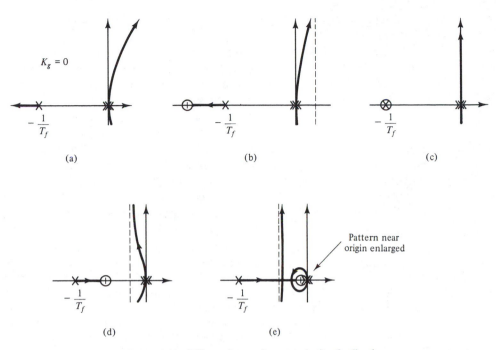

Figure 6.13 Effect of zero due to velocity feedback.

longer adequate and the computer aids of Appendix B are required. But the sketches easily showed that P control is inadequate, and quickly homed in on the suitable range of parameters of the compensator, without any numerical values.

6.6 ROOT LOCI FOR PHASE-LEAD AND PHASE-LAG COMPENSATION

Passive electrical networks to realize these very important forms of compensation were given in Fig. 2.8, and active realizations in Section 5.5. Including a gain, the transfer functions can be written as follows:

$$G_c(s) = \frac{K_c(s + z)}{s + p} \tag{6.17}$$

For a lead, $z < p$, and for a lag, $z > p$. Figure 6.14 shows the pole–zero patterns. Phase-lead compensation is an approximation to PD control, and is often preferable

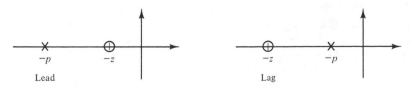

Figure 6.14 Phase lead and phase lag.

to reduce the effect of signal noise. In PD control, compensator output and input are related by $m = K_c e + K_d \dot{e}$. Here $\dot{e}$ is the slope of a plot of e versus time, and this term tends to amplify any irregularities in e due to noise. To reduce this effect, the ratio p/z should not be made larger than necessary to serve the purpose. Like PD control, this purpose is usually to improve stability. Phase-lag compensation is commonly used, like PI control, to improve accuracy.

However, phase lead may also improve accuracy, and phase lag may improve stability. The frequency response methods of the next two chapters provide greater insight into these effects, including how accuracy under dynamic conditions may be expressed, and improved. Design is in fact often carried out more conveniently and with greater insight in this frequency domain.

The examples below illustrate the approach for design by root loci, and are also useful to interpret the effects of the compensators on the loci.

Example 6.6.1   Phase-Lead Compensation

In the system of Fig. 6.8, the P control of Example 6.4.2 is to be replaced by phase lead, intended to "pull" the locus branches for P control, shown as dashed curves in Fig. 6.15, to the left by means of the added zero. Ignoring for now the weaker effect of the added pole, which is often placed at 10 times the distance to the origin, the zero is chosen by arguments similar to those used for PI control in Fig. 6.11. The choice -1.3 in Fig. 6.15 implies a closed-loop pole on the branch between -2 and -1.3, but this is better than one between the origin and the zero if it were chosen to the right of -1. With the pole at -13, the asymptotes intersect the axis at $(-1 - 2 - 13 + 1.3)/3 = -4.9$, instead of at -1 as for P control. The resulting loci show the desired stabilizing effect.

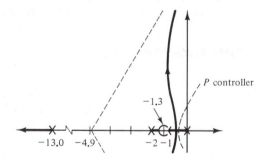

Figure 6.15   Example 6.6.1: phase-lead compensation.

More formally, a desired closed-loop pole position can be chosen, and the position of the phase-lead pole can then be selected such that the net sum of the vector angles at the desired position will be an odd multiple of $\pm 180°$. This ensures that it will lie on the locus.

Example 6.6.2   Phase-Lag Compensation

The dashed curves in Fig. 6.16 again show the loci for P control of the preceding process. Now a pole–zero pair is added close to the origin, much closer than pictured in Fig. 6.16 to enable the shape of the loci near the origin to be indicated. As suggested by the vectors to a point on the dashed locus in Fig. 6.16, for such

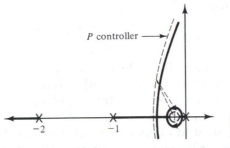

P controller

Figure 6.16 Example 6.6.2: phase-lag compensation.

a pair the net contribution to the vector angles at the point is small. Therefore, the main branches often change only little, as indicated in Fig. 6.16. The picture near the origin is of the type shown in Fig. 6.10(c). The zero pulls the loci back to the axis, where one branch moves right to the zero and the other left to the breakaway point of the main branches.

To see why this compensation is used even if the effect on the main branches is quite small, consider the loop gain function

$$G_cG = K_c \frac{s + z}{s(s + 1)(s + 2)(s + p)} = \frac{K_c z}{2p} \frac{1 + s/z}{s(s + 1)(0.5s + 1)(1 + s/p)} \tag{6.18}$$

The vectors $(s + z)$ and $(s + p)$ are not much different for points of interest on the main branches. Therefore, K_c will have a value similar to that for P control. However, the second form of (6.18) shows that the gain contains the factor z/p. For a choice $z = 0.2$, $p = 0.04$, this equals 5. Hence, even allowing for a reduction of K_c, steady-state errors will be improved by a considerable factor. A closed-loop pole is present on the real axis branch that turns right to the zero at $-z$. But, from $C/R = G_cG/(1 + G_cG)$, this zero is also a closed-loop zero, and is usually close enough to the pole to ensure an acceptably small residue at this pole.

6.7 OTHER USES OF ROOT LOCI AND CONCLUSION

The root locus technique has been discussed, and its application to analysis and design. Both series and feedback compensation were considered. Examples demonstrated how differently shaped loci may describe the same system, depending on whether they picture the effect of a controller parameter or the controller gain on the closed-loop poles. The loci are used to select parameters to achieve a specified damping ratio or time constant and to find the corresponding closed-loop poles. The denominator of C/R is then available in factored form, and Example 6.5.3 illustrates how such results can be used to find the step response by the methods of Chapter 3.

The root locus technique applies also to find the roots of polynomials generally. For example, the equation

$$a_5 s^5 + a_4 s^4 + a_3 s^3 + a_2 s^2 + a_1 s + a_0 = 0 \tag{6.19}$$

can be written as follows:

$$(a_5s^5 + a_4s^4 + a_3s^3)\left[1 + \frac{a_2}{a_5}\frac{s^2 + (a_1/a_2)s + a_0/a_2}{s^3(s^2 + (a_4/a_5)s + a_3/a_5)}\right] \qquad (6.20)$$

The last term can be considered as a loop gain function, with root locus gain a_2/a_5, for which loci can be constructed. Other forms are possible in (6.20), but the one shown has the advantage that the open-loop poles and zeros can be found from the solution of quadratics.

Root loci are also used to examine the sensitivity of the closed-loop pole positions to plant parameter variations. As an example, let in the system of Fig. 6.8, with open-loop poles at 0, -1, -2 and with gain $K = 0.65$, the position of the pole at -1 be uncertain. Loci can be used to show the effect of variations δ of this pole on the closed-loop system poles. The system characteristic equation

$$s(s + 1 + \delta)(s + 2) + 0.65 = s^3 + (3 + \delta)s^2 + (2 + 2\delta)s + 0.65 = 0 \quad (6.21)$$

is rearranged to isolate δ, and written in the form of (6.20):

$$(s^3 + 3s^2 + 2s + 0.65) + \delta s(s + 2)$$

$$= (s^3 + 3s^2 + 2s + 0.65)\left[1 + \delta\frac{s(s + 2)}{s^3 + 3s^2 + 2s + 0.65}\right] \qquad (6.22)$$

The last term is the loop gain function of interest, with δ as the root locus gain. Figure 6.17(a) shows the loci for $\delta > 0$. The open-loop zeros are 0 and -2, and the open-loop poles are the roots of (6.9) for $K = 0.65$, given by the denominator of (6.12). For $\delta < 0$, the root locus gain is negative, and the $0°$ *locus* introduced at the end of Section 6.2 is needed. Because the minus sign represents 180°, the angle condition now requires the net sum of the vector angles to be an even multiple of $\pm 180°$. This implies that directions of asymptotes change by 180°, and angles of departure or arrival at complex poles and zeros also. Further, parts of the real axis to the left of an even number of poles and zeros on this axis, and to the right of all of these, belong to the loci.

Figure 6.17(b) shows the locus sketch for the example. Evidently, negative variations δ are destabilizing, as expected because the pole -1 is then moving closer to the origin. The magnitude condition or the Routh–Hurwitz criterion can be used to determine the value of δ at which stability is lost.

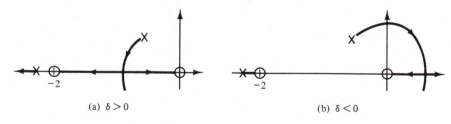

(a) $\delta > 0$ (b) $\delta < 0$

Figure 6.17 Loci for parameter sensitivity.

PROBLEMS

6.1. Plot the loci of the closed-loop system poles for varying K of systems with the following loop gain functions.

(a) $\dfrac{K}{s}$ (b) $\dfrac{K}{s + 1}$ (c) $\dfrac{K}{s - 1}$

(d) $\dfrac{K}{s^2}$ (e) $\dfrac{K}{s^2 + 4}$ (f) $\dfrac{K}{s^2 - 4}$

(g) $\dfrac{K}{s^2 + 2s + 2}$ (h) $\dfrac{K(s + 2)}{s(s + 3)}$

6.2. Sketch the general shape of the loci of the system poles for systems with the following loop gain functions.

(a) $\dfrac{K(s + 1)}{s^2}$. Use the angle condition to show that the locus contains a circle centered at -1 with radius 1.

(b) $\dfrac{K(s + 2)}{s(s + 1)}$

(c) $\dfrac{K}{(s^2 + 2s + 2)(s^2 + 6s + 10)}$

(d) $\dfrac{K(s + 2)}{(s + 1)(s^2 + 6s + 11.25)}$

6.3. For the system in Fig. P6.3 with $H = 1$, $G = [K(s + 1)]/[s(s + 2)]$:
(a) Draw the root loci.
(b) Use them to find K for a 2-sec time constant of the dominating closed-loop pole.
(c) Find the other system pole, by root locus methods (not analytically, although for this problem root loci are not at all required).
(d) Use the roots of the system characteristic equation found in parts (b) and (c) to write the closed-loop transfer function $C(s)/R(s)$ and plot its pole–zero pattern.
(e) Add the pole–zero pattern of a step input $R(s) = 1/s$ to obtain that of the corresponding $C(s)$.
(f) Calculate $c(t)$, evaluating the residues graphically, and also find the steady-state errors for unit step and unit ramp inputs.

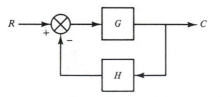

Figure P6.3

6.4. In Fig. P6.4 with $G(s) = 1/[s(s + 1)]$, to evaluate the effect of adding a pole or a zero to the open-loop pole–zero pattern, sketch and compare the loci for

(a) $G_c = K$ (b) $G_c = K/(s + 2)$ (c) $G_c = K(s + 2)$

How is relative stability affected?

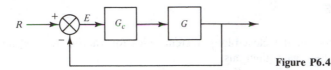

Figure P6.4

6.5. For a unity feedback hydraulic position servo with loop gain function

$$G = \frac{K}{s(s^2 + 2s + 5)}$$

(a) Sketch the loci of the closed-loop poles for varying K.
(b) Find reasonably accurately the value of K for which the time constant of the dominating pair of closed-loop poles is 2 sec.
(c) Determine the position of the third pole.
(d) Calculate the unit step response.

6.6. In Fig. P6.4 with a simple lag plant $G(s) = 1/(0.25s + 1)$, PI control $G_c(s) = K(s + 2)/s$ is used to reduce steady-state errors.
(a) Plot the loci and use them to find K for a dominating system time constant of 1 sec.
(b) For this K, find the second pole on the loci, and determine the unit step response from the pole–zero pattern of the output $C(s)$.

6.7. In Fig. P6.3 with

$$G = \frac{4K}{(0.5s + 3)(2s^2 + 8s + 12)} \qquad H(s) = \frac{0.1}{0.05s + 0.4}$$

(a) Sketch the loci of the closed-loop poles for varying K.
(b) Find the approximate value of K so that the damping ratio of the dominating closed-loop poles will be 0.5.

6.8. For a system with loop gain function

$$G = \frac{K(s + 3)}{s^2 - 2s + 10}$$

(a) Plot the open-loop pole–zero pattern and determine whether the system is open-loop stable.
(b) Sketch the root loci for varying K.
(c) Find the range of K for which the system is stable.
(d) Find K for a damping ratio of about 0.7.

6.9. In Fig. P6.4 with a plant $G = 1/[(s + 1)(s + 3)]$ consisting of two simple lags, and PI control $G_c = K(s + 2)/s$:
(a) Sketch the loci of the closed-loop system poles for varying K.
(b) Find, reasonably accurately, the value of K for a damping ratio 0.5 for the dominating pair of poles.

6.10. For a fluid power position servo with loop gain function

$$G = \frac{K}{s(s^2 + 6s + 13)}$$

(a) Sketch the loci of the system poles for varying K.
(b) Find K for a damping ratio 0.707 of the dominating poles.
(c) Where is the third pole for this K?

6.11. Figure P6.11 shows root loci for

$$GH = \frac{K}{s(0.1s^2 + 0.4s + 0.5)}$$

Note that in contrast with Problems 6.5 and 6.10 the loci from the complex pair of open-loop poles dip down to the real axis. This change of configuration may occur if the pair lies relatively far to the left.

(a) How does the nature of the transient response, in terms of time constants and damping ratios of its components, change as K is increased?

(b) What is the limiting value of K for stability?

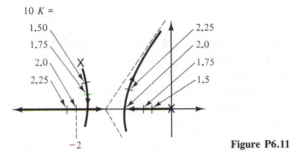

Figure P6.11

6.12. For a system with loop gain function

$$G = \frac{K(s^2 + 4s + 8)}{s^2(s - 1)}$$

sketch the loci and find the range of values of K for which the system is stable.

6.13. For an open-loop unstable system with loop gain function

$$GH = \frac{K(s + 1)}{s(s - 1)}$$

sketch the loci of the system poles and find K for a system time constant of 1 sec from the loci.

6.14. Plot the root loci for a system with the loop gain function

$$\frac{K}{s(s + 6)(s^2 + 6s + 13)}$$

(*Hint:* Courageous use of the rules and the angle condition will yield loci consisting only of straight lines.)

6.15. Repeat Problem 6.14 for

$$\frac{K}{(s + 2)(s + 6)(s^2 + 8s + 20)}$$

and find the limiting value of K for stability.

6.16. Sketch the loci of the closed-loop system poles for varying K of a system with loop gain function

$$\frac{K}{s(0.5s^2 + s + 1)}$$

and find the limiting value of K for stability.

6.17. From Problem 6.2(a), the loci of $K(s + 2)/s^2$ contain a circle of radius 2 centered at -2. Use these loci to find K for a damping ratio 0.707.

6.18. If Fig. P6.4 models a motor position servo with $G(s) = 1/[s(s + 1)]$ and $G_c = K$:
   (a) Plot the loci and use them to find K for a damping ratio of about 0.7.
   (b) Calculate the unit step response, using the system poles as found from the loci.

6.19. Plot the loci for a loop gain function $GH = K/[s(s + 4)]$ and use them to find the lowest value of K that will minimize system settling time. Also, is there a change of nature of the transient response as K is increased, and where does it occur?

6.20. Sketch the loci of

$$GH = \frac{K}{s(s + 1)(s + 3)(s + 4)}$$

and determine the limiting value of K for stability by use of the characteristic equation and the condition $s = j\omega$.

6.21. If Fig. P6.3 with $G = K/[s(0.1s + 1)]$ and $H = 1/(0.02s + 1)$ represents a position servo with appreciable sensor time constant:
   (a) Plot loci and find K for a damping ratio 0.5 for unity feedback (i.e., assuming an ideal feedback sensor).
   (b) If the actual sensor has in fact a significant time constant, what will be the actual damping ratio if the value of K found in part (a) is used? [Use the angle condition for some points along a horizontal through the "design point" of part (a) to sketch the section of the actual loci which is of interest more accurately.] An approximate answer will suffice. Compare it with the design value in part (a).

6.22. For a unity feedback hydraulic position servo with loop gain function

$$G = \frac{K}{s(s^2 + 2s + 2)}$$

   (a) Sketch the loci of the closed-loop system poles for varying K and find K for a damping ratio 0.5 for the complex poles.
   (b) Calculate the unit step response and the unit ramp steady-state following error for the design of part (a).

6.23. For a motor position servo with loop gain function

$$G = \frac{K}{s(0.25s + 1)(0.1s + 1)}$$

sketch the loci of the closed-loop system poles for varying K, and find K for a damping ratio 0.5 of the dominating pair.

6.24. In Fig. P6.4 with $G(s) = 1/[(s + 1)(s + 4)]$, design a physically realizable controller $G_c(s)$ such that:

   1. The steady-state error for step inputs is zero.
   2. The dominating time constant is 0.5 sec.
   3. The damping ratio is about 0.7.

6.25. In Fig. P6.4 with $G(s) = 1/[(0.1s + 1)(0.02s + 1)]$ a system damping ratio 0.707 is required.

    (a) For $G_c(s) = K_c$, find K_c and the corresponding steady-state error after a unit step.

    (b) Design PD control $G_c = K + K_d s$ to halve the steady-state error in part (a). Solve this analytically after verifying that root locus design would involve considerable trial and error.

6.26. **(a)** In Fig. P6.4 with $G(s) = 1/[(s + 1)(0.5s + 1)]$, sketch root loci for PI control $G_c = K_p + K_i/s$ if (i) K_i/K_p large, (ii) $K_i/K_p = 2$, (iii) $K_i/K_p = 1$, and (iv) $K_i/K_p = 0.1$.

    (b) Which of conditions (i) to (iii) is preferred, and why?

    (c) For condition (iv), will the locus branches at considerable distance to the origin differ much from those for P control $G_c = K_p$? If not, why not?

6.27. In Problem 6.26 it is desired to compare condition (iv) with P control $G_c = K_p$.

    (a) For $G_c = K_p$, find K_p and the system poles for a damping ratio 0.707.

    (b) Since condition (iv) with the same K_p will have little effect on these poles (why not?), use K_p as in part (a) and assume that the poles in part (a) are also system poles for condition (iv). Use the magnitude condition to determine the system pole near the origin for the value of K_p, from the open-loop pole–zero pattern.

    (c) Calculate the initial magnitude of the slowly decaying exponential transient due to this pole. Why is it rather small?

    (d) Which of conditions (i) to (iv) would you choose if fast convergence to a value near steady state is important and slow convergence to zero steady-state error is acceptable?

6.28. Fig. P6.4 with $G(s) = 3/[s^2(s + 3)]$ can represent the pitch control system of a missile.

    (a) Sketch loci to determine whether $G_c(s)$ could be chosen to be a simple gain.

    (b) Choose an idealized controller which would stabilize the system and for which the system time constant at high gains will be 1 sec.

    (c) What are the steady-state errors for unit step and unit ramp inputs?

6.29. In Fig. P6.4 a plant transfer function $G(s) = 1/[s(s - 2)]$ may represent a tall rocket. Such a plant is open-loop unstable (i.e., unstable without feedback control), like a pencil standing on its end.

    (a) Plot the loci to determine whether the system can be stabilized by P control $G_c = K_c$.

    (b) If not, could the unstable pole of G be canceled by a zero of G_c to stabilize the system, and if not, why not?

    (c) Choose an idealized controller that can stabilize the system, and find the corresponding range of gains for stability.

6.30. In order to appreciate that the solution in Problem 6.29(c) is not unique, sketch root loci for the following controllers:

    (a) Phase-lead compensation $G_c = K(s + 1)/(s + 8)$

    (b) Idealized quadratic lead $G_c = K(s^2 + 4s + 5)$

    Note how rough locus sketches can give a quick idea of the potential of contemplated forms of compensators.

6.31. Using root locus sketches:

    (a) Investigate the stability of the minor loop in Fig. P6.31.

    (b) Determine whether the overall system can be stable if G_c is just a gain.

(c) What form of idealized controller G_c would you propose, and how would one find its parameters?

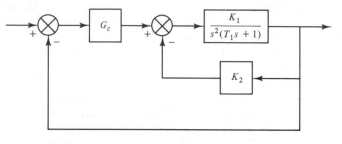

Figure P6.31

6.32. The feedback $H = (1 + K_t s)$ in Fig. P6.3 can represent a parallel combination of direct feedback and minor loop rate feedback. If $G(s) = K/(s^2 + 2s + 3.25)$ represents a spring–mass–damper system with a position output and a force input:

(a) Find the constraints on K and/or K_t for a steady-state error of 10% following step inputs.

(b) Write the characteristic polynomial with the constraints of part (a), and rearrange it to obtain the loop gain function for the construction of loci of the closed-loop poles for varying K_t.

(c) Calculate the value of K_t for a damping ratio 0.707 from the quadratic characteristic equation, and use the corresponding roots in sketching the loci for varying K_t.

6.33. In the system with rate feedback shown in Fig. P6.33:

(a) Sketch the loci and find K for a system damping ratio 0.5 for the dominating poles.

(b) Find the steady-state errors for step and ramp inputs for K of part (a).

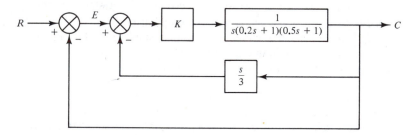

Figure P6.33

6.34. If Fig. P6.4 models a position servo with $G(s) = 1/[s(s + 3)]$, choose p and K in the controller $G_c = K(s + 1)/(s + p)$ such that:

1. The steady-state unit ramp following error is zero.

2. The damped natural frequency of oscillatory components in the transient response is 4 rad/sec.

6.35. In Fig. P6.4, if the open-loop unstable plant $G(s) = 1/[s(s - 4)]$ represents a tall rocket, use root loci to determine the range of K for which the system will be stable if G_c is a phase-lead network

$$G_c(s) = \frac{K(s + 2)}{s + 20}$$

6.36. In Fig. P6.4, let $G(s)$ be a motor plus load transfer function $1/[s(Js + B)]$ in which damping B is negligible, so that, for $J = 1$, G can be approximated by $G = 1/s^2$. Sketch loci for $G_c = K_p$ and for phase-lead compensation $G_c = K(s + 2)/(s + 8)$. Note the stabilizing effect of the lead and give the system time constant for high gains. Also, sketch the different general shape of the loci if the zero is moved close enough to the origin.

6.37. For Fig. P6.4 with $G(s) = 1/[s(s + 1)]$, choose the pole(s) and zero(s) of a controller G_c, and sketch the corresponding loci, such that no closed-loop poles will have a time constant above 0.5 sec or a damping ratio below 0.5 for some value of gain. How does the general shape of the loci change if the pole of G_c is moved relatively far away?

6.38. In Fig. P6.4, let G be a hydraulic servo with transfer function $G(s) = \omega_n^2/(s^2 + \omega_n^2)$.
(a) Can a compensator $G_c = K$ be used?
(b) Use root loci to investigate the use of a compensator $G_c = K(s + a)$, $a > 0$.
(c) Similarly, examine the use of $G_c = K(s^2 + a^2)$ with a both larger and smaller than ω_n.

6.39. In Fig. P6.4, $G(s) = K_h/(ms^2 + bs + k)$ represents a lightly damped spring–mass–damper system controlled by a hydraulic servo of which the transfer function can be approximated by a constant K_h over the frequency range of interest. It is specified that the system must follow ramp inputs with zero steady-state error.
(a) Sketch loci to determine whether the simplest possible controller which will meet the error specification can be used.
(b) Sketch the general shapes of the loci for $G_c = K(s + a)/s^2$ for some positions of the zero $-a$ from far left to close to the origin.

6.40. In Fig. P6.4 with $G = 1/[(s + 1)(s + 2)(s + 3)]$:
(a) Sketch the loci for $G_c = K$ and determine the limiting value of K for stability.
(b) Repeat for the lag compensator $G_c = (K/5)(s + 2.5)/(s + 0.5)$ and compare the steady-state errors for step inputs with part (a).

6.41. In Fig. P6.4 with $G = K/[s(s + 1)]$:
(a) With $G_c = 1$ find K for a damping ratio 0.707 and the steady-state error for a unit ramp input.
(b) Sketch the loci and repeat the above for $G_c = (10s + 1)/(40s + 1)$. How does this phase-lag compensator affect behavior?

6.42. In Fig. P6.4 with $G = 1/(s^2 + s + 5/4)$, sketch the loci for a phase-lead compensator $G_c(s) = K(s + 0.5)/(s + 5)$.

6.43. Use root locus sketches to determine the stability of systems with the following characteristic equations.
(a) $s^3 + 3s^2 + 2s + 1 = 0$
(b) $s^3 + 3s^2 + 2s + 8 = 0$

6.44. Repeat Problem 6.43 for:

(a) $s^4 + 5s^3 + 6s^2 + 2s + 1 = 0$
(b) $s^5 + 5s^4 + 6s^3 + 2s^2 + 4s + 4 = 0$

6.45. For a system with loop gain function

$$\frac{K(s + 5)}{s(s + 2)(s + 3)}$$

(a) Sketch the loci of the closed-loop poles for varying K. It may be shown that for $K = 8$ the closed-loop poles are located at -4, $-0.5 \pm j\,3.12$.
(b) Sketch loci to show the effect of variations δ of the open-loop pole at -2 on the closed-loop poles, for $K = 8$. Which direction of variation is dangerous?

7

FREQUENCY RESPONSE METHODS

7.1 INTRODUCTION

In frequency response methods, system behavior is evaluated from the steady-forced response to a sinusoidal input

$$r(t) = A \sin \omega t \tag{7.1}$$

These methods for analysis and design are very popular, and frequency response testing, over a range of frequencies ω, is often the most convenient method for measurement of systems dynamics. Also, the techniques of the preceding chapters are not well adapted to certain frequency response concepts of great practical importance. For example, if high-frequency "noise" is superimposed on the input, how can the system be designed to respond well to the input but "filter out" the noise? Or, for audio systems, instrumentation, control systems, and so on, how does one design filters that pass only input signal components in a selected range of frequencies? This relates to the problem of designing systems for a specified *bandwidth,* that is, range of frequencies over which it responds well. Further, a maximum percentage error may be specified for input signals in a normal operating range of frequencies, thus extending accuracy specifications from static to dynamic conditions.

Compensator design is often more convenient in the frequency domain, and can be carried out, for the general configuration in Fig. 7.1, by the use of frequency response plots of the loop gain function G_cGH. Polar plots, Nyquist diagrams, and M and N circles will be used to develop frequency-domain criteria for absolute and relative stability. Frequency-domain criteria for speed of response are also discussed. Bode plots and the Nichols chart are alternative forms of

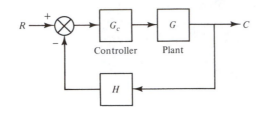

Figure 7.1   Feedback system.

plots which are much more convenient for practical analysis and design. Design by frequency response methods is therefore delayed until Chapter 8, where these types of plots are discussed.

7.2 FREQUENCY RESPONSE FUNCTIONS AND POLAR PLOTS

Figure 7.2 shows a system with transfer function $G(s)$ and a sinusoidal input

$$r(t) = A \sin \omega t \tag{7.2}$$

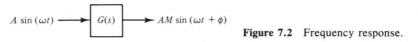

Figure 7.2   Frequency response.

Using the transform of $r(t)$ given in Table 1.6.1, the transform $C(s)$ of the system output is

$$C(s) = \frac{A\omega G(s)}{s^2 + \omega^2} = \frac{K_1}{s + j\omega} + \frac{K_2}{s - j\omega} + \cdots \tag{7.3}$$

Of the partial fraction expansion, only the terms due to the roots of the denominator of $R(s)$ are shown. As discussed below (3.24), these give the steady-forced part of the solution needed in frequency response methods. As in Section 3.3, the residues are

$$K_1 = \frac{A\omega G(s)}{s - j\omega}\bigg|_{s = -j\omega} = \frac{AG(-j\omega)}{-2j} \qquad K_2 = \frac{AG(j\omega)}{2j}$$

and the forced response is

$$c_s(t) = \frac{A[-G(-j\omega)e^{-j\omega t} + G(j\omega)e^{j\omega t}]}{2j} \tag{7.4}$$

$G(j\omega)$ is a complex variable, which can be represented in the alternative ways indicated in Fig. 7.3, either as a sum of a real part $a(\omega)$ and an imaginary part $jb(\omega)$, or as a vector of length $M(\omega)$ and phase angle $\phi(\omega)$:

$$G(j\omega) = a(\omega) + jb(\omega) = M(\omega)e^{j\phi(\omega)}$$
$$M(\omega) = |G(j\omega)| = \sqrt{a^2(\omega) + b^2(\omega)} \tag{7.5}$$
$$\phi(\omega) = \tan^{-1}\frac{b(\omega)}{a(\omega)}$$

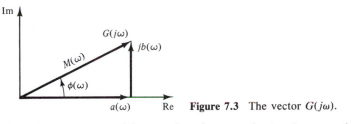

Figure 7.3   The vector $G(j\omega)$.

For $G(-j\omega)$, the imaginary part, and hence the phase angle ϕ, changes sign:

$$G(-j\omega) = M(\omega)e^{-j\phi(\omega)} \tag{7.6}$$

Substituting (7.5) and (7.6) into (7.4) yields

$$c_s(t) = \frac{AM(\omega)(e^{j(\omega t + \phi)} - e^{-j(\omega t + \phi)})}{2j}$$

or

$$c_s(t) = AM(\omega) \sin[\omega t + \phi(\omega)] \tag{7.7}$$

This result is indicated in Fig. 7.2, and shows the following:

1. For a sinusoidal input, the forced response is also sinusoidal, and of the same frequency.
2. The magnitude M of the frequency response function $G(j\omega)$, obtained by replacing s by $j\omega$ in the transfer function $G(s)$, equals the ratio of output amplitude to input amplitude.
3. The phase angle ϕ of $G(j\omega)$ is the phase angle of the output relative to that of the input.

The Laplace operator $s = \sigma + j\omega$ in $G(s)$ is replaced by $s = j\omega$, so for $G(j\omega)$ only values of s along the imaginary axis ($\sigma = 0$) are considered.

*****

$G(j\omega)$ can be plotted on a complex plane or polar plot as a vector of length $M(\omega)$ and phase angle $\phi(\omega)$, positive counterclockwise from the positive real axis. As ω varies, the end point of the vector describes the polar plot.

For a simple lag

$$G(j\omega) = \frac{K}{j\omega T + 1} \tag{7.8}$$

$$M(\omega) = K/\sqrt{1 + (\omega T)^2} \qquad \phi(\omega) = -\tan^{-1} \omega T$$

M and ϕ are verified by inspection of Fig. 7.4(a) and yield the polar plot in Fig. 7.4(b).

Rapid sketching of the general form of such plots is frequently useful. To this end, $\omega \to 0$ is considered first. The diagram in Fig. 7.4(a) or the equation for $G(j\omega)$ shows that the lag then approaches K, on the positive real axis. Next, for $\omega \to \infty$, $M \to 0$ and Fig. 7.4(a) shows that the phase of the denominator of G approaches $+90°$, so that of G itself $-90°$. Thus for $\omega \to \infty$ the plot approaches

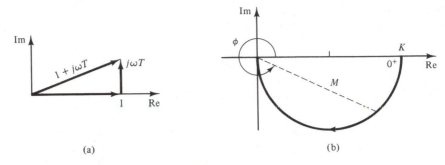

(a)   (b)

Figure 7.4 Polar plot of simple lag.

the origin from the bottom. Alternatively, for $\omega \to \infty$, $G(j\omega) \to K/(j\omega T)$, and each factor j in a denominator represents $-90°$ since

$$\frac{1}{j} = \frac{1}{j}\frac{j}{j} = \frac{j}{j^2} = -j \qquad (\text{i.e., } -90°)$$

The diagram of Fig. 7.4(a) is also very useful to show the behavior between 0^+ and $+\infty$. Both the magnitude and phase of the denominator of $G(j\omega)$ are seen to increase continuously with ω, because the imaginary part $j\omega T$ increases with ω. Therefore, for $G(j\omega)$ itself the phase must increase continuously in negative direction, from $0°$ to $-90°$, while the distance to the origin decreases continuously.

*****

Figure 7.5 shows similarly derived sketches for a number of common functions. In Fig. 7.5(a), for $\omega \to 0^+$ the function values are positive and real, and for

$$P_i \equiv j\omega T_i + 1; \quad Q \equiv (j\omega/\omega_n)^2 + 2\zeta(j\omega/\omega_n) + 1$$

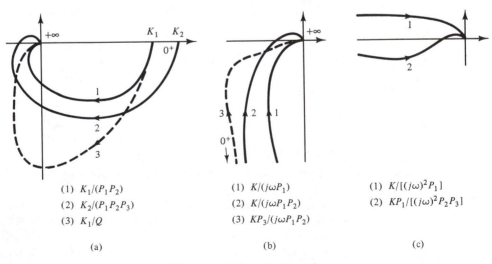

(1) $K_1/(P_1 P_2)$ (1) $K/(j\omega P_1)$ (1) $K/[(j\omega)^2 P_1]$

(2) $K_2/(P_1 P_2 P_3)$ (2) $K/(j\omega P_1 P_2)$ (2) $KP_1/[(j\omega)^2 P_2 P_3]$

(3) K_1/Q (3) $KP_3/(j\omega P_1 P_2)$

(a)   (b)   (c)

Figure 7.5 Polar plot examples.

$\omega \to +\infty$ all tend to the origin. For (a1) and (a3) the functions approach a constant divided by $(j\omega)^2$ as $\omega \to \infty$, and for (a2) a constant divided by $(j\omega)^3$. Each denominator factor $j\omega$ represents an angle $-90°$, so (a1) and (a3) approach the origin from $180°$, and (a2) from $-270°$ (or $+90°$). The diagram in Fig. 7.4(a) helps to show that for (a1) and (a2) the vector from the origin must rotate clockwise and shorten as ω increases. This is also true for (b1), (b2), and (c1) in Fig. 7.5(b) and (c). Note that for $\omega \to \infty$, (b3) approaches a constant divided by $(j\omega)^2$, and (c2) a constant divided by $(j\omega)^3$. So (b3) approaches the origin from the left, and (c2) from the top. For $\omega \to 0^+$, the functions (b) approach $K/(j\omega)$, and (c) approach $K/(j\omega)^2$. Therefore the plots (b) start far out along the negative imaginary axis, and (c) far out along the negative real axis. In (b3) and (c2) the numerator factors contribute positive phase angles (lead), so counterclockwise rotation. The equations in Fig. 7.6 can be used to calculate points, if needed. For example, for case (b3),

$$M = \frac{KM_3}{\omega M_1 M_2} \qquad \phi = -90° + \phi_3 - \phi_1 - \phi_2 \qquad (7.9)$$

Appendix B gives a program for computer construction. It is design oriented in that for a given plant the effect of selected dynamic compensators on the polar plot can be studied.

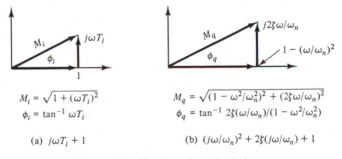

$$M_i = \sqrt{1 + (\omega T_i)^2}$$
$$\phi_i = \tan^{-1} \omega T_i$$

(a) $j\omega T_i + 1$

$$M_q = \sqrt{(1 - \omega^2/\omega_n^2)^2 + (2\zeta\omega/\omega_n)^2}$$
$$\phi_q = \tan^{-1} 2\zeta(\omega/\omega_n)/(1 - \omega^2/\omega_n^2)$$

(b) $(j\omega/\omega_n)^2 + 2\zeta(j\omega/\omega_n) + 1$

Figure 7.6 Simple and quadratic lags.

7.3 NYQUIST STABILITY CRITERION

In Fig. 7.1, the closed-loop system poles are the roots of the characteristic equation $1 + G_cGH = 0$, so can be called the zeros of $(1 + G_cGH)$. The poles of $(1 + G_cGH)$ (i.e., the roots of its denominator) equal those of the loop gain function G_cGH, the open-loop poles. Let

$$1 + G_cGH = K\frac{(s + z_1)(s + z_2) \cdots}{(s + p_1)(s + p_2) \cdots} \qquad (7.10)$$

The poles $-p_1$, $-p_2$, ... of $(1 + G_cGH)$ are usually known, but the zeros $-z_1$, $-z_2$, ... are not. If they were, stability analysis would be unnecessary. Let the pole–zero pattern be as shown in Fig. 7.7, where $-z_1$ and $-z_2$ are

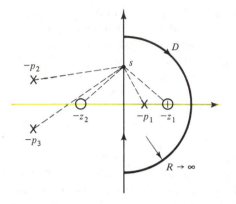

Figure 7.7 Nyquist contour D.

not known. To prove stability, it is necessary and sufficient to show that no zeros $-z_i$ are inside the *Nyquist contour* D which encloses the entire right half of the s-plane. D consists of the imaginary axis from $-j\infty$ to $+j\infty$ and a semicircle of radius $R \rightarrow \infty$. In principle, stability analysis is based on plotting $(1 + G_cGH)$ in a complex plane as s travels once clockwise around the closed contour D. As discussed in Section 3.2 and elsewhere, the factors $(s + z_i)$ and $(s + p_i)$ are vectors from $-z_i$ and $-p_i$ to s, and for any value of s the magnitude and phase of $(1 + G_cGH)$ in (7.10) can be determined graphically by measuring the vector lengths and angles in Fig. 7.7, if the z_i were known.

Note that on the imaginary axis $s = j\omega$. Thus the plot of $(1 + G_cGH)$ for s traveling up the imaginary axis from $\omega = 0^+$ to $\omega \rightarrow +\infty$ is in effect just the polar plot of the frequency response function $(1 + G_cGH)$. Hence frequency response functions can be found graphically, by measurement from the pole–zero pattern.

Figure 7.7 shows that if s moves once clockwise around D, vectors $(s + z_i)$ and $(s + p_i)$ rotate 360° clockwise for each pole and zero inside D, and undergo no net rotation for poles and zeros outside D.

From (7.10), if the vector $(s + z_1)$ in the numerator rotates 360° clockwise, this will contribute a 360° clockwise rotation of the vector $(1 + G_cGH)$ in the complex plane in which it is plotted. If vector $(s + p_1)$ in the denominator rotates 360° clockwise, this will contribute a 360° counterclockwise revolution of $(1 + G_cGH)$. Poles and zeros outside D do not contribute any net rotation. The result can be expressed as follows:

Principle of the Argument. If $(1 + G_cGH)$ has Z zeros and P poles inside the Nyquist contour D, a plot of $(1 + G_cGH)$ as s travels once clockwise around D will encircle the origin of the complex plane in which it is plotted $N = Z - P$ times in clockwise direction.

In principle, this completes the attributes needed for stability analysis. For stability the condition $Z = 0$, so $N = -P$, is necessary and sufficient. Apparently, then, if $(1 + G_cGH)$ is plotted from (7.10), but with its numerator given in polynomial rather than factored form, as s travels once around D, the system will be stable if and only if the plot encircles the origin P times in counterclockwise direction.

The following observation allows this result to be stated in a more convenient form:

The encirclements of a plot of $(1 + G_cGH)$ around the origin equal the encirclements of a plot of G_cGH around the -1 point, on the negative real axis.

With this, the following has been proved.

Nyquist stability criterion.   A feedback system is stable if and only if the number of counterclockwise encirclements of a plot of the loop gain function G_cGH about the -1 point is equal to the number of poles of G_cGH inside the right-half plane, called open-loop unstable poles.

Usually, systems are open-loop stable, that is, $P = 0$. In this case, the criterion becomes:

An open-loop stable feedback system is stable if and only if a plot of G_cGH, as s travels once around the Nyquist contour, does not encircle the -1 point.

In the marginal case where G_cGH has poles on the imaginary axis, these will be excluded from the Nyquist contour by semicircular indentations of infinitesimal radius around them. This is shown in Fig. 7.8 for the common case of a pole at the origin. The plot of G_cGH as s travels once around D is called a *Nyquist diagram* and is needed to use the criterion.

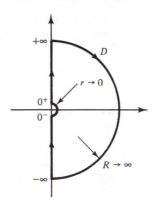

Figure 7.8 Indented Nyquist contour D.

7.4 NYQUIST DIAGRAMS AND STABILITY

In Section 7.2 the sketching of polar plots of frequency response functions was discussed. These are also part of Nyquist diagrams, for values of s on the imaginary axis from 0^+ to $+\infty$. From (7.6), the mirror image of the polar plot relative to the real axis is the part of the Nyquist diagram corresponding to the section of the Nyquist contour between 0^- and $-\infty$ on the imaginary axis.

For the part of the Nyquist contour along the semicircle of radius $R \to \infty$ the value of G_cGH is constant, usually zero. This is because, as discussed in Section 6.1, physical realizability prevents the order of the numerator of G_cGH from being larger than that of the denominator, and in practice it is usually smaller. Hence, for $|s| \to \infty$, G_cGH generally tends to zero.

It remains to consider the case of poles on the imaginary axis. The form of a loop gain function with n poles at the origin is

$$G_cGH = \frac{K}{s^n} \frac{(s + z_1)(s + z_2) \cdots}{(s + p_1)(s + p_2) \cdots} \tag{7.11}$$

Just as $(s + p_1)$ is a vector from the pole $-p_1$ to s, so each factor s in the denominator is a vector from a pole at the origin to points s. In Fig. 7.8, as s travels around the small semicircle of radius $r \to 0$, each of these vectors rotates $180°$ counterclockwise, at a constant $|s| \to 0$, between the points for frequencies 0^- and 0^+ on the imaginary axis. This means that $|G_cGH| \to \infty$ and G_cGH rotates $180°$ clockwise between frequencies 0^- and 0^+ for each pole of G_cGH at the origin.

Figure 7.9 shows some of the polar plots in Fig. 7.5 completed into Nyquist diagrams. In Fig. 7.9(a), only the mirror image need be added to the plot (a2) in Fig. 7.5. If the numerical values are such that the -1 point is located where shown, the system is stable, because the system is open-loop stable, and the -1 point is not encircled. However, the effect of an increase of gain K is that each point on the plot is moved radially outward proportionally, since the phase angle of G_cGH is unaffected but the magnitude is proportional to K. Beyond a certain gain, the -1 point will be encircled twice in the clockwise direction as ω increases from $-\infty$ to $+\infty$, and the system will be unstable. Note that for

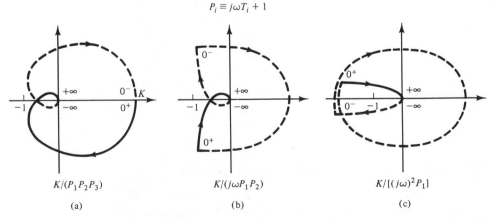

$$P_i \equiv j\omega T_i + 1$$

$K/(P_1 P_2 P_3)$ $K/(j\omega P_1 P_2)$ $K/[(j\omega)^2 P_1]$

(a)   (b)   (c)

Figure 7.9 Nyquist diagram examples.

(a1) and (a3) in Fig. 7.5 the -1 point will be approached for high gains, but cannot be encircled, so these open-loop stable systems cannot lose stability. Root locus sketches readily confirm these results.

In Fig. 7.9(b), after completing the mirror image of the polar plot, the ends 0^- and 0^+ are connected by a clockwise 180° rotation at large radius due to the pole at the origin. The system is stable, but would become unstable if radially expanded by an increase of gain such that -1 is encircled.

In Fig. 7.9(c) there must be a 360° clockwise rotation at large radius between 0^- and 0^+, due to the double pole at the origin. Following the curve from $-\infty$ to $+\infty$ shows that the -1 point is encircled twice in clockwise direction, so the system is unstable for any value of gain K.

The Nyquist diagram in Fig. 7.10(a), which incorporates the polar plot (c2) in Fig. 7.5, is useful to show that care is necessary when deciding how many encirclements are present. It is convenient to visualize an elastic with one end fixed at -1 and the other moved along the closed contour from $-\infty$ to $+\infty$. The appearance to the contrary, this will show that there is no net encirclement of -1. Therefore, because there are no right-half-plane poles, the system shown is stable. This, incidentally, shows the stabilizing effect of phase-lead compensation $(j\omega T_1 + 1)/(j\omega T_3 + 1)$ on the system in Fig. 7.9(c). Figure 6.15 showed a root locus interpretation, while on the Nyquist diagram the effect is seen to be an introduction of positive phase angles which deform the curve in a counterclockwise direction, causing it to pass on the other side of the -1 point.

Figure 7.10(b) shows an open-loop unstable case. For $\omega \to 0^+$ the function approaches $-K/(j\omega)$. With the minus sign equivalent to $\pm 180°$, and with the factor $j\omega$ in the denominator contributing $-90°$, the polar plot "starts" far out along the positive imaginary axis. For $\omega \to +\infty$, it tends to a constant divided by $(j\omega)^2$, so approaches the origin from the left. The pole at the origin again

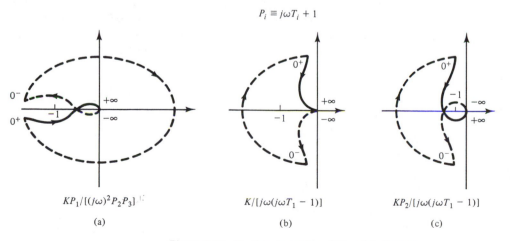

Figure 7.10 Further examples of Nyquist diagrams.

causes a 180° clockwise rotation at large radius from 0^- to 0^+. There is one clockwise encirclement of the -1 point. Hence the system is unstable, for all gains K, since with $P = 1$ stability requires one counterclockwise encirclement.

In Fig. 7.10(c) a zero has been added, and the system is now stable, since there is one counterclockwise encirclement of the -1 point. In this example a reduction of gain below the value at which the locus will still cross the negative real axis to the left of -1 would cause instability.

7.5 RELATIVE STABILITY: GAIN MARGIN AND PHASE MARGIN

Most practical systems are not open-loop unstable, so that stability requires zero encirclements of the -1 point. To determine this, it is in fact not necessary to plot the complete Nyquist diagram; the polar plot, for ω increasing from 0^+ to $+\infty$, is sufficient. Figures 7.9 and 7.10 help to verify that the following is equivalent to zero encirclements.

Simplified Nyquist Criterion. If G_cGH does not have poles in the right-half s-plane, the closed-loop system is stable if and only if the -1 point lies to the left of the polar plot when moving along this plot in the direction of increasing ω.

For example, the polar plot of a loop gain function shown in Fig. 7.11 indicates a stable system. If the curve passes through -1, the system is on the verge of instability. For adequate relative stability it is reasonable that the curve should not come too close to the -1 point. Gain margin and phase margin are two common design criteria, which specify the distance of a selected point of the polar plot to -1. Both are defined in Fig. 7.11:

1. Gain margin $= 1/0C$.   (7.12a)
2. Phase margin $\phi_m = 180°$ plus the phase angle of G_cGH at the crossover frequency ω_c at which $|G_cGH| = 1$. It is also the negative phase shift (i.e., clockwise rotation) of G_cGH which will make the curve pass through -1.   (7.12b)

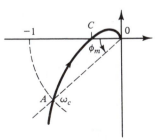

Figure 7.11 Phase margin and gain margin.

Each specifies the distance to -1 of only one point on the curve, so misleading indications are possible. Phase margin is used very extensively in practice.

Bode plots, discussed in Chapter 8, are much more convenient, but these criteria can also be calculated from polar plots. For example, Fig. 7.11 shows that gain margin corresponds to a zero imaginary part of G_cGH.

Example 7.5.1

Consider Fig. 7.10(a):

$$G_cGH = \frac{K(1 + j\omega T_1)}{(j\omega)^2(1 + j\omega T_2)(1 + j\omega T_3)} \tag{7.13}$$

The imaginary part can be identified by multiplying the numerator and denominator by $(1 - j\omega T_2)(1 - j\omega T_3)$. Then

$$G_cGH = K\frac{1 - \omega^2(T_2T_3 - T_1T_2 - T_1T_3) + j\omega(T_1 - T_2 - T_3 - \omega^2 T_1T_2T_3)}{-\omega^2(1 + \omega^2 T_2^2)(1 + \omega^2 T_3^2)}$$

The imaginary part is zero if $\omega^2 = (T_1 - T_2 - T_3)/(T_1T_2T_3)$, and substituting this yields $|G_cGH| = 0C$. Note that only one intersection can occur here. If

$$G_cGH = \frac{K(j\omega T_1 + 1)(j\omega T_4 + 1)}{(j\omega)^2(j\omega T_2 + 1)(j\omega T_3 + 1)(j\omega T_5 + 1)} \tag{7.14}$$

then two intersections could occur, as in Fig. 7.12, which represents a stable system since the -1 point lies left of the curve.

Example 7.5.2

Consider Fig. 7.9(b):

$$G_cGH = \frac{K}{j\omega(j\omega T_1 + 1)(j\omega T_2 + 1)} \tag{7.15}$$

This simpler example can be treated as the preceding one, but it may also be recognized that since the factor $j\omega$ provides $-90°$ of the $-180°$ required along the negative real axis, the product

$$(j\omega T_1 + 1)(j\omega T_2 + 1) = 1 - \omega^2 T_1T_2 + j\omega(T_1 + T_2)$$

must yield the remaining $-90°$. This requires that $\omega^2 = 1/(T_1T_2)$, for which

$$G_cGH = \frac{K}{(j\omega)^2(T_1 + T_2)} = \frac{-K}{\omega^2(T_1 + T_2)} = \frac{-KT_1T_2}{T_1 + T_2}$$

Hence the gain margin is $(T_1 + T_2)/(KT_1T_2)$.

Correlations with time-domain relative stability measures will be considered shortly.

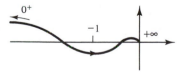

0^+ -1 $+\infty$

Figure 7.12   Polar plot for (7.14).

7.6 *CLOSED-LOOP FREQUENCY RESPONSE AND M CIRCLES*

By means of the Nyquist stability criterion, it has been possible to determine the stability of the closed-loop system from a frequency response plot of its loop gain function. It is now shown how for systems with unity feedback, the closed-loop frequency response may also be found from these plots. With input R, output C, and loop gain function $G(j\omega)$ in the forward loop, the closed-loop frequency response is given by

$$\frac{C(j\omega)}{R(j\omega)} = \frac{G(j\omega)}{1 + G(j\omega)} \qquad (7.16)$$

Consider now the polar plot of $G(j\omega)$ in Fig. 7.13(a). By definition, $G(j\omega)$ is the vector from 0 to points A, with length $0A$ and phase angle ϕ_a. But also, $1 + G(j\omega)$ is the vector from B to A, because it is the vector sum of $G(j\omega)$ and the vector 1, which is the vector from B to 0. Therefore, the magnitude $M(\omega)$ and phase angle $\phi(\omega)$ of the closed-loop frequency response function C/R are given by

$$M(\omega) = \left|\frac{C}{R}\right| = \frac{0A}{BA} \qquad (7.17)$$

$$\phi(\omega) = \text{phase angle } \frac{C}{R} = \phi_a - \phi_b$$

By measuring the vector lengths as A moves along the curve, the magnitude M can be plotted versus ω, as shown in Fig. 7.13(b) for a typical case. Low relative stability means that the curve comes close to -1, so that BA will be small over a range of frequencies. Thus a low phase margin will be reflected in severe resonance peaking in $M(\omega)$.

*****

For any given point A, $0A$, BA, ϕ_a, and ϕ_b have given values, and therefore M and ϕ in (7.17) also. This permits loci to be constructed for constant values of M and of ϕ. For example, along a vertical midway between 0 and -1, $0A = BA$, so this is the locus for $M = 1$. The real axis to the left of -1 and to the right of 0 is the locus for $\phi = 0$, since $\phi_a = \phi_b$, and this axis between -1 and 0 is the locus for $\phi = \pm 180°$.

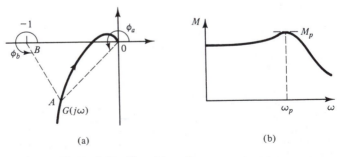

(a)   (b)

Figure 7.13   Closed-loop frequency response.

It can be shown that, with the exceptions noted above, the loci for constant M and for constant ϕ are circles, called *M circles* and *N circles*, respectively. Some M circles are shown in Fig. 7.14. The centers are on the real axis, and their locations x and the circle radii r are given by

$$x = \frac{M^2}{1 - M^2} \qquad r = \left| \frac{M}{1 - M^2} \right| \qquad (7.18)$$

These loci are not used nearly as often for system analysis and design as is their translation to another type of plot, the Nichols chart, discussed in Chapter 8.

In Fig. 7.14, if $G(j\omega)$ is a polar plot superimposed on a graph on which the M and N circles are printed, M and ϕ for the values of frequency ω along $G(j\omega)$ can be read off from the intersections with these loci. Thus M can be plotted versus ω as in Fig. 7.13(b), and ϕ also.

In particular, in Fig. 7.14 the maximum value of M occurs at the point of tangency of $G(j\omega)$ with the locus for $M = 2$. Thus in this case the resonant peak $M_p = 2$ and occurs at the resonant frequency ω_p given by $G(j\omega)$ at the point of tangency.

Note that in addition to resonant peaking M_p, the phase margin and gain margin can also be read in Fig. 7.14. Of the three, the tangency circle M_p is a closed-loop performance measure, and the other two are criteria applied to the open-loop gain function.

7.7 FREQUENCY RESPONSE PERFORMANCE CRITERIA

In Sections 3.5 and 3.6 transient performance was discussed by examination of the quadratic lag (closed-loop) system transfer function.

$$\frac{C(s)}{R(s)} = \frac{\omega_n^2}{s^2 + 2\zeta\omega_n s + \omega_n^2} \qquad (7.19)$$

This was based on the fact that the performance of many systems is dominated

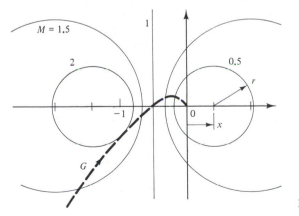

Figure 7.14 *M* circles.

by one complex conjugate pair of poles. It is appropriate, therefore, to consider the frequency response of the system (7.19):

$$\frac{C(j\omega)}{R(j\omega)} = \frac{1}{(j\omega/\omega_n)^2 + 2\zeta(j\omega/\omega_n) + 1} \tag{7.20}$$

From Fig. 7.6:

$$M = \left|\frac{C}{R}\right| = \frac{1}{\sqrt{(1 - \omega^2/\omega_n^2)^2 + (2\zeta\omega/\omega_n)^2}} \tag{7.21}$$

The typical frequency response plot is shown in Fig. 7.15. The *resonant peak* M_p at the *resonant frequency* ω_p is found by setting the derivative of M to zero.

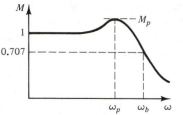

Figure 7.15 Frequency response second-order system.

This yields the following results for ω_p/ω_n and M_p, where M_p is obtained by substituting ω_p/ω_n into (7.21):

$$\frac{\omega_p}{\omega_n} = \sqrt{1 - 2\zeta^2} \qquad M_p = \frac{1}{2\zeta\sqrt{1 - \zeta^2}} \qquad \zeta \leqslant 0.707 \tag{7.22}$$

These equations are plotted in Fig. 7.16(a).

The resonant peaking, as indicated in Section 7.6, is a key performance measure of the closed-loop response. Figure 7.16(a) links it directly to the damping ratio ζ, the relative stability measure for transient response. Low damping ratio ζ means both severe step response overshoot and severe resonance peaking.

In transient response, ζ and ω_n corresponding to the dominating poles determine speed of response and overshoot. The real part of the poles determines settling time, and the imaginary part the frequency of transient oscillations and hence the rise time. Increasing ω_n for given ζ reduces both settling time and rise time, improving the speed of response. In frequency response, a very common measure of speed of response is the system *bandwidth* ω_b, identified in Fig. 7.15 as the range of frequencies over which M equals at least 0.707 ($= \sqrt{2}/2$) of its value at $\omega = 0$:

$$M(\omega_b) = 0.707M(0) \tag{7.23}$$

Substituting 0.707 for M in (7.21), squaring both sides, and solving the resulting quadratic equation for ω_b/ω_n yields

$$\frac{\omega_b}{\omega_n} = [1 - 2\zeta^2 + \sqrt{2 - 4\zeta^2(1 - \zeta^2)}]^{1/2} \tag{7.24}$$

This is shown plotted in Fig. 7.16(b).

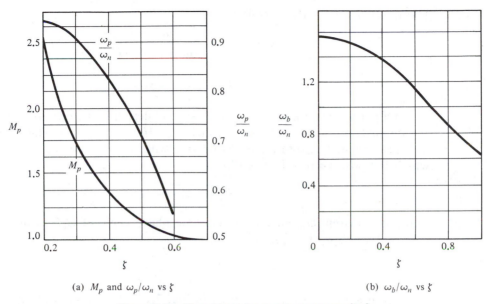

Figure 7.16   Closed-loop frequency response criteria.

When $\zeta = 0.707$, the bandwidth ω_b equals the undamped natural frequency ω_n. The bandwidth is the range of frequencies over which the response is considered to be satisfactory. Beyond ω_b, in particular for lower values of ζ, the response drops off rapidly. For given ζ, ω_b is proportional to ω_n, and a measure of speed of response. Raising ω_b reduces settling time and rise time of the step response.

Open-Loop Frequency Response Performance Measures

Resonance peaking and bandwidth are performance specifications for the closed-loop system. However, the polar plots discussed earlier and the Bode plots in Chapter 8 are plots of the loop gain function. The question is, therefore: What requirements should be imposed on these plots to satisfy specifications on closed-loop bandwidth ω_b and relative stability ζ or M_p?

To correlate open-loop and closed-loop criteria, it is noted that C/R of (7.19) can be considered to be realized by a unity feedback ($H = 1$) system with loop gain function.

$$G_c G = \frac{\omega_n^2}{s(s + 2\zeta\omega_n)} \tag{7.25}$$

The frequency response function is

$$G_c G(j\omega) = \frac{\omega_n^2}{-\omega^2 + 2j\zeta\omega_n\omega}$$ (7.26)

$$|G_c G| = \frac{\omega_n^2}{\sqrt{\omega^4 + (2\zeta\omega_n\omega)^2}}$$

Of particular importance for analysis and design is the *crossover frequency* ω_c, defined in (7.12b) as that for which $|G_c GH| = 1$, and at which the phase margin is defined. From (7.26), $|G_c G| = 1$ when $\omega_c^4 + (2\zeta\omega_n\omega_c)^2 - \omega_n^4 = 0$, of which the positive root is given by

$$\frac{\omega_c}{\omega_n} = (\sqrt{4\zeta^4 + 1} - 2\zeta^2)^{1/2}$$ (7.27)

The phase margin ϕ_m equals 180° plus the phase angle of $G_c G$ at ω_c, so $\phi_m = 180° - \tan^{-1} 2\zeta\omega_n/(-\omega_c)$, or

$$\phi_m = \tan^{-1} \frac{2\zeta}{(\sqrt{4\zeta^4 + 1} - 2\zeta^2)^{1/2}}$$ (7.28)

Both ω_c/ω_n and ϕ_m are plotted in Fig. 7.17. The ϕ_m–ζ plot represents an important and useful correlation between frequency response and transient response measures of relative stability. It is also noted that with increasing ζ the crossover frequency of the loop gain function lies farther below the undamped natural frequency of the closed-loop system.

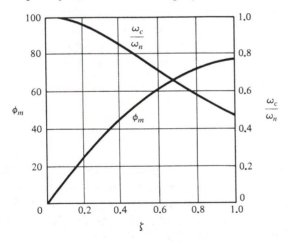

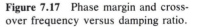

Figure 7.17 Phase margin and crossover frequency versus damping ratio.

Table 7.7.1 shows the ratio of crossover to bandwidth frequency for the range of values of ζ or ϕ_m, found from Figs. 7.16(b) and 7.17. For a specified phase margin this table shows how the crossover frequency of the loop gain function should be chosen to meet a given specification on the closed-loop system bandwidth. It is interesting to note that the ratio is rather constant.

The important Bode plot design technique in Chapter 8 is based on the

TABLE 7.7.1 BANDWIDTH–CROSSOVER FREQUENCY CORRELATION

ζ	0	0.1	0.2	0.3	0.4	0.5	0.6	0.707	0.8	0.9	1.0
ω_c/ω_b	0.64			0.63	0.62		0.63	0.64	0.68	0.71	0.77
ϕ_m	0	12	24	35	45	53	61	67	72	75	78

loop gain function. The table is useful to choose the required crossover frequency. It is also evident that raising the crossover frequency increases the speed of response.

7.8 CONCLUSION

Frequency response methods and frequency response plots have been discussed. Polar plots, Nyquist diagrams, and M circles were used to develop criteria for absolute and relative stability. Frequency response performance criteria for stability and speed of response were correlated with those for transient analysis.

Nyquist diagrams (or root loci) should be used for open-loop unstable plants. However, for most systems, Bode plot techniques (Chapter 8) are much more convenient for analysis and design and allow the clearest visualization of the effects of compensation on performance. The Nichols chart, used for the same purpose as M circles, is related to Bode plot techniques, and will be applied as well.

Construction of accurate polar plots is laborious, and computer aids are essential for their application in detailed analysis and design. The interactive program with graphics in Appendix B is oriented toward the latter in that for a given plant polar plots can be obtained for selected dynamic compensators.

PROBLEMS

7.1. Determine the steady-forced response of a system with transfer function

$$T(s) = \frac{1}{(s + 1)(0.1s + 1)}$$

to the following sinusoidal inputs.
(a) $r(t) = 2 \sin 0.5t$
(b) $r(t) = 2 \sin 5t$
Note the effect of frequency on output amplitude and phase shift.

7.2. In Problem 7.1, determine the frequency at which the output lags the input by 90°, and find the ratio of output to input amplitude at this frequency. [*Hint: T(s) =* $1/(0.1s^2 + 1.1s + 1)$, and the phase shift is $-90°$ when $T(j\omega)$ plots along the negative imaginary axis.]

7.3. For $T(s) = 1/[s(s + 1)(0.1s + 1)]$, find the intersection of $T(j\omega)$ with the real axis of the complex plane.

7.4. To relate the frequency response function and the s-plane, determine the magnitude and phase angle of the frequency response function

$$G(j\omega) = \frac{10(j\omega + 1)}{j\omega + 5}$$

at $\omega = 1$, $\omega = 5$, and $\omega = 25$ from measurement on the pole–zero pattern of $G(s)$.

7.5. Figure P7.5 shows the poles of a quadratic lag transfer function with root locus gain $(1 + 16 = \omega_n^2 =)$ 17. To show that low damping (i.e., poles at an angle relatively close to the imaginary axis) is equivalent to high resonance peaking in a plot of the magnitude of the frequency response function versus frequency, obtain an adequate number of points for such a plot by measurements on the pole–zero pattern. Note that the high peak is due to a short vector in the denominator.

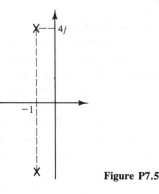

Figure P7.5

7.6. In Problem 7.5, also use measurement on the pole–zero pattern to determine and plot the phase angle of the frequency response function versus frequency, at $\omega = 1, 3, 4.123 \ (= \sqrt{17})$, and 5.

7.7. Sketch polar plots of (a) $2/(s + 1)$ and (b) $2/(0.1s + 1)$ based on calculation of at least three points. How do the plots differ? Prove that the shape is circular.

7.8. Construct polar plots for

(a) $G(s) = \dfrac{10}{(s + 1)(0.1s + 1)}$

(b) $G(s) = \dfrac{10}{(s + 1)(0.5s + 1)}$

from calculations at frequencies 0.5, 1, 2, and 6. Comment on the nature of the plot if the second time constant is quite small relative to the dominating one of 1 sec.

7.9. Construct and compare polar plots of $G(s) = \omega_n^2/(s^2 + 2\zeta\omega_n s + \omega_n^2)$ for $\omega_n = 1$ and (a) $\zeta = 0.75$, (b) $\zeta = 0.25$ from calculated points for $\omega = 0.5, 1$, and 2.

7.10. Sketch polar plots of

(a) $G(s) = \dfrac{1}{s(s + 1)}$

(b) $G(s) = \dfrac{1}{s(s + 1)(0.5s + 1)}$

(c) $G(s) = \dfrac{1}{s^2(s + 1)}$

based on the behavior at low and high frequencies and points calculated for $\omega = 0.5, 1,$ and 2.

7.11. Extend the polar plots of Problem 7.10 into Nyquist diagrams and use these to determine stability and the number of unstable poles of systems of which these are the loop gain functions.

7.12. Sketch Nyquist diagrams for the loop gain functions

(a) $\dfrac{K}{s - 1}$

(b) $\dfrac{1}{s(s - 1)}$

and in part (a) give the range of K for which the system is stable. Verify the results by root locus sketches.

7.13. Figure P7.13 shows the polar plot of an open-loop stable system for gain $K = 1$.
   (a) Determine the limits on K for stability.
   (b) What are the system phase and gain margins?
   (c) Sketch the Nyquist diagram corresponding to Fig. P7.13 and verify the conclusion on stability by considering encirclements.

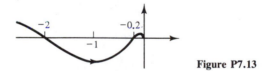

Figure P7.13

7.14. Sketch the polar plot for a system with loop gain function

$$G(s) = \dfrac{10}{s(1 + s/4)(1 + s/16)}$$

and determine the gain and phase margins.

7.15. Gain margin and phase margin should be measures of the distance of the polar plot to -1. Accordingly, determine in each of Fig. P7.15(a) and (b) which measure would and which would not provide a misleading idea of the degree of relative stability.

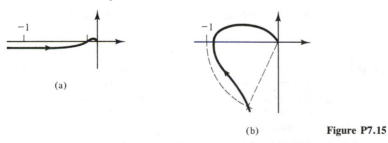

(a)

(b)   **Figure P7.15**

7.16. Plot the closed-loop frequency response function magnitude characteristic of a system versus frequency if the system has unity feedback and the polar plot of its loop gain function is as shown in Fig. P7.16. Note how nearness of this plot to −1 translates into a high resonance peak.

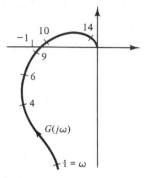

Figure P7.16

7.17. Sketch the polar plot for a loop gain function

$$G(s) = \frac{7}{s(0.1s + 1)(0.02s + 1)}$$

based on calculated points at $\omega = 5$, 10, and 15, and the intersection with the real axis, and determine the gain margin and the approximate phase margin.

7.18. In Fig. P7.18, let

$$G(s) = \frac{5}{s(s + 1)(0.5s + 1)}$$

Based on calculations at $\omega = 0.5$, 1, and 1.5:

(a) Construct the polar plot of $G_c G$ for $G_c = 1$ and determine stability.

(b) Construct a polar plot and find phase and gain margins to determine the effect of a phase-lag network $G_c = (10s + 1)/(50s + 1)$.

(c) Has the compensation affected the steady-state errors for step and ramp inputs?

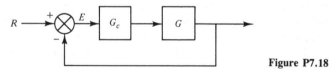

Figure P7.18

7.19. In Problem 7.18, replace the phase-lag by phase-lead compensation

$$G_c = \frac{2s + 1}{0.2s + 1}$$

Sketch the polar plot of $G_c G$ based on the frequencies $\omega = 2$ and 4 to estimate gain margin and phase margin. What is the effect on steady-state errors? Note that both phase lag and phase lead bend the plot around the other side of −1. Compare how each achieves this (i.e., how each moves a point on G to change the plot in the desired direction).

7.20. A system with unity feedback has the loop gain function

$$G(s) = \frac{K}{(s + 1)(s^2 + 2s + 2)}$$

(a) Find the limiting value of K for stability.
(b) Sketch the polar plot for $K = 2.5$ based on points calculated at $\omega = 1$, 1.5, 1.8, and 2.
(c) How does doubling K to $K = 5$ affect this plot?
(d) For $K = 5$, estimate the maximum peaking of the closed-loop frequency response, and plot the corresponding M-circle to verify the estimate.

7.21. In Problem 7.20, use M-circles to find the value of K for which the closed-loop frequency response peaking is about 20%.

7.22. Determine the required positions of the dominating pair of poles of a system if the resonance peak is specified to be 1.15 and the bandwidth 10 rad/sec.

7.23. Experimental measurements yield a plot of the magnitude of the frequency response function with a resonance peak 1.35 at a frequency of 10 rad/sec.
(a) Estimate ζ and ω_n of the dominating system poles.
(b) Estimate the bandwidth.
(c) Estimate the percentage overshoot in response to step inputs.

7.24. Estimate the phase margin and crossover frequency specifications to which a loop gain function must be designed if the closed-loop system must meet one of the following sets of requirements:

1. An effective damping ratio $\zeta = 0.6$ and a bandwidth of $\omega_b = 10$ rad/sec.
2. A resonance peak 1.15 occurring at a frequency of 10 rad/sec.
3. A step response with an overshoot of 20% and a settling time of 1 sec.

FREQUENCY RESPONSE
ANALYSIS AND DESIGN

8.1 INTRODUCTION

In this chapter the emphasis will be primarily on the use of Bode plots and the Nichols chart for open-loop stable and minimum-phase systems. For open-loop-unstable plants (i.e., with right-half-plane poles), Nyquist diagrams or root loci should be used. For nonminimum-phase plants there are zeros in the right-half plane. This is discussed separately in this chapter. Systems with transport lag, due to pipelines for example, are also considered, as are certain particular control configurations, such as feedforward control for reducing the effect of disturbances.

8.2 BODE PLOTS

Consider the general loop gain function

$$G_c GH = \frac{K}{(j\omega)^n} \frac{S_1 S_2 \cdots Q_1 Q_2 \cdots}{S_{k+1} S_{k+2} \cdots Q_{l+1} Q_{l+2} \cdots} \tag{8.1}$$

where

$$S_i \equiv j\omega T_i + 1 \qquad Q_i \equiv \left(\frac{j\omega}{\omega_{ni}}\right)^2 + 2\zeta_i \frac{j\omega}{\omega_{ni}} + 1$$

K is the gain defined in Section 4.5, since the constant terms are unity. T_i, ω_{ni}, and ζ_i are all positive, so no poles and, for now, no zeros lie in the right-half plane.

G_cGH of (8.1) is the product of only four types of *elementary factors:*

1. Gain K
2. Integrators $1/(j\omega)^n$ or differentiators $(j\omega)^n$
3. Simple lag $1/S_i$ or simple lead S_i
4. Quadratic lag $1/Q_i$ or quadratic lead Q_i

For a product $G_cGH = M_1 e^{j\phi_1} M_2 e^{j\phi_2} \cdots = M e^{j\phi}$, $M = M_1 M_2 \cdots$ and $\phi = \phi_1 + \phi_2 + \cdots$. The phase angle ϕ is expressed as a sum. The magnitude M will also be expressed as a sum, by using decibels (dB) as units:

$$M \text{ in dB} = M_{dB} = 20 \log_{10} M \qquad (8.2)$$

$$20 \log M = 20 \log M_1 + 20 \log M_2 + \cdots$$

A decibel conversion table is given in Table 8.2.1. For a value outside the table, say $M^1(10)^n$, where M^1 is a value in the table, the conversion is $20 \log M^1(10)^n = 20 \log M^1 + 20 \log (10)^n = 20 \log M^1 + 20n$.

In Bode plots, the magnitude M in dB and the phase angle ϕ in degrees are plotted against ω on semilog paper. The development has shown the following:

Bode magnitude and phase-angle plots of G_cGH are obtained by summing those of its elementary factors.

TABLE 8.2.1 DECIBEL CONVERSION: $m = 20 \log_{10} M$

M	0	1	2	3	4	5	6	7	8	9
0.0	$m =$	−40.00	−33.98	−30.46	−27.96	−26.02	−24.44	−23.10	−21.94	−20.92
0.1	−20.00	−19.17	−18.42	−17.72	−17.08	−16.48	−15.92	−15.39	−14.89	−14.42
0.2	−13.98	−13.56	−13.15	−12.77	−12.40	−12.04	−11.70	−11.37	−11.06	−10.75
0.3	−10.46	−10.17	−9.90	−9.63	−9.37	−9.12	−8.87	−8.64	−8.40	−8.18
0.4	−7.96	−7.74	−7.54	−7.33	−7.13	−6.94	−6.74	−6.56	−6.38	−6.20
0.5	−6.02	−5.85	−5.68	−5.51	−5.35	−5.19	−5.04	−4.88	−4.73	−4.58
0.6	−4.44	−4.29	−4.15	−4.01	−3.88	−3.74	−3.61	−3.48	−3.35	−3.22
0.7	−3.10	−2.97	−2.85	−2.73	−2.62	−2.50	−2.38	−2.27	−2.16	−2.05
0.8	−1.94	−1.83	−1.72	−1.62	−1.51	−1.41	−1.31	−1.21	−1.11	−1.01
0.9	−0.92	−0.82	−0.72	−0.63	−0.54	−0.45	−0.35	−0.26	−0.18	−0.09
1.0	0.00	0.09	0.17	0.26	0.34	0.42	0.51	0.59	0.67	0.75
1.1	0.83	0.91	0.98	1.06	1.14	1.21	1.29	1.36	1.44	1.51
1.2	1.58	1.66	1.73	1.80	1.87	1.94	2.01	2.08	2.14	2.21
1.3	2.28	2.35	2.41	2.48	2.54	2.61	2.67	2.73	2.80	2.86
1.4	2.92	2.98	3.05	3.11	3.17	3.23	3.29	3.35	3.41	3.46
1.5	3.52	3.58	3.64	3.69	3.75	3.81	3.86	3.92	3.97	4.03
1.6	4.08	4.14	4.19	4.24	4.30	4.35	4.40	4.45	4.51	4.56
1.7	4.61	4.66	4.71	4.76	4.81	4.86	4.91	4.96	5.01	5.06
1.8	5.11	5.15	5.20	5.25	5.30	5.34	5.39	5.44	5.48	5.53
1.9	5.58	5.62	5.67	5.71	5.76	5.80	5.85	5.89	5.93	5.98
2.	6.02	6.44	6.85	7.23	7.60	7.96	8.30	8.63	8.94	9.25
3.	9.54	9.83	10.10	10.37	10.63	10.88	11.13	11.36	11.60	11.82
4.	12.04	12.26	12.46	12.67	12.87	13.06	13.26	13.44	13.62	13.80
5.	13.98	14.15	14.32	14.49	14.65	14.81	14.96	15.12	15.27	15.42
6.	15.56	15.71	15.85	15.99	16.12	16.26	16.39	16.52	16.65	16.78
7.	16.90	17.03	17.15	17.27	17.38	17.50	17.62	17.73	17.84	17.95
8.	18.06	18.17	18.28	18.38	18.49	18.59	18.69	18.79	18.89	18.99
9.	19.08	19.18	19.28	19.37	19.46	19.55	19.65	19.74	19.82	19.91

These plots are much easier to make than polar plots or Nyquist diagrams, and can readily be interpreted in terms of different aspects of system performance. Bode plots of the four factors will be obtained first.

1. *Gain $K > 0$* [Fig. 8.1(a)]:
   $M_{dB} = 20 \log K$, $\phi = 0$, both independent of ω.

2. *Integrators $1/(j\omega)^n$* [Fig. 8.1(b)]:
$$M_{dB} = 20 \log|j\omega|^{-n} = 20 \log \omega^{-n} = -20\,n \log \omega \qquad (8.3)$$

At $\omega = 1$, $M_{dB} = 0$, and at $\omega = 10$, one decade (dec) away from $\omega = 1$, $M_{dB} = -20n$. Hence on a log scale the magnitude plot is a straight line crossing the 0 dB axis at $\omega = 1$ at a slope of $-20n$ dB/dec. Phase angle $\phi = -n\,90°$ and is independent of frequency.

For *differentiators* $(j\omega)^n$, the plots are the mirror images relative to the 0 dB and 0° axes. This is also true for the *leads* corresponding to the simple and quadratic lag below.

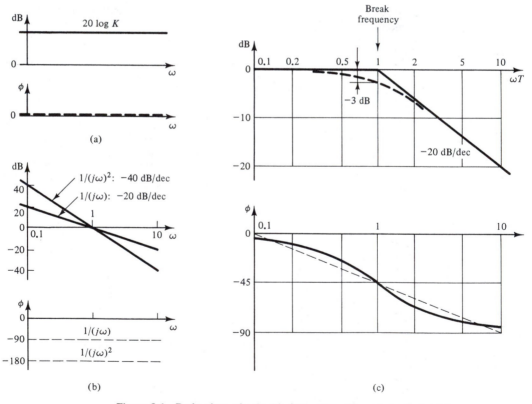

Figure 8.1 Bode plots of gain (a), integrators (b), and simple lag (c).

3. *Simple lag* $1/(j\omega T + 1)$ [Fig. 8.1(c), or half the values for $\zeta = 1.0$ in Fig. 8.2]: By inspection or using Fig. 7.6(a):

$$M_{dB} = 20 \log 1/\sqrt{1 + (\omega T)^2} \qquad \phi = -\tan^{-1} \omega T \qquad (8.4)$$

$$\omega T \ll 1: M_{dB} \longrightarrow 20 \log 1 = 0 \text{ dB} \qquad \phi \longrightarrow 0°$$

So the 0 dB axis is the low-frequency asymptote.

$$\omega T \gg 1: M_{dB} \longrightarrow 20 \log (\omega T)^{-1} = -20 \log \omega T \qquad \phi \longrightarrow -90°$$

Analogous to the integrator, this high-frequency asymptote crosses the 0 dB axis at $\omega T = 1$, at a slope of -20 dB/dec. This is the *asymptotic approximation*. The asymptotes meet at the *break frequency* or *corner frequency* given by $\omega T = 1$ on the normalized plot. On a log ω scale it is at $\omega = 1/T$.

Closer to $\omega T = 1$, the actual values can be calculated from (8.4). The deviations from the asymptotes and the phase angles ϕ at several frequencies are shown in Table 8.2.2. At $\omega T = 1$, the deviation is -3 dB and the phase $-45°$. At $\omega T = 0.5$ and $\omega T = 2$ the deviation is -1 dB. Note that at 0.1 of its break frequency the simple lag contributes $-5.7°$ phase angle, and at 10 times the break frequency $-84.3°$, $5.7°$ away from $-90°$. Figure 8.1(c) shows by a dashed line between $(\omega T = 0.1, \phi = 0)$ and $(\omega T = 10, \phi = -90°)$ an asymptotic approximation to the phase-angle curve. Its error is below $6°$ at all frequencies in this range.

TABLE 8.2.2   SIMPLE LAG $1/(j\omega T + 1)$

ωT	0.1	0.2	0.5	1	2	5	10
Deviation from asymptotes (dB)	-0.04	-0.2	$-1.$	$-3.$	$-1.$	-0.2	-0.04
Phase angle ϕ (deg)	-5.7	-11.3	-26.6	-45	-63.4	-78.7	-84.3

4. *Quadratic lag* (Fig. 8.2):

$$\frac{1}{(j\omega/\omega_n)^2 + 2\zeta(j\omega/\omega_n) + 1}$$

From Fig. 7.6(b):

$$M_{dB} = 20 \log \left[\left(1 - \frac{\omega^2}{\omega_n^2} \right)^2 + \left(\frac{2\zeta\omega}{\omega_n} \right)^2 \right]^{-1/2} \qquad (8.5)$$

$$\phi = -\tan^{-1} \frac{2\zeta\omega/\omega_n}{1 - \omega^2/\omega_n^2}$$

$\omega/\omega_n \ll 1: M_{dB} \to 20 \log 1 = 0$ dB; $\phi \to 0°$. So the 0 dB axis is again the low-frequency asymptote.

$\omega/\omega_n \gg 1: M_{dB} \to 20 \log (\omega/\omega_n)^{-2} = -40 \log (\omega/\omega_n)$; $\phi \to -180°$.

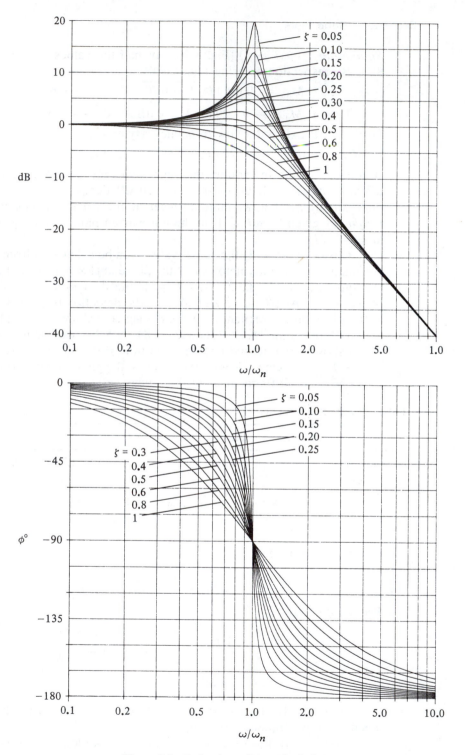

Figure 8.2 Bode plots of a quadratic lag.

Analogous to integrators, this high-frequency asymptote crosses the 0 dB axis at $\omega/\omega_n = 1$, at a slope of -40 dB/dec. Closer to $\omega/\omega_n = 1$, (8.5) gives the actual curves. At $\omega/\omega_n = 1$, $M_{\text{dB}} = 20 \log 0.5/\zeta$; $\phi = -90°$. Smaller damping ratios ζ cause more severe peaking of M_{dB} and more abrupt change of ϕ near $\omega/\omega_n = 1$. The Bode plots for a quadratic lead are again mirror images relative to the 0 dB and 0° axes.

Before showing examples of the construction of Bode plots by use of those of the elementary factors, the translation of the concepts of phase margin and gain margin to the Bode plot is considered. From Fig. 7.11, the phase margin ϕ_m is the sum of 180° and the phase angle at the frequency where $|G_c GH| = 1$ (i.e., 0 dB). Hence, as shown by the partial plots in Fig. 8.3, the phase margin ϕ_m is the distance of the phase-angle curve above $-180°$ at the *crossover frequency* ω_c, where the magnitude plot crosses the 0 dB axis. Similarly, from Fig. 7.11 the gain margin equals 1 divided by the magnitude at the frequency where the phase angle is $-180°$. GM_{dB}, the gain margin in dB, is therefore the distance of the magnitude below 0 dB at this frequency, as shown in Fig. 8.3. The following examples illustrate the use of these criteria.

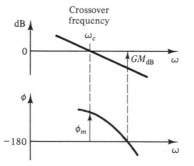

Figure 8.3 Phase margin and gain margin.

8.3 EXAMPLES OF BODE PLOTS

As discussed in Section 8.2, Bode plots are obtained by summing those of the elementary factors present.

Example 8.3.1 Asymptotic Magnitude Plots

In each case in Fig. 8.4 the Bode plots of the elementary factors, obtained in Section 8.2, are shown by dashed lines. Summation then gives the desired plots.

In practice, the dashed lines are usually not shown, and the final plot is made immediately. This is done by determining first the low-frequency asymptote. From (8.1), as $\omega \to 0$,

$$\text{low-frequency asymptote} = \frac{K}{(j\omega)^n} \tag{8.6}$$

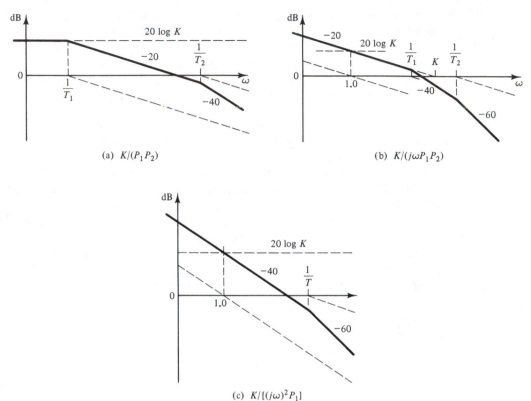

(a) $K/(P_1 P_2)$ (b) $K/(j\omega P_1 P_2)$

(c) $K/[(j\omega)^2 P_1]$

Where
$P_i \equiv j\omega T_i + 1$

Figure 8.4 Asymptotic magnitude plots.

This consists of a gain and integrators, and crosses the 0 dB axis at ω given by

$$\left| \frac{K}{(j\omega)^n} \right| = \frac{K}{\omega^n} = 1: n = 1: \omega = K \tag{8.7}$$

$$n = 2: \omega = \sqrt{K} \tag{8.8}$$

For $n = 0$ (Fig. 8.4a) the low-frequency asymptote has zero slope, for $n = 1$ (Fig. 8.4b) the slope is -20 dB/dec, and for $n = 2$ (Fig. 8.4c) -40 dB/dec. With the low-frequency asymptotes known, the other factors are considered in the order of increasing break frequencies. For example, in Fig. 8.4(b), by (8.7) the low-frequency asymptote—in this case its extension—intersects the 0 dB axis at $\omega = K$. Moving to the right along this asymptote, the first break frequency encountered, at $1/T_1$, is due to a simple lag. It contributes a change of slope of -20 dB/dec, so that at $1/T_1$ the slope changes from -20 to -40 dB/dec. The next break, at $1/T_2$, is again due to a simple lag, so here the slope changes from -40 to -60 dB/dec.

If G_cGH does not have right-half-plane poles or zeros, there is a unique relation between the magnitude plot and the phase-angle curve. Recalling that a factor $j\omega$ in the denominator means $-90°$, it is evident from (8.6) that a zero slope of the low-frequency asymptote means $0°$ phase angle at low frequencies, -20 dB/dec means $-90°$, and -40 dB/dec means $-180°$. Similarly, if the slope of the high-frequency asymptote is $-20\ m$ dB/dec, the phase angle at high frequencies is $-m.90°$.

Example 8.3.2   Dynamic Compensators

Figure 8.5 shows Bode plots of common compensators. A phase-lag compensator [Fig. 8.5(b)] contributes negative phase angles to a loop gain function, and a phase lead contributes positive phase angles, as in Fig. 8.5(c). In Fig. 8.5(b) and (c), the

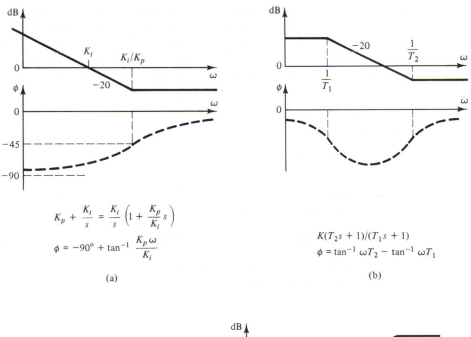

$$K_p + \frac{K_i}{s} = \frac{K_i}{s}\left(1 + \frac{K_p}{K_i}s\right)$$

$$\phi = -90° + \tan^{-1}\frac{K_p\omega}{K_i}$$

(a)

$$K(T_2s + 1)/(T_1s + 1)$$

$$\phi = \tan^{-1}\omega T_2 - \tan^{-1}\omega T_1$$

(b)

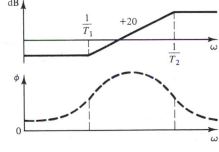

$$K(T_1s + 1)/(T_2s + 1)$$

$$\phi = \tan^{-1}\omega T_1 - \tan^{-1}\omega T_2$$

(c)

Figure 8.5 Dynamic compensators: (a) PI control; (b) phase-lag compensation; (c) phase-lead compensation.

frequency and phase angle at the maximum deviation from $0°$ can be shown to be

$$\omega_m = \sqrt{\frac{1}{T_1}\frac{1}{T_2}} \qquad \phi_m = \pm\sin^{-1}\frac{1 - T_2/T_1}{1 + T_2/T_1} \qquad (8.9)$$

For $T_2/T_1 = 0.1$, $\phi_m = \pm55°$. The phase-angle curves can be plotted by calculation of ϕ at several frequencies from the equations in Fig. 8.5. Alternatively, straight-line approximations of the angles for the simple lags and leads can be drawn as in Fig. 8.1(c), and used to find approximate angles graphically. The discussion of an approach that is often preferable follows.

Appendix B gives a design-oriented interactive program for asymptotic magnitude plots and for phase-angle curves. But manual plotting is much less objectionable than for polar plots. The asymptotic magnitude plot is drawn first. Very often the actual magnitude curve and the phase-angle curve are not required over the entire frequency range, and it is sufficient to obtain these values only where needed to solve the particular problem, say to determine phase margin or gain margin. An approach that is often convenient is, therefore, to calculate phase angles and deviations from asymptotic magnitude plots at selected frequencies. This is done by adding the contributions of the elementary factors, as in the following example.

Example 8.3.3

$$G_c GH = \frac{0.325}{s(s + 1)(0.5s + 1)} \qquad (8.10)$$

Figure 6.8 shows a root locus design for this example. Determine the phase margin and gain margin.

From (8.7), the low-frequency asymptote crosses the 0 dB axis at $\omega_c = 0.325$, at a slope of -20 dB/dec, as shown in Fig. 8.6. From Fig. 8.1(c), Table 8.2.2, or by using half the values for $\zeta = 1$ in Fig. 8.2, the simple lag with break frequency 1 contributes at 0.325 a deviation of less than 0.5 dB between asymptote and true curve, and the simple lag at 2 contributes even less. So $\omega_c = 0.325$ can be taken to be the crossover frequency for phase margin calculation. The integrator contributes

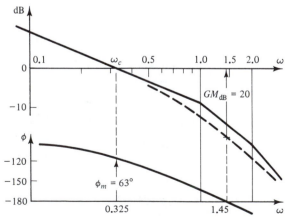

Figure 8.6 Example 8.3.3.

$- 90°$ at all frequencies. The simple lag break at 1 contributes at 0.325 an angle $-\tan^{-1} (0.325/1) = -18°$, and that at 2 contributes $-\tan^{-1} (0.325/2) = -9°$. Hence

$$\phi(0.325) = -90 - 18 - 9 = -117° \qquad \phi_m = 180 - 117 = +63° \qquad (8.11)$$

Determination of the gain margin requires the frequency at which the phase is $-180°$. From the slopes in Fig. 8.6, the angle is $-90°$ at low and $-270°$ at high frequencies. An asymptotic phase-angle plot could be drawn, but inspection suggests that $-180°$ will occur between 1 and 2 rad/sec.

$$\begin{aligned} \phi(1) &= -90 - 45 - 26.6 = -161.7 \\ \phi(1.5) &= -90 - 56.3 - 36.9 = -183.2 \end{aligned} \qquad (8.12)$$

A curve through the points of (8.11) and (8.12) yields

$$\phi(1.45) \approx -180 \qquad \text{gain margin frequency} = 1.45 \qquad (8.13)$$

Figure 8.6 suggests considerable deviation of the actual magnitude curve at this frequency. At $\omega = 1$, the simple lag break at this frequency contributes -3 dB, and that at 2 rad/sec contributes -1 dB at $\omega = 1$ (Table 8.2.2), so the total deviation at $\omega = 1$ is -4 dB. This is also the total deviation at $\omega = 2$. The actual curve in Fig. 8.6 is sketched through these two points and yields at $\omega = 1.45$:

$$GM_{dB} = 20 \text{ dB} \qquad GM = 10 \qquad (8.14)$$

8.4 CLOSED-LOOP FREQUENCY RESPONSE AND THE NICHOLS CHART

Example 8.4.1   Example 8.3.3 Continued

The numerical values for this example were obtained in a root locus design (Fig. 6.8) in which a damping ratio 0.707 was specified for the dominating poles. Based on the correlations for a quadratic lag in Fig. 7.17 or Table 7.7.1, this suggests a phase margin of 67°, not far from the actual value $\phi_m = 63°$ of (8.11) for the third-order system. Further, with a crossover frequency $\omega_c = 0.325$, Table 7.7.1 suggests a closed-loop system bandwidth ω_b of

$$\omega_b \approx \frac{0.325}{0.64} = 0.5 \text{ rad/sec} \qquad (8.15)$$

Also, Fig. 7.16(a) indicates that $M_p \approx 1$, so the closed-loop frequency response should show no peaking.

The foregoing predictions of ω_b and M_p of the closed-loop response are based on Bode plots of the loop gain function and the correlations in Section 7.7. If it is desired to verify them, the closed-loop response should be plotted.

In Section 7.6, closed-loop frequency response was considered on the polar plot. For unity feedback systems, the magnitude and phase angle of the closed-loop frequency response function can be found from the intersections of the polar plot with M and N circle loci. However, just as Bode plots are more convenient than polar plots, the Nichols chart is more convenient than the use of M circles and N circles.

The Nichols chart is shown in Fig. 8.7(a). The magnitude of the loop gain

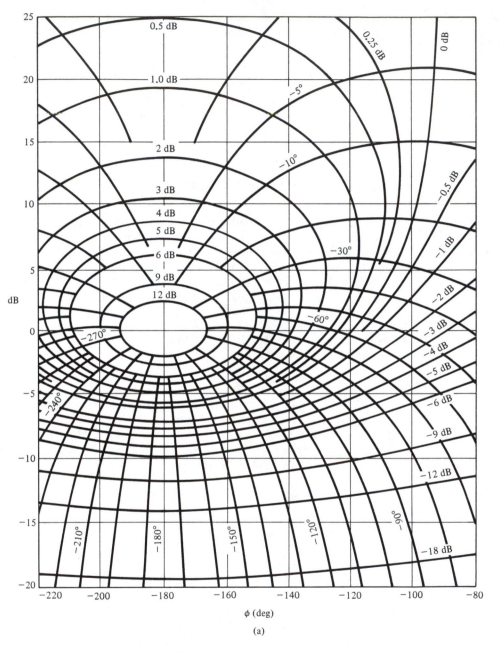

(a)

Figure 8.7(a) Nichols chart.

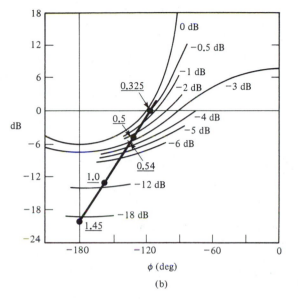

Figure 8.7(b)   Example 8.4.2.

function in dB is plotted against its phase angle, in degrees. This is a dB magnitude versus phase-angle plot and is often obtained most easily by transferring these data from Bode plots. The plot becomes a Nichols chart when the M and N circles are translated to the loci for constant magnitude M and phase angle ϕ of the closed-loop frequency response function. The loci for M are identified by the magnitude in dB. In principle, the M loci could be obtained from the M circles in Fig. 7.14 by plotting the distance of points to the origin vertically, in dB, and the phase angle horizontally. The Nichols chart is used extensively for analysis and design. With the loop gain function plotted on this graph, the resonant peaking M_p, in dB, and the resonant frequency ω_p are found from the point of tangency with the locus for maximum M, as for M circles. The detail of the plot shown in Fig. 8.8 indicates how phase margin ϕ_m and gain margin GM_{dB} are found. GM_{dB} is the distance of the magnitude below 0 dB along the

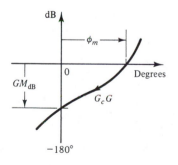

Figure 8.8 Phase and gain margin on a Nichols chart.

$-180°$ vertical, and ϕ_m is the distance of the phase angle to the right of $-180°$ along the 0 dB line.

*****

The correlations represented in Fig. 7.16, Fig. 7.17, and Table 7.7.1 allow predictions of closed-loop frequency response peaking M_p, resonant frequency ω_p, and bandwidth ω_b to be made from Bode plot data. Transferring the Bode plots to the Nichols chart will give the actual values of these performance measures. These may verify the predictions, or suggest a design iteration on the Bode plots, say, to increase the crossover frequency ω_c. Since the system bandwidth ω_b is the frequency at which M falls below $0.707 = -3$ dB:

> On the Nichols chart, the bandwidth is the frequency at which the loop gain function plot crosses the $M = -3$ dB locus.

Example 8.4.2 Resonance and Bandwidth for Fig. 8.6

In Example 8.4.1, open-loop to closed-loop correlations were used to predict from the Bode plot that the closed-loop frequency response will not show a resonant peak, and will have a bandwidth $\omega_b = 0.5$ [equation (8.15)].

To check these predictions, the Bode plots in Fig. 8.6 have been transferred to the Nichols chart in Fig. 8.7(b). The plot verifies the absence of a resonance peak; only at low frequencies, where the angle approaches $-90°$, does M approach 0 dB. The plot intersects the $M = -3$ dB locus at the bandwidth $\omega_b \approx 0.54$ rad/sec, not far from the predicted value 0.5 rad/sec.

8.5 PERFORMANCE SPECIFICATIONS ON THE BODE PLOT

An important reason for the widespread use of Bode plots is the ease of interpreting performance specifications on the asymptotic magnitude plot. Several are considered in turn, for the unity feedback configuration with series compensation $G_c(s)$ shown in Fig. 8.9.

1. *Relative stability:* G_cG must have adequate length of not more than a -20 dB/dec slope at or near crossover frequency ω_c.

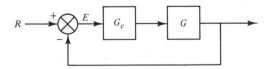

Figure 8.9   Series compensation.

In Fig. 8.10, if other break frequencies are far removed from ω_c, as suggested by the dashed lines, (a), (b), and (c) approximate the functions and phase margins indicated. Thus, if inspection of a Bode plot reveals little or no length of not more than a -20 dB/dec slope at or near ω_c, it is immediately evident that ϕ_m will be inadequate.

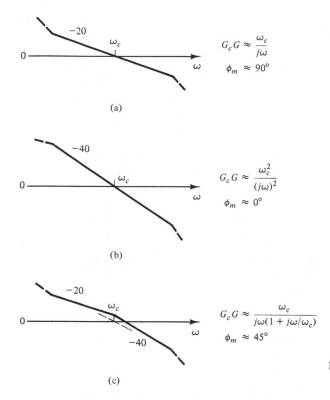

$$G_c G \approx \frac{\omega_c}{j\omega}$$

$$\phi_m \approx 90°$$

(a)

$$G_c G \approx \frac{\omega_c^2}{(j\omega)^2}$$

$$\phi_m \approx 0°$$

(b)

$$G_c G \approx \frac{\omega_c}{j\omega(1 + j\omega/\omega_c)}$$

$$\phi_m \approx 45°$$

(c)

Figure 8.10 Phase margin versus Bode plot slopes near 0 dB.

2. *Steady-state accuracy:* To improve steady-state accuracy, the low-frequency asymptote must be raised or its slope changed.

The low-frequency asymptote is $K/(j\omega)^n$. From Table 4.5.1, for $n = 0$ (type 0) the steady-state error after a unit step is $1/(1 + K)$, so reduces if the zero slope low frequency asymptote $20 \log K$ is raised. For zero steady-state error after a step, the system must be at least type 1 ($n = 1$), so the low-frequency asymptote slope must be at least -20 dB/dec. If a Bode plot shows a -40 dB/dec low-frequency asymptote, the system is type 2, so has zero steady-state errors for both steps and ramps. Steady-state errors for $n = 1$ and 2 also reduce as K is increased, so as the low-frequency asymptotes are raised, since, from (8.7) and (8.8), these intersect the 0 dB axis at $\omega = K$ and $\omega = \sqrt{K}$, respectively.

  Note that instead of from these intersections, the asymptotes can also be plotted by calculating the magnitude of $K/(j\omega)^n$ at any convenient value of ω, where K may be determined by a specification on steady-state errors. Say that the steady-state ramp following error $1/K$ ($n = 1$) may be 10%. Then $1/K = 0.1$, so $K = 10$ and the dB magnitude is $20 \log (10/\omega)$. At, say, $\omega = 0.1$, this is 40 dB, so the -20 dB/dec asymptote must pass $\omega = 0.1$ at a level of 40 dB.

3. *Accuracy in the operating range:* To ensure specified accuracy over a

normal range of frequencies, the plot may not fall below a given level over this range. To improve accuracy, this level must be raised.

In Fig. 8.9, $E/R = 1/(1 + G_cG)$. If up to 10% error is permitted up to a certain frequency, $|G_cG| \geq 10 = 20$ dB is approximately the minimum allowed level up to this frequency. Thus the notion of accuracy is extended to dynamic operation. Raising low-frequency asymptotes not only improves steady-state accuracy but also that at other frequencies where this has raised the plot.

4. *Crossover frequency and bandwidth:* Crossover frequency ω_c is a measure of bandwidth ω_b, so of speed of response, according to $\omega_c \approx 0.63\omega_b$.

See Table 7.7.1, for $\phi_m \leq 67°$. Roughly, if $|G_cG| \gg 1$, then $C/R = G_cG/(1 + G_cG) \approx 1$, so output follows input almost perfectly. For $|G_cG| \ll 1$, $C/R \approx G_cG \ll 1$. The crossover $|G_cG| = 1 = 0$ dB separates these two ranges.

5. *Noise rejection:* To ensure a specified attenuation (reduction) of noise components in the input above a certain frequency, the plot above that frequency should be below a certain level.

Noise at high frequencies, above the bandwidth, is often present in input signals. To cause the system to act as a filter, it may be specified that noise amplitudes shall be reduced to, say, 10% above a certain frequency. Sufficiently above the bandwidth, this requires that

$$\left|\frac{C}{R}\right| \approx |G_cG| \leq 0.1 = -20 \text{ dB}$$

so the plot must be below -20 dB. Noise rejection considerations are one reason why the bandwidth of a system should not be made larger than necessary.

These criteria show how different aspects of performance are reflected in individual features of the plot and permit specifications to be translated into requirements on the Bode plot. The task of system design is to derive the compensators that will meet these requirements.

8.6 BODE PLOT DESIGN

The Bode plot design examples below will be interpreted also by polar plot sketches, and Nichols charts will be used on occasion to verify closed-loop response predictions. This will provide an indication of the design techniques available for these plots.

Example 8.6.1   Choice of Gain for $G_c = K$

$$G_cG = \frac{K}{s(s + 1)(0.5s + 1)} \tag{8.16}$$

is shown in Fig. 8.6, 8.12, and partially in Fig. 8.11, for $K = 0.325$. Changing K will not affect the phase-angle curve and will just raise or lower the magnitude plot. On polar plots, as discussed earlier, it moves points radially, and on the Nichols chart the curve is moved parallel to the magnitude axis. For $K = 0.325$, $\phi_m = 63°$, but to reduce the steady-state error $1/K$, $\phi_m = 53°$ will be accepted. As indicated in Fig. 8.11, $\phi_m = 53°$ will result if the actual curve crosses 0 dB at 0.43 rad/sec. To realize this, the plot must be raised by 3 dB; that is, the gain 0.325 must be raised by a factor α given by 20 log $\alpha = 3$. Hence $\alpha = 1.41$ and the new gain is $K = 1.41 \times 0.325 = 0.45$. For polar plot or Nichols chart design, from Fig. 7.17 $\phi_m = 53°$ corresponds to $\zeta \approx 0.5$, and Fig. 7.16(a) shows that $\zeta = 0.5$ is about equivalent to a closed-loop resonance peaking $M_p = 1.15$. So the gain would be adjusted so that G_cG becomes tangent to the corresponding M locus.

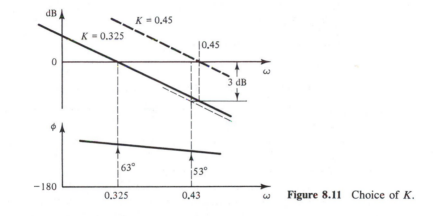

Figure 8.11 Choice of K.

Example 8.6.2   Phase-Lag Compensation in Example 8.6.1

If the error $1/K$ in Example 8.6.1 may not exceed 40%, $K = 2.5$ would be required. The Bode plot for $G_c = K = 2.5$ in Fig. 8.12 shows crossover about midway between the -20- and -60 dB/dec slope sections, so that the phase margin is probably near zero. What is desired is a way to raise the plot for $G_c = K = 0.325$ only at low frequencies, and leave it unchanged over a wide enough range around its crossover frequency 0.325 that the phase margin changes little. Phase-lag compensation, repeated as G_c/K in Fig. 8.12 from Fig. 8.5(b), can meet this requirement. Since G_c and G are in series, their Bode plots, or those of G_c/K and KG, add. If $G_c/0.325$ is chosen to be 0 dB, asymptotically, above 0.0325 rad/sec, the asymptotes of G_cG and $0.325G$ will coincide above this frequency. If $G_c/0.325 = K_c(T_1s + 1)/(T_2s + 1)$, the highest break frequency $1/T_1 = 0.0325$ is chosen a factor of 10 below 0.325. Then G_c will not contribute more than about $-5°$ phase lag at 0.325, so will not reduce the phase margin by more than this amount. The value $1/T_2 = 0.00422$ is then found graphically where the -40 dB/dec asymptote of G_cG starting at 0.0325 intersects the desired level of the low-frequency asymptote, for which the gain is 2.5. Since the gain of G is 1, it follows that the desired compensator is

$$G_c = \frac{2.5(1 + s/0.0325)}{1 + s/0.00422} \tag{8.17}$$

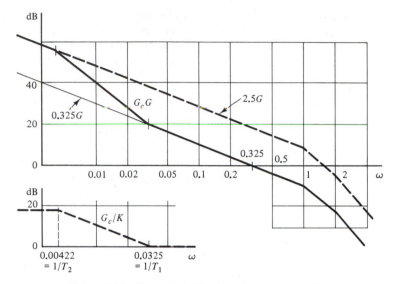

Figure 8.12 Example 8.6.2: phase-lag compensation.

Note that since $K_c = 2.5/0.325$ and the high-frequency gain $K_c T_1/T_2$ of $G_c/0.325$ is 1, T_2 may be found analytically from $T_2 = K_c T_1$.

Since $|G_c G| > 100$ up to 0.01 rad/sec, the errors for inputs up to this frequency will not exceed 1%. Above 2 rad/sec, $|G_c G| < 0.1$, so less than 10% of noise at higher frequencies will appear in the output.

Low-frequency accuracy has been improved without loss of relative stability. The example also shows how phase-lag compensation may improve stability without loss of low-frequency accuracy. If the gain 2.5 were part of the plant, G_c of (8.17), but with the gain 2.5 replaced by 1, would serve to lower the higher-frequency portion of the plot for $K = 2.5$ and would realize an adequate length of -20 dB/dec slope at crossover. The crossover frequency would be found from the phase-angle curve, calculated over the range where the desired phase margin is expected to occur.

Figure 8.13 shows a polar plot to interpret the stabilizing effect possible with a phase lag. G_c moves a point A on G to B on $G_c G$. G_c introduces a negative phase shift (i.e., clockwise rotation) ϕ_c and a gain reduction. The benefit derives from the latter, because by shortening $0A$ to $0B$ it causes $G_c G$ to pass on the other side of -1.

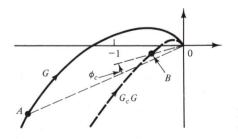

Figure 8.13 Phase-lag compensation on polar plot.

Example 8.6.3   Phase-Lead Compensation for $G = 1/s^2$

Figure 8.14 shows the Bode plot of K/s^2. Raising or lowering it will not produce -20 dB/dec slopes, nor will phase-lag compensation. But the phase-lead characteristic, repeated as G_c/K in Fig. 8.14 from Fig. 8.5(c), when added to the plot of KG, can produce the desired -20 dB/dec at the crossover of $G_c G$.

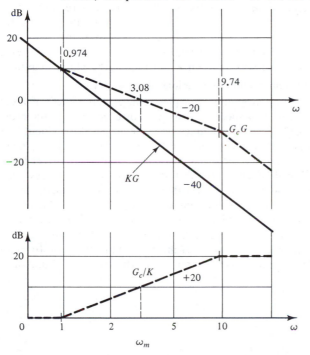

Figure 8.14   Example 8.6.3: phase-lead compensation.

Let a damping ratio $\zeta = 0.5$ be specified. From Fig. 7.17, this suggests a phase margin $\phi_m = 53°$. Since the phase of G is $-180°$, the lead compensator $G_c/K = (T_1 s + 1)/(T_2 s + 1)$ (which has a gain of $1 = 0$ dB at low frequencies) must contribute at least $+53°$. From (8.9), $T_2/T_1 = 0.1$ will provide $+55°$ at $\omega_m = 1/\sqrt{T_1 T_2}$. This is satisfactory provided that the crossover frequency ω_c of $G_c G$ is made to coincide with ω_m: $\omega_c = \omega_m = 1/\sqrt{T_1 T_2}$. To make this happen, it is noted that on a log scale ω_m lies midway between the break frequencies $1/T_1$ and $1/T_2$. Since between these frequencies the asymptotic plot of G_c/K contributes a change of magnitude of 20 log (T_1/T_2), at the midpoint ω_m the lead contributes 10 log $(T_1/T_2) = 10$ log $10 = 10$ dB. Thus ω_c should occur where KG is 10 dB below the 0 dB axis, since then its sum with G_c/K crosses at ω_c. If K is known from a steady-state error specification, $KG = K/s^2$ crosses the 0 dB axis at the known frequency $\sqrt{K}$ [equation (8.8)]. On the plot this frequency is 1.75, so $K = 1.75^2$, and KG is -10 dB at 3.08 rad/sec:

$$\omega_c = \omega_m = 3.08 = \frac{1}{\sqrt{T_1 T_2}} = \frac{1}{T_2 \sqrt{10}}$$

$$T_2 = 0.103 \quad \left(\frac{1}{T_2} = 9.74\right) \qquad T_1 = 1.03 \quad \left(\frac{1}{T_1} = 0.974\right)$$

If K is not set by a steady-state error specification, but the bandwidth ω_b is specified instead, ω_c is estimated from Table 7.7.1 as $\omega_c = 0.62\omega_b$, and KG must pass at 10 dB below the 0 dB axis at this frequency. The value $\omega_c = 3.08$ corresponds to $\omega_b = 5$.

The present design was derived from a third alternative, namely, a specified settling time $T_s = 2$ sec. Since $T_s = 4/(\zeta\omega_n)$ and $\zeta = 0.5$, this requires that $\omega_n = 4$. From Fig. 7.17, for $\zeta = 0.5$, $\omega_c/\omega_n = 0.77$, so $T_s = 2$ translates to an estimated requirement of $\omega_c = 4 \times 0.77 = 3.08$. Observe, finally, from Fig. 8.14 that both lowering and raising the Bode plot, by changing K, will move the crossover closer to the -40 dB/dec slope sections and hence will reduce the phase margin.

Example 8.6.4   Phase-Lead Compensation in Examples 8.6.1 and 8.6.2.

Figure 8.15 shows again the magnitude and phase-angle plots of KG of Examples 8.6.1 and 8.6.2. Let the system speed of response corresponding to the crossover frequencies 0.43 and 0.325 in these examples be inadequate. It is evident from the phase-angle characteristic of KG that phase lead is necessary if a phase margin of 53° is to exist over about 0.5 rad/sec. The maximum possible crossover frequency depends on the phase lead added, which in turn depends on T_2/T_1. As discussed in Section 6.6, noise considerations suggest that this ratio not be made larger than necessary. Say that $T_2/T_1 = 0.1$ is acceptable. This provides a 55° lead at $\omega_m = 1/\sqrt{T_1 T_2} = 1/(T_2\sqrt{10})$. Since the phase of G is $-180°$ at 1.45, it can make $\omega_c = 1.45$ provided that $\omega_c = \omega_m = 1.45 = 1/(T_2\sqrt{10})$, or $1/T_2 = 4.59$ and $1/T_1 = 0.459$. As in Example 8.6.3, to make $\omega_c = \omega_m$, KG must be 10 log 10 = 10 dB below the 0 dB axis at 1.45. The low-frequency asymptote of the plot of KG drawn in Fig.

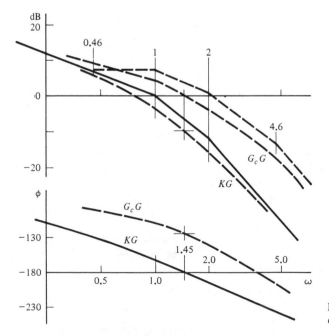

Figure 8.15   Example 8.6.4: phase-lead compensation.

8.15 to meet this requirement intersects the 0 dB axis at $\omega = 1$, so that $K = 1$ and

$$G_c = \frac{1 + s/0.46}{1 + s/4.6} \qquad (8.18)$$

The following observations are made:

1. *Closed-loop performance:* From Fig. 7.17, a phase margin of 53° corresponds to a closed-loop damping ratio $\zeta = 0.5$, for which Fig. 7.16(a) indicates a closed-loop frequency response peaking $M_p = 1.15 = 1.2$ dB, and Table 7.7.1 suggests a bandwidth $\omega_b = \omega_c/0.62 = 1.45/0.62 = 2.34$. To check these predictions, the Bode plots of G_cG in Fig. 8.15 have been transferred to the Nichols chart in Fig. 8.16. The curve intersects the -3 dB locus at $\omega_b = 2.4$, close to the predicted bandwidth of 2.34 based on Bode plot design and Table 7.7.1, and it is about tangent to the $M = 1$ dB locus, close to the predicted peaking of 1.2 dB.

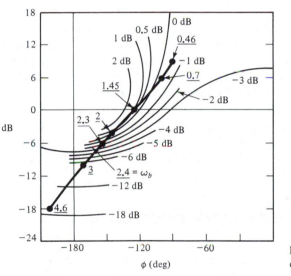

Figure 8.16   Example 8.6.4: Nichols chart.

2. The phase lead extends the -20 dB/dec slope to higher frequencies and hence increases bandwidth without loss of phase margin.

3. The phase margin is about 53° in both Examples 8.6.1 and 8.6.4, but the phase lead allows an increase of gain from 0.45 to 1, and thus reduces steady-state errors.

4. The steady-state error is larger than that for the phase-lag compensated system in Example 8.6.2, where $K = 2.5$. If the specifications should require both $\omega_c = 1.45$ and $K = 2.5$, phase lag can be added to the lead-compensated system in Fig. 8.15 to raise the low-frequency asymptote.

5. Figure 8.17 interprets phase-lead compensation on a polar plot. G_c moves

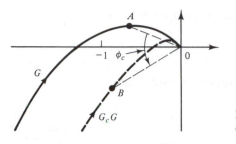

Figure 8.17 Phase-lead compensation on polar plot.

a point A on G to B on G_cG. The benefit derives from the positive phase shift (i.e., counterclockwise rotation) ϕ_c introduced by G_c. Although $0B > 0A$, this causes the plot to pass on the other side of the -1 point. To obtain maximum phase shift near -1, the break frequencies of the lead must be in the range of those near -1. This contrasts with the phase lag in Fig. 8.13. Its break frequencies must be in a lower range, to realize the magnitude reduction near the -1 point on which its use relies.

6. For the PID controller discussed at the end of Section 5.4, the I and D actions do not interfere with each other because the I action is made to occur much slower than the D action. This corresponds to present results in observations 4 and 5, with phase-lag action, which approximates I control, occurring at much lower frequencies than phase-lead action, which approximates D control.

7. It is useful to observe in Example 8.6.4 that if the break frequencies of the plant were, say, 1 and 10 rad/sec instead of 1 and 2, a smaller ratio of the break frequencies of the lead, that is, closer to 1, would have sufficed to achieve $\omega_c = 1.45$ with $\phi_m = 53°$.

8.7 OPEN-LOOP UNSTABLE OR NONMINIMUM-PHASE PLANTS

If the plant G is open-loop unstable, that is, has poles in the right-half s-plane, then, as mentioned earlier, Nyquist diagrams or root loci should be used to determine system stability. If this shows stability, Bode plots can be applied to find phase margin, as well as loop gain characteristics, accuracy, and crossover frequency, or to choose gains for a specified crossover frequency. Bode plots are always valid for the latter uses.

For nonminimum-phase plants, that is, with right-half-plane zeros, Bode plots are valid for stability analysis, but care is necessary because, as with right-half-plane poles of G, there is no longer a unique relation between the magnitude and phase-angle plots. Thus the very useful design guide which aims for an adequate length of -20 dB/dec slope near crossover is no longer valid. Figure 8.18 shows vector diagrams and Bode diagrams for $(1 + Ts)$ and $(1 - Ts)$. For $(1 + Ts)$ the angle increases from 0 to $+90°$, and for $(1 - Ts)$ it decreases from

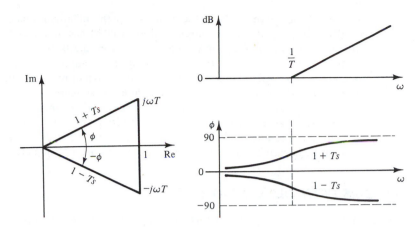

Figure 8.18 Plots of $(1 + j\omega T)$ and $(1 - j\omega T)$.

0 to $-90°$ as ω increases from 0 to $+\infty$. The magnitude $|1 + j\omega T| = \sqrt{1 + (\omega T)^2}$ is the same for both, so its plot does not reflect the difference of phase angles.

Example 8.7.1 Nonminimum-Phase System

To sketch root loci for the example in Fig. 8.19, the loop gain function is first rewritten to extract the root locus gain:

$$\frac{(-KT_n/T)(s - 1/T_n)}{s(s + 1/T)}$$

For $K > 0$, the root locus gain is negative, so that, as for the example in Fig. 6.17, the asymptote is at $0°$ instead of $180°$, because the angle condition is satisfied in this direction.

The asymptotic Bode magnitude plot in Fig. 8.19 is the same as for a numerator factor $(1 + T_n s)$. Therefore, right-half-plane zeros and poles are identified by small circles. To predict stability from the magnitude plot, note that the phase of a function $G_a = G(1 - T_n s)$ is the same as that of $G_b = G/(1 + T_n s)$. Thus, in Fig.

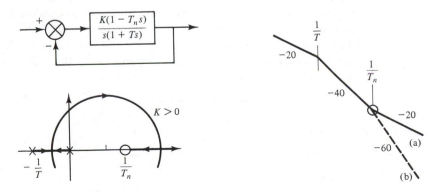

Figure 8.19 Example 8.7.1: nonminimum-phase system.

8.19 the angle curve for (a) is the same as for the minimum-phase function shown by (b), and adequate phase margin will exist if (b) has an adequate length of -20 dB/dec slope at the frequency where (a) crosses the 0 dB axis. These ideas are useful to estimate the severity of any stability problems visually from the magnitude plot (a), without actually plotting (b). But for design of nonminimum-phase systems it is usually recommended that the complete phase-angle curve be drawn as well.

Due to the -60 dB/dec slope of (b) beyond $1/T_n$, nonminimum-phase zeros can impose severe restrictions on crossover frequency and bandwidth. However, if their break frequencies occur far above the crossover frequency, they contribute only small negative angles at this frequency and are not objectionable.

8.8 SYSTEMS WITH TRANSPORT LAG

The terms *transport lag, transportation lag, dead time,* or *delay time* have been used to describe elements with the transfer function e^{-Ts}. From (1.17) or Table 1.6.1, if $F(s)$ is the transform of a function $f(t)$, then $e^{-Ts}F(s)$ is the transform of $f(t - T)$, the function $f(t)$ delayed by T, as illustrated in Fig. 8.20(a). Transport lag is often needed to model the effect of flow through long fluid lines, as indicated in Fig. 8.20(b), and occurs in other contexts as well. If the fluid line has length L and the fluid velocity is v, the dead time during which the process at the outlet is unaware of a change of $f(t)$ at the inlet is $T = L/v$. Long pipelines between subsystems in power plants and other processes can have a profound effect on systems dynamics. In the schematic diagram of a level control system in Fig. 4.1, long pneumatic lines may connect the controller to the actuator, and the pipeline between the control valve and the tank may introduce a considerable delay. Lumping the delays in a loop into a single transport lag T, the loop gain function in the block diagram of Fig. 4.2 can be written in the following form, for P control:

$$G_cG = \frac{Ke^{-Ts}}{(T_1s + 1)(T_2s + 1)} \tag{8.19}$$

The series expansion of e^{-Ts} implies that G_cG theoretically has an infinite number of zeros, complicating root locus analysis. An approximation useful to show the general effect is to use the first two terms of the series: $e^{-Ts} \approx 1 - Ts$. This indicates a nonminimum-phase system, and serves, from Section 8.7, to emphasize the potentially strong effect, depending on the value of T compared

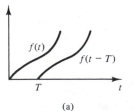

(a)

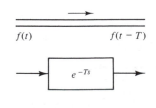

(b)   **Figure 8.20** Transport lag.

to T_1 and T_2. Another approximation often used for initial analysis is $e^{-Ts} \approx$ $1/(1 + Ts)$. Unless T is small compared to T_1 and T_2, this also shows the bandwidth-limiting effect of transport lag.

Analysis in the frequency domain is both exact and more convenient.

$$|e^{-j\omega T}| = |\cos \omega T - j \sin \omega T| = 1 \qquad \phi(e^{-j\omega T}) = -\omega T \qquad (8.20)$$

Thus transport lag does not affect the magnitude plot of $K/[(T_1 s + 1)(T_2 s + 1)]$ shown in Fig. 8.21, but changes its phase-angle curve by $-\omega T$. The curve identified by ϕ_2 is for larger T than ϕ_1. If the horizontal at angle ϕ_m above $-180°$ indicates the desired phase margin, it is evident that larger transport lag T reduces the crossover frequency possible without phase-lead compensation. This shows that if a system schematic suggests appreciable transport lag, careful analysis is necessary to ensure that adequate stability is maintained.

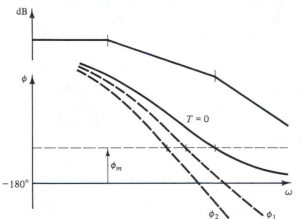

Figure 8.21   Design with transport lag.

8.9 FEEDFORWARD CONTROL

Feedforward control is used extensively in practice to reduce the effect on the system output of measurable disturbance inputs, that is, disturbances which via a sensor can be made available as signals, such as water, oil, or pneumatic supply pressure variations. In many cases it can give a dramatic reduction of output deviations from the desired value.

Figure 8.22, without the dashed link, is similar to the disturbance input system model in Fig. 4.13 of Section 4.4. While the feedback loop acts to reduce the effect of disturbances D on C, it can do so only after C has already been affected by D. The idea of the feedforward link G_f is in effect to counteract the disturbance before it changes C. In Fig. 8.22, if $R = 0$ and without the feedback loop,

$$M = (L - G_f G_c G_1)D \qquad (8.21)$$

If G_f could be made equal to $L/(G_c G_1)$, the effect of D on M, and on C, would be eliminated.

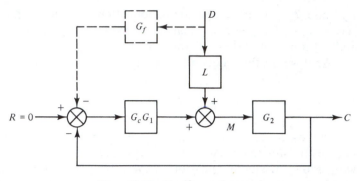

Figure 8.22   Feedforward control.

However, usually this form of G_f has more zeros than poles and hence is not physically realizable. But it is evident, and can be explored by the use of Bode plots, that simple forms of G_f can often greatly reduce the size of the effective disturbance input

$$D^1 = (L - G_f G_c G_1)D \tag{8.22}$$

to the feedback system. This is particularly advantageous if the block G_2 is relatively slow acting compared to the others. Then the steady-state gains of these other blocks are most important, and choosing G_f to be a constant gain can greatly reduce disturbance effects.

8.10 CONCLUSION

Although other forms of plots were discussed, the heavy emphasis in this chapter has been on analysis and design by means of Bode plots. With the performance measures of Sections 8.5 and 7.7, these combine ease of plotting and ready identification of features of the performance. The particular cases of nonminimum-phase plants and systems with transport lag were considered, as was the use of feedforward control to reduce the effect of measurable disturbances.

For routine analysis and design, Appendix B gives an interactive program for constructing asymptotic magnitude plots and actual phase-angle curves. It is design oriented in that for a given plant the effect of selected dynamic compensators can be shown.

PROBLEMS

8.1. Construct asymptotic Bode magnitude plots for the following transfer functions.

    (a) $\dfrac{4}{s+2}$ **(b)** $\dfrac{4}{(0.4s+1)(s+1)}$ **(c)** $\dfrac{8}{s(1.25s+1)(s+2)}$

8.2. Construct asymptotic Bode magnitude plots for the following transfer functions. In the case of underdamped quadratics, always sketch the true magnitude curve locally based on the peak value alone.

(a) $\dfrac{5(s + 0.6)}{s(2.5s + 1)(s + 2)(0.25s + 1)}$ (b) $\dfrac{3.125}{s(s^2 + 0.625s + 1.5625)}$

(c) $\dfrac{1.6}{(s + 0.4)(s + 0.8)(s + 1)}$

8.3. Construct asymptotic Bode magnitude plots for the following loop gain functions, and determine by inspection, noting the contributions of the elementary factors, whether or not the corresponding systems are stable.

(a) $\dfrac{72}{s(s^2 + 2s + 9)}$ (b) $\dfrac{45}{s^2(s + 5)}$

8.4. For a system with loop gain function

$$G(s) = \frac{20}{(s + 5)(0.1s + 1)(0.025s + 1)}$$

(a) Plot the asymptotic Bode magnitude plot.
(b) At two convenient frequencies, 10 and 20, find the deviation of the true magnitude curve from the asymptotes, and sketch a section of this curve for use in part (c).
(c) Determine the phase margin and gain margin.

8.5. In Fig. P8.5, $G_c = K$ and $G = 2000/[s(s + 20)(0.01s + 1)]$.
(a) Construct the asymptotic Bode magnitude plot and find the approximate deviation with the true curve for a frequency near crossover to get a better approximation of the crossover frequency, for $K = 1$.
(b) Determine the phase margin for $K = 1$.
(c) How does a change of gain K affect Bode magnitude and phase-angle plots? What is the effect of reducing K by a factor m?
(d) Find K to get a phase margin of 50°.

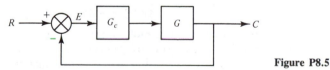

Figure P8.5

8.6. In Fig. P8.5 with $G = 4/[s(s + 4)(0.0625s + 1)]$ and $G_c = K$, find K to obtain a phase margin of $\phi_m = 54°$.

8.7. In Fig. P8.5, G is the transfer function of a field-controlled motor in a position control system:

$$G = \frac{1}{s(0.1s + 1)(0.02s + 1)}$$

If $G_c = K$, find K for a phase margin of 45° and determine the corresponding gain margin.

8.8. In Fig. P8.5, G is the plant in a two-tank level control system:

$$G = \frac{1}{(0.1s + 1)(0.02s + 1)}$$

(a) If $G_c = K$, find K for $\phi_m = 55°$ and determine the corresponding steady-state error for unit step inputs.

(b) To reduce this error to zero, introduce PI control by making $G_c = K(1 + 5/s)$, with K as in part (a). Modify the Bode plot to account for this, and verify that the phase margin at ω_c of part (a) is reduced by no more than 6°, which will be taken to be acceptable in this case.

(c) How would the phase margin be affected if the parameter 5 in G_c were increased?

8.9. In Fig. P8.5 with $G = 1/[s(0.1s + 1)]$, $G_c = K$:

(a) Draw the asymptotic Bode magnitude plot without establishing a 0 dB axis level, and plot the phase-angle curve from points calculated at $\omega = 4, 5, 6, 8, 10, 15, 20,$ and 30.

(b) Determine the crossover frequencies ω_c required for phase margins ϕ_m of 65°, 55°, 45°, and 35°.

(c) At the frequencies ω_c of part (b) indicate the approximate deviations of the true curve to the asymptotic magnitude plot.

(d) Draw the 0 dB axis levels associated with the values ϕ_m specified in part (b) and find the values of K needed to realize these phase margins.

(e) Predict the closed-loop resonance peaking M_p and bandwidth ω_b for each of the designs from the open-loop response data.

8.10. It is desired to compare the predictions of ω_b and M_p in Problem 8.9 with closed-loop results obtained using the Nichols chart.

(a) Transfer the Bode plot data for the lowest value of K in Problem 8.9 to the chart.

(b) How does a curve on the Nichols chart change if K is changed from K_1 to K_2?

(c) Plot the Nichols chart curves for all values of K in Problem 8.9.

(d) From the plots, find ω_b and M_p for each case and compare them with the predictions.

8.11. In Fig. P8.5 with $G_c = K$ and

$$G = \frac{1}{s(0.1s + 1)(0.01s + 1)}$$

(a) Draw the asymptotic Bode magnitude plot with undetermined 0 dB axis level.

(b) Calculate the phase angle at a number of frequencies, plot the phase-angle curve, and find the crossover frequency ω_c if the desired phase margin is 45°.

(c) Use ω_c to set the 0 dB axis level on the Bode magnitude plot, corrected from the asymptotic plot at several frequencies.

(d) Replot the results on the Nichols chart and compare bandwidth ω_b and resonance peaking M_p with those predicted from the Bode plot data.

8.12. Figure P8.12 shows the Bode plots for the loop gain function of a unity feedback system.

(a) What is the loop gain function?

(b) Predict the closed-loop system resonance peaking and bandwidth.

8.13. In Problem 8.12, replot the Bode plot data on the Nichols chart and:

(a) Compare ω_b and M_p obtained from the Nichols chart with the predictions of Problem 8.12(b).

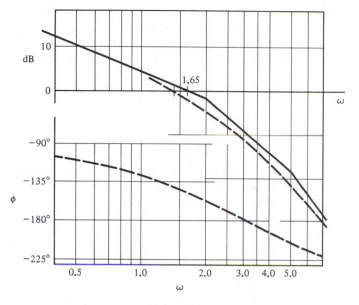

Figure P8.12

(b) Plot the closed-loop frequency response function magnitude versus frequency, with magnitude in dB, on semilog graph paper.

8.14. Determine a desired Bode plot and corresponding loop gain function of a system with unity feedback which satisfies the following specifications:

1. The steady-state error for step inputs must be zero.
2. The steady-state error following unit ramps may not exceed 2%.
3. The error in response to sine inputs up to 10 rad/sec may not exceed 10%.
4. To limit the effect of high-frequency noise, the output in response to sine inputs above 250 rad/sec may not exceed 10% of the input.
5. To ensure adequate bandwidth, the crossover frequency must be about 50 rad/sec.
6. The phase margin must be at least about 55°.

8.15. In Fig. P8.5 with

$$G(s) = \frac{1}{s[(s^2/50^2) + 2 \times 0.5 \times (s/50) + 1]}$$

(a) For $G_c = K$, find K to achieve a phase margin of 50°, and plot the Bode magnitude plot and the phase-angle curve.
(b) Design phase-lag compensation to reduce steady-state errors for ramp inputs by a factor of 10 while reducing the phase margin by no more than 6°.

8.16. In Problem 8.15:
(a) If G_c is a phase-lead compensator with a ratio 10 of the break frequencies, find the maximum crossover frequency that can be achieved for a 45° phase margin.

(b) Determine the gain and break frequencies of this compensator and construct the Bode magnitude plot of G_cG.

(c) Plot the phase-angle curves of G and G_cG to verify that this compensation is not very effective in raising crossover frequency and bandwidth for underdamped quadratics, because of the relatively fast change of phase angle near the undamped natural frequency.

(d) What is the effect on steady-state errors relative to $G_c = K$?

8.17. In Fig. P8.5 with

$$G = \frac{1}{s(0.1s + 1)(0.01s + 1)}$$

design a series compensator G_c to achieve a crossover frequency of 30 rad/sec and a phase margin of about 45°.

8.18. In Fig. P8.5 with $G = 1/[(0.1s + 1)s]$, design series compensation to meet the following specifications:

1. The steady-state error following ramp inputs may not exceed 2%.
2. The error in response to sinusoidal inputs up to 5 rad/sec should not exceed about 5%.
3. The crossover frequency should be about 50 rad/sec to meet bandwidth requirements while limiting the response to high-frequency noise.
4. The ratio of the break frequencies of G_c should not exceed 5, to limit noise effects.
5. The phase margin should be about 50°.

8.19. In Fig. P8.5 with a plant consisting of three simple lags in series

$$G = \frac{1}{(s + 1)(0.25s + 1)(0.1s + 1)}$$

design a PI controller if a crossover frequency of about 2 rad/sec and a phase margin of about 50° are desired.

8.20. For the plant of Problem 8.19, design a controller G_c to satisfy the following specifications:

1. The steady-state error for step inputs may not exceed 10%.
2. The crossover frequency of G_cG should be at least 7 rad/sec.
3. The phase margin should be about 45°.

8.21. In Fig. P8.5 with $G = 1/[s(0.5s + 1)]$, design phase-lag compensation to meet the following specifications:

1. The steady-state unit ramp following error may not exceed 10%.
2. The phase margin should be at least 45°.
3. The error in response to sinusoidal inputs up to 0.1 rad/sec may not exceed about 4%.

8.22. In Fig. P8.5 with $G = 1/[(s + 1)(0.25s + 1)^2]$:
  (a) If $G_c = K$, find K for a phase margin of about 48°.
  (b) Design phase-lag compensation G_c to reduce the steady-state error of part (a) by a factor of 10, for a phase margin of about 42°.

8.23. In Fig. P8.5 with $G = 1/[s(0.2s + 1)(0.05s + 1)]$:
  (a) If $G_c = K$, find K for a 45° phase margin and the corresponding crossover frequency.
  (b) Design phase-lead compensation with a ratio 10 of the break frequencies to maximize the system bandwidth, maintaining about 45° phase margin.

8.24. In Fig. P8.5 with $G = 100/[s(s + 10)^2]$:
  (a) If $G_c = K$, find K for a unit ramp steady-state following error of 5% and construct the corresponding Bode plot.
  (b) Design phase-lag compensation G_c to achieve at least 55° phase margin without loss of the steady-state accuracy of part (a).

8.25. In Fig. P8.5 with $G = 1/[s(0.1s + 1)(0.02s + 1)]$, design a phase-lag compensator to satisfy the following specifications:

  1. The steady-state unit ramp following error may not exceed 2%.
  2. The phase margin should be at least 50°.

8.26. In Fig. P8.5 with $G = 1/[s(0.05s + 1)(0.01s + 1)]$, design phase-lag compensation to meet the following specifications:

  1. The steady-state ramp input following error may not exceed 5%.
  2. The phase margin must be at least 50°.

8.27. In Fig. P8.5 with $G = 1/[s(0.1s + 1)(0.001s + 1)]$:
  (a) For $G_c = K$ find K and plot KG for 0.1% steady-state error following ramp inputs. Determine the stability.
  (b) With K as above, determine the effect of a lead network

$$G_c = K\left[\frac{(s/50) + 1}{(s/400) + 1}\right]$$

  on steady-state accuracy and stability and on the bandwidth.
  (c) Could phase-lag compensation have been used if no loss of bandwidth relative to part (a) is permitted?

8.28. In Fig. P8.5 with $G = 1/[(2s + 1)(0.2s + 1)]$:
  (a) If $G_c = K$, find K for 50° phase margin and construct the Bode plot for this gain.
  (b) Design PI control to improve steady-state errors without loss of bandwidth, for a 45° phase margin.

8.29. In Fig. P8.5 with $G = 1/[(s + 1)(0.1s + 1)]$, design a controller G_c to meet the following specifications:

  1. Zero steady-state error for step inputs.

2. The steady-state unit ramp following error may not exceed 2%.

3. The error in response to sinusoidal inputs up to 5 rad/sec may not exceed 10%.

4. The phase margin should not be below about 63°.

Note that a combination of two types of controllers, in series, can be necessary to meet a set of specifications.

8.30. In Fig. P8.5 with $G = 1/[s(0.1s + 1)(0.2s + 1)]$:
  (a) If $G_c = K$, find K and construct the Bode plot for a steady-state unit ramp following error of 3.3% and determine system stability.
  (b) Design phase-lag compensation to stabilize the system, with about 50° phase margin, without loss of steady-state accuracy.

8.31. In Fig. P8.31, choose K and b to meet the following specifications:

  1. The unit ramp steady-state following error should not exceed 10%.

  2. The phase margin should be about 65°.

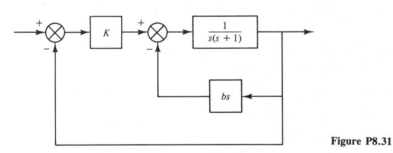

Figure P8.31

8.32. In Fig. P8.5 with $G = (1 - 0.1s)/[s(1 + s)]$:
  (a) If $G_c = K$, find K for a phase margin of about 55°.
  (b) Design a phase-lead network G_c with a ratio 10 of the break frequencies to achieve maximum improvement of crossover frequency of part (a), without loss of phase margin.
  (c) Compare the steady-state errors of parts (a) and (b).

8.33. In Fig. P8.5 with $G = (1 - s)/[s(1 + s)]$:
  (a) Construct the Bode magnitude plot for $G_c = K$, with K chosen for a phase margin of about 55°.
  (b) What is the maximum crossover frequency achievable with phase-lead compensation with ratio 10 of the break frequencies if the phase margin is to equal that of part (a)?

8.34. In Fig. P8.5 with $G_c = K$ and $G = e^{-T_{ds}}/(s + 1)$, where G is a transfer function frequently used to approximate the dynamic behavior of processes:
  (a) Construct Bode magnitude and phase-angle plots for $T_d = 0, 0.1$, and 0.5 sec, for $K = 1$.
  (b) For $T_d = 0.1$ and 0.5, find the values of K to achieve 50° phase margin and the corresponding crossover frequencies.

8.35. If the combination of a transport lag and a simple lag as in Problem 8.34 cannot model a process adequately, an additional simple lag may often do so. Let

$$G = \frac{e^{-T_d s}}{(s + 1)(0.2s + 1)} \qquad T_d = 0.2 \qquad G_c = K$$

Construct Bode magnitude and phase-angle curves both with and without the transport lag, and find and compare the values of K and the corresponding crossover frequencies for a phase margin of about 55°.

8.36. In Fig. P8.36:
  (a) Find K for a phase margin of about 55° and the corresponding steady-state error following a unit step change of disturbance input D in the absence of feedforward control G_f.
  (b) Construct the asymptotic Bode magnitude plot of C/D for part (a).
  (c) Use Bode plots to choose a constant-gain feedforward control G_f which counterbalances the direct effect of the disturbance over a wide frequency range. Construct the resulting C/D plot, and compare with that of part (b) to judge the effect of feedforward on steady-state and dynamic disturbance response.

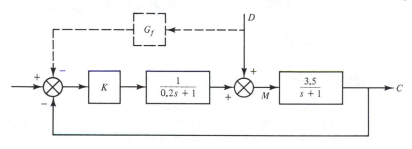

Figure P8.36

8.37. Repeat Problem 8.36 if the time constants, but not the gains, of the two simple lag blocks are interchanged. Compare the effectiveness of feedforward control, chosen for best results at low frequencies, with that in Problem 8.36.

DIGITAL CONTROL SYSTEMS

9.1 INTRODUCTION

A very strong trend to digital computer control is evident in most areas of application. Large process applications are found in power plants, paper mills, refineries, chemical plants, steel mills, and so on. Aircraft control, manufacturing, machine tools and robots, environmental control, and transportation are but a few of an expanding list of applications.

The introduction of minicomputers after about 1965 and of microcomputers since about 1975, and the greatly increased power, reduced cost, and improved reliability of computer hardware have been mainly responsible for this. Microprocessor-based digital single-loop controllers (i.e., the equivalent of pneumatic or electronic process controllers) have become available, and computer control is now competitive in every way even for simple single-loop systems.

Digital control offers important advantages in flexibility of modifying controller characteristics, or of adapting the controller if plant dynamics change with operating conditions. In multivariable systems, with more than one input and output, modern techniques for optimizing system performance or reducing interactions between feedback loops can be implemented.

It should be emphasized that feedback control in the sense of earlier chapters is only one of the functions of the computer. In fact, most of the information transfer between process and computer is of an on-off nature and exploits the logical decision-making capacity of the computer. A controls practice example will be discussed to illustrate this and to broaden the focus of attention beyond strict feedback control.

To enable the computer to meet the variety of demands imposed on it, it is time-shared among its tasks. This inherently requires sampling of the variables of interest. The sampling of an output with a sampling interval of T seconds may be visualized by assuming the presence of a relay in the feedback path which closes momentarily every T seconds. Thus feedback is not continuous but only intermittent, at the sampling instants. In the course of the sampling interval all outputs are sampled, and the computer uses the sample values to calculate actuating signals to the system actuators according to control algorithms. As will be discussed, frequently these are digital implementations of PID control, separate for the individual feedback loops in the process.

Sampling is a fundamental departure from the continuous systems considered thus far, and its effect on system dynamics will require careful attention. A large sampling interval reduces computing and sampling requirements, but too large an interval will not adequately represent actual signal variations and could cause instability.

9.2 COMPONENTS IN A PROCESS CONTROL CONFIGURATION

Large processes with many feedback loops have conventionally frequently been controlled by separate PID controllers for each loop, which are also expected to handle moderate interactions between the loops and under static conditions make each output equal to the corresponding input.

In *supervisory computer control* the computer is used to adjust the controller set points, the desired values. This permits simultaneous and rapid adjustment of numerous inputs to change operating points or product composition, or to reduce interactions among the loops.

Direct digital control was a logical next step. The PID controllers are eliminated, and the computer performs, for each loop in turn, their calculations of P, I, and D actions to produce the signals to the actuators. Originally, when economic considerations effectively limited computer control to large processes, control was generally based on a relatively large central computer. Later developments in the configurations of process control are discussed subsequently, but to introduce the typical system and its components it is convenient to assume centralized control.

Figure 9.1 shows the schematic diagram for a system of which a number of analog outputs must be controlled. *Sensors* measure the system outputs. Those which provide a digital signal are preferable, such as shaft position encoders or turbine flow meters, but usually the sensor outputs are continuous.

Instrumentation can refer to several possibilities. If the sensor output is not electrical, a transducer (TDR) is used to change it to a proportional electrical signal. The sensor or TDR may be followed by a transmitter (TMR) if needed because of the distance to the computer or the quality and strength of the signal.

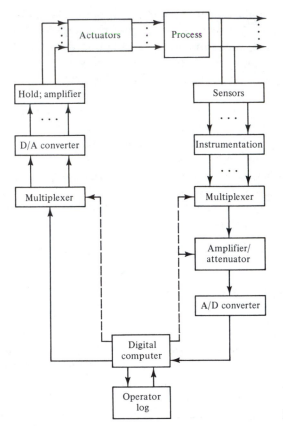

Figure 9.1 Computer control system.

Signal conditioning filters may be present to improve the signals. Simple lag filters to remove high-frequency noise from the sensor outputs are very common and will be found to have special significance in connection with the effects of sampling.

The *multiplexer* connects each signal in turn, as selected by the computer, to a single *analog-to-digital (A/D) converter*. The *amplifier/attenuator,* its scaling selected by the computer for each signal, may be present to obtain voltage ranges suitable for conversion.

The *computer* must apply a scaling inverse to that above, and convert the values to engineering units from calibration data of the sensors and other instrumentation. The system inputs and stored values of present and past output samples are used to calculate signals to the actuators during each sampling interval, on the basis of the computer control algorithm selected. Such algorithms are discussed later, and often implement PID control for each loop.

In addition to these cyclic programs, the computer must handle emergencies and respond to irregular demands for action generated by the operator and the program which use the logical decision-making capacity of the computer. This activity is discussed later. The *priority interrupt* feature is the key in allowing the computer to meet these requirements in organized fashion. The inputs to the computer have assigned priority levels. Unless interrupt is inhibited, which could be the case while cyclic programs are executed, an input will interrupt action on an input of lower priority. The latter is resumed later at the point of interruption.

The *output multiplexer,* on the output side of the computer, directs each digital output to the proper *actuator.* Again, digital actuators are desirable. A very important one is the stepping motor, which advances a certain number of degrees for each pulse it receives. However, most actuators are of the analog type, such as motors, pneumatic valve actuators, or hydraulic cylinders, and require *digital-to-analog conversion (D/A).*

The output of a converter is a sequence of voltage pulses, while the signal to the actuator must be a continuous signal, preferably that of which the pulses are the samples. The *zero-order hold* (ZOH) is a very common technique for approximating this signal reconstruction. This simply holds the last sample voltage constant until the next sample, and produces a staircase approximation to the desired signal. Power *amplification* is also necessary, as is an electrical-to-pneumatic transducer for pneumatic actuators.

9.3 FEATURES AND CONFIGURATIONS OF COMPUTER CONTROL

Section 9.2 focused on the control of a number of analog outputs. Some of the many other activities of the computer are process monitoring and data logging; alarming and taking appropriate actions when variables exceed permissible limits; sequencing of multiple parallel actuators; process startup and shutdown; sensing the status of contacts, indicating whether on-off valves are open or closed; switching on a motor; or opening a valve. Others, relating more closely to the feedback loops of Section 9.2, are:

1. Provisions for "bumpless transfer" between manual and automatic control, to avoid potentially severe switchover transients
2. Provisions to limit, for reasons of safety, the commanded change of an actuator position in one sampling interval to a given percentage of its value

Before initiating any action, the computer must make sure that it is safe to do so, in that variables are inside safe limits, preliminary commands have been obeyed, and the computer is not in a manual operating mode.

Pending a somewhat detailed example in the next section showing the extent to which straight engineering considerations of controls practice can complicate a level control, a power plant turbine startup procedure is outlined briefly. When not running, the rotor is rotated slowly by a turning gear driven by an electric motor. This is to prevent thermal sag of the rotor, which would cause destructive unbalance forces. Before rolling the rotor off this gear, the computer checks steam and bearing lubricant pressures. If a specified minimum speed is not reached in a specified time, this phase is repeated after a specified delay. Otherwise, the rotor is accelerated according to a recommended curve that avoids excessive thermal stress. The bearing vibrations are interrogated at given intervals, and if they exceed safe limits, the rotor is held at safe speeds, away from critical values, for a specified delay.

The last few years have seen important developments in the system configurations of process control. A strong trend is evident toward distributed control and away from the centralized configuration, and its large and impressive central control room, implied in Section 9.2.

Briefly, the following system types have been distinguished:

Type 1 or centralized configuration: All controllers are in a central control room, with long individual lines to and from actuators and sensors. The controllers in this configuration may also be analog, including pneumatic.

Type 2 or star configuration: A central supervisory computer is connected to several unit control rooms, of which one is indicated in Fig. 9.2(a). This configuration greatly reduces wiring and installation costs for large installations because the controllers in the unit control rooms are much closer to the actuators and sensors. Noise pickup and susceptibility to damage are reduced correspondingly. The controllers can continue to operate if the link to the supervisory computer, through which the operator can access the outputs and adjust set points, fails. These links are "data highways", in which multiplexing techniques are used to time-share a pair of wires between many digital signals.

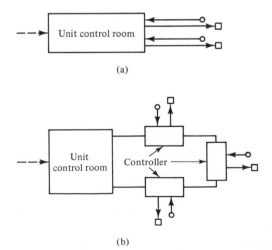

(a)

(b)

Figure 9.2 Type 2 (a) and type 3 (b) configurations.

Type 3 or distributed configuration: This type is illustrated in Fig. 9.2(b) and represents true distributed control. A data highway in a loop configuration permits bidirectional communication with a supervisory computer in the unit control room. Communication still exists with all controllers if the link is damaged at one point. The controller for each loop may be mounted adjacent to the corresponding actuator.

This distributed control has been made possible by the availability of general-purpose digital single-loop controllers, microprocessor based, since about 1979. These permit adjustment of controller set points and of the parameters of the control algorithms by a remote supervisory computer. Controllers are available which also allow a choice among a number of different control algorithms, such as:

1. Several forms of PID algorithms.
2. Ratio control, to control, say, the ratio of pulverized coal and airflow for a power plant.
3. Feedforward control, discussed in Section 8.9.
4. Cascade control: Say that to increase a temperature the controller increases a fuel valve opening. The expected larger fuel flow may not result if the fuel source pressure has dropped due to disturbances. Therefore, the controller output is used instead to provide the setpoint of a flow controller, which acts quickly to counteract the disturbances and supply the desired flow.

Microprocessor-based *Programmable Controllers* should also be mentioned. These are very widely applied, and are available at many levels of sophistication, ranging from control of a simple sequence of operations, and including sequences of which the steps may require closed-loop control of several variables to track variable inputs.

9.4 CONTROL PRACTICE: A LEVEL CONTROL EXAMPLE

Figure 9.3(a) shows a typical schematic diagram of a large feedwater level control system for power plants and Fig. 9.3(b) the logic diagram for control valve operation. Although somewhat simplified, the system provides a good illustration of the many considerations that may affect the implementation of a control loop.

A level sensor and transmitter (TMR) provide a feedback signal in the range 4 to 20 mA. The controller may be analog or digital, or both may be present with the former acting as a backup for the latter. A malfunction of the digital system would then cause automatic switchover to the backup, but manual selection would also be possible. Most controllers have provisions for "bumpless transfer," to avoid severe switchover transients.

The E/P converter changes the electrical controller output into a proportional

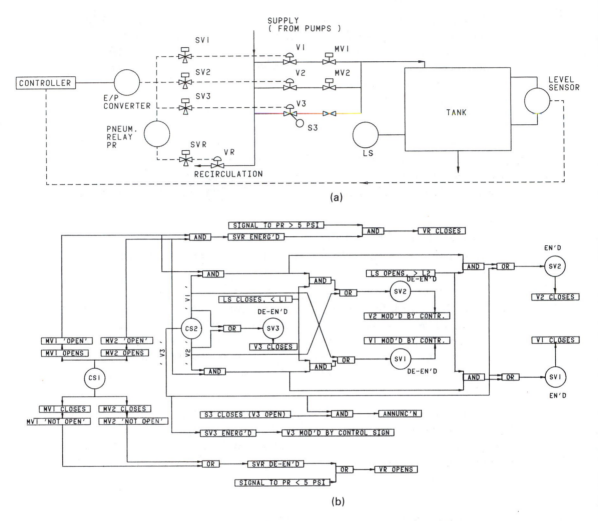

Figure 9.3   Level control example: controls practice.

3 to 15 psi pneumatic signal needed for operation of the control valves. The valve system is rather more complicated than the pneumatically actuated valve of smaller systems and is discussed below. A recirculation loop is provided to protect the pumps that supply the water, by ensuring that the flow does not fall below a minimum value. LS represents one or more level switches, which give warning signals if the tank level passes set values.

Control Valve Operation

The valves V1 and V2 are pneumatically actuated control valves of full (100%) capacity. That is, each can supply the maximum tank flow. A small, 5% capacity valve V3 is needed to control the small flows during system startup and shutdown. The large valves cannot do this adequately. For control-loop linearity, it is desirable to select valves with a linear installed flow characteristic (i.e., which provide a linear relation between valve opening and valve flow during operation). To approximate this, the drop in supply pump discharge pressure as pump flow increases must be taken into account.

When only the small valve V3 is in operation, the recirculation valve Vr is open to protect the pumps and the discharge pressure is almost constant. Hence a valve with a linear "inherent" flow characteristic will also operate linearly after installation. But with the large valves V1 and V2, pump pressure drops as a valve opens to pass more flow. To counteract this, the valves selected should have an inherent equal-percentage flow characteristic; that is, for a constant pressure drop across the valve the flow should increase stronger than linearly with valve opening. The large valves are arranged to close on loss of control air pressure or electrical power, to avoid system damage if the valves failed open. But the small valve is arranged to open under these conditions, to ensure the minimum flow needed for safety.

The valves MV1 and MV2 downstream of V1 and V2 are motorized isolating valves which are either fully open or fully closed and permit V1 and/or V2 to be effectively removed from the system. It is noted that the control valves V1 and V2 in their closed positions may have very significant leakage flow rates.

Valve Operating Logic

Figure 9.3(b) shows a logic diagram for valve operation. It is discussed to illustrate this aspect of computer control, and the extent to which straightforward engineering considerations may complicate a basic level control system.

CS1 and CS2 are the main control switches. For startup, CS1 is closed, closing motorized valves MV1 and MV2, and CS2 is switched to V3. This energizes the solenoid valve SV3, admitting control air pressure to V3, and allowing it to modulate the flow in response to the control signal. During this stage of the startup the solenoid valve SVr is deenergized and the recirculation valve Vr is open. When V3 is fully open, the switch S3 closes, giving an annunciation to the operator.

The operator then switches CS1 to "open," causing MV1 and MV2 to open. When their "fully open" positions, sensed by the closure of limit switches, are annunciated to the operator, he switches CS2 to, say, V1. This deenergizes both SV3 and SV1, closing V3 and passing the control air pressure signal to V1. V1 now controls flow to maintain the level set point. While the flow is still small, it is desired to keep the recirculation valve Vr open. This is the function of the pneumatic relay PR. If its input, the control air pressure, is below 5 psi, its output is 0 psi, and Vr stays open. But if the input is above 5 psi the output is 15 psi, and Vr closes if SVr is energized.

It may be that during system overload or valve malfunction V1 cannot maintain the set-point level. If the level should drop below a certain value 1_1, a current relay LS is closed by a level sensor/transmitter. As indicated in Fig. 9.3(b), provided that MV1 is fully open, this deenergizes SV2 and allows V2 to operate in parallel with V1. When the level recovers to 1_2, SV2 is again energized and V2 closes.

Figure 9.3(b) is apparently based on straightforward engineering considerations. However, to combine these into a minimal logic diagram such as that shown may require experience and an adequate knowledge of logic circuits.

9.5 CONTROL ALGORITHMS AND FINITE DIFFERENCES

The preceding sections have been concerned with a description of computer control systems and their components. Before turning to the dynamic analysis of such systems, it is desirable to round off the present picture by a preliminary discussion of the implementation of controller transfer functions by digital *control algorithms*. As will be seen in Chapter 10, there are several techniques for this purpose. This section is restricted to the use of finite-difference approximations, with emphasis on the very common case of PI and PID control. The transfer function $D(s)$ for PID control and the corresponding time-domain relation between input $e(t)$ and output $u(t)$ are

$$D(s) = \frac{U(s)}{E(s)} = K_p + \frac{K_i}{s} + K_d s \tag{9.1a}$$

$$u(t) = K_p e(t) + K_i \int_0^t e \, dt + K_d \dot{e}$$

Numerous variations of algorithms exist, depending on the finite-difference forms used to approximate the terms. Such approximations may also be based on the differential form of (9.1a):

$$\dot{u} = K_i e + K_p \dot{e} + K_d \ddot{e} \tag{9.1b}$$

The notations $e_k = e(kT)$ and $u_k = u(kT)$ will be used to denote the variables at time $t = kT$, as indicated in Fig. 9.4. In the present context, the sequence $e_k, k = 0, 1, 2, \ldots$, represents the sample values of the error, and u_k the signal to the actuator, say a valve.

Control algorithms are in general of two kinds:

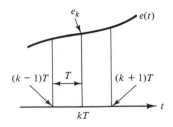

Figure 9.4   Finite differences.

1. *Position algorithm:* Here the output of the algorithm is u_k, the desired valve position.
2. *Velocity algorithm:* The algorithm supplies $\Delta u_k = u_k - u_{k-1}$, the desired change of valve position between $(k - 1)T$ and kT.

Velocity algorithms are often preferred because they have the fail-safe feature of leaving the valve where it is if computer malfunction causes a zero output. They also facilitate bumpless transfer between manual and automatic control, and adapt naturally to actuators such as stepping motors which require such incremental input signals.

Finite-Difference Approximations of Derivatives

In Fig. 9.4, the following approximations, called *first differences,* can be used for $\dot{e}(t)$, the slope of the $e(t)$ curve at $t = kT$:

$$\frac{e_{k+1} - e_k}{T} \qquad \text{forward difference}$$

$$\frac{e_k - e_{k-1}}{T} \qquad \text{backward difference} \qquad (9.2)$$

$$\frac{e_{k+1} - e_{k-1}}{2T} \qquad \text{central difference}$$

If e_{k+1} is available, the last tends to give the best approximation. *Second differences* to approximate $\ddot{e}$ may be derived as the first difference of first differences:

$$\frac{e_{k+2} - 2e_{k+1} + e_k}{T^2} \qquad \text{forward difference}$$

$$\frac{e_k - 2e_{k-1} + e_{k-2}}{T^2} \qquad \text{backward difference} \qquad (9.3)$$

$$\frac{e_{k+1} - 2e_k + e_{k-1}}{T^2} \qquad \text{central difference}$$

The last, for example, may be derived as

$$\frac{[(e_{k+1} - e_k)/T] - [(e_k - e_{k-1})/T]}{T}$$

Example 9.5.1   PI and PID Control Algorithms

   (a) *PI control:* From (9.1b) with $K_d = 0$, using backward differences, $(u_k - u_{k-1})/T = K_i e_k + K_p(e_k - e_{k-1})/T$, or

$$u_k = u_{k-1} + (K_p + K_i T)e_k - K_p e_{k-1} \qquad (9.4a)$$

This is the algorithm for calculating the new signal u_k based on past values of u and present and past values of e.

(b) *PID control:* Using (9.1b) with backward differences gives

$$\frac{u_k - u_{k-1}}{T} = K_p \frac{e_k - e_{k-1}}{T} + K_i e_k + K_d \frac{e_k - 2e_{k-1} + e_{k-2}}{T^2}$$

Rearranging, this becomes

$$u_k = u_{k-1} + \left(K_p + K_i T + \frac{K_d}{T} \right) e_k - \left(K_p + \frac{2K_d}{T} \right) e_{k-1} + \frac{K_d}{T} e_{k-2} \qquad (9.4b)$$

The position algorithms (9.4a) and (9.4b) can be written as velocity algorithms $\Delta u_k = u_k - u_{k-1}$.

Example 9.5.2   **Algorithms for Phase-Lead or Phase-Lag Compensation with Transfer Function**

$$G_c(s) = \frac{U(s)}{E(s)} = \frac{K(s + b)}{s + a} \qquad (9.5a)$$

Cross-multiplying and inverting yields

$$\dot{u} + au = K\dot{e} + Kbe \qquad (9.5b)$$

(a) *Backward differences:*

$$\frac{1}{T}(u_k - u_{k-1}) + au_k = \frac{K}{T}(e_k - e_{k-1}) + Kbe_k$$

or

$$u_k = \frac{u_{k-1} + K(1 + bT)e_k - Ke_{k-1}}{1 + aT} \qquad (9.5c)$$

(b) *Forward differences:*

$$\frac{1}{T}(u_{k+1} - u_k) + au_k = \frac{K}{T}(e_{k+1} - e_k) + Kbe_k$$

Solving for u_{k+1}, and then reducing all subscripts by one since u_k is the value to be found, yields a different algorithm than (9.5c):

$$u_k = (1 - aT)u_{k-1} + Ke_k - K(1 - bT)e_{k-1} \qquad (9.5d)$$

Example 9.5.3   **Numerical Solution of Differential Equations**

(a) $\dot{y} + ay = x$: Backward, forward, and central differences, respectively, yield the algorithms

$$\frac{1}{T}(y_k - y_{k-1}) + ay_k = x_k \qquad y_k = \frac{y_{k-1} + Tx_k}{1 + aT}$$

$$\frac{1}{T}(y_{k+1} - y_k) + ay_k = x_k \qquad y_{k+1} = (1 - aT)y_k + Tx_k$$

$$\frac{1}{2T}(y_{k+1} - y_{k-1}) + ay_k = x_k \qquad y_{k+1} = -2aTy_k + y_{k-1} + 2Tx_k$$

Of course, for calculation of y_k the subscripts in the last two algorithms can be reduced by one.

(b) $\ddot{y} + a\dot{y} + by = x$: Using central differences:

$$\frac{1}{T^2}(y_{k+1} - 2y_k + y_{k-1}) + \frac{a}{2T}(y_{k+1} - y_{k-1}) + by_k = x_k$$

or

$$y_{k+1} = \frac{(2 - bT^2)y_k + (0.5aT - 1)y_{k-1} + T^2x_k}{1 + 0.5aT}$$

Finite-Difference Approximations of Integrals

For the integral term in (9.1a), if in Fig. 9.4 v_{k-1} approximates $\int e\,dt$, the area under the curve, up to $(k - 1)T$, then the approximation v_k up to kT may be written alternatively as

$$v_k = v_{k-1} + Te_{k-1} \qquad \text{forward rectangular rule}$$

$$v_k = v_{k-1} + Te_k \qquad \text{backward rectangular rule} \qquad (9.6)$$

$$v_k = v_{k-1} + \frac{T}{2}(e_{k-1} + e_k) \qquad \text{trapezoidal rule}$$

Figure 9.4 shows that the second terms on the right are different approximations to the area between $(k - 1)T$ and kT, the last being the most accurate.

Example 9.5.4   PI and PID Control Algorithms

(a) *PI control:*

$$\frac{U(s)}{E(s)} = K_p + \frac{K_i}{s} \qquad u = K_p e + K_i \int e\,dt \qquad (9.7a)$$

Using the backward rectangular rule:

$$u_k = K_p e_k + v_k \qquad v_k = v_{k-1} + K_i Te_k \qquad (9.7b)$$

In the form of a velocity algorithm:

$$\Delta u_k = u_k - u_{k-1} = K_p(e_k - e_{k-1}) + K_i Te_k$$

$$= (K_p + K_i T)e_k - K_p e_{k-1} \qquad (9.7c)$$

(b) *PI control:* Using the trapezoidal rule,

$$u_k = K_p e_k + v_k \qquad v_k = v_{k-1} + 0.5K_i T(e_{k-1} + e_k)$$

$$u_k = u_{k-1} + K_p(e_k - e_{k-1}) + 0.5K_i T(e_{k-1} + e_k) \qquad (9.7d)$$

$$= u_{k-1} + (K_p + 0.5K_i T)e_k + (0.5K_i T - K_p)e_{k-1}$$

(c) *PID control:* With trapezoidal integration and backward differences for the derivative,

$$u_k = u_{k-1} + K_p(e_k - e_{k-1}) + \frac{K_d}{T}(e_k - 2e_{k-1} + e_{k-2}) + 0.5K_i T(e_{k-1} + e_k)$$

$$= u_{k-1} + \left(K_p + \frac{K_i T}{2} + \frac{K_d}{T}\right)e_k - \left(K_p + \frac{2K_d}{T} - \frac{K_i T}{2}\right)e_{k-1} + \frac{K_d}{T}e_{k-2} \qquad (9.8)$$

These algorithms can be compared with those in Example 9.5.1. The better algorithm (9.8) results if in the equation from which (9.4b) was derived the term $K_i e_k$ is replaced by $0.5K_i(e_{k-1} + e_k)$.

Example 9.5.5   Lead or Lag Compensation

$$G_c(s) = \frac{U(s)}{E(s)} = \frac{K(s + b)}{s + a} \qquad (9.9a)$$

or solution of the differential equation

$$\dot{u} + au = K\dot{e} + Kbe \tag{9.9b}$$

using trapezoidal integration. Rearrange the equation to $\dot{u} - K\dot{e} = -au + Kbe$ and write it in the integral form

$$u - Ke = \int_0^t (-au + Kbe) \, d\tau \tag{9.9c}$$

Using the last of (9.6), this gives

$$u_k - Ke_k = u_{k-1} - Ke_{k-1} + 0.5T(-au_{k-1} + Kbe_{k-1} - au_k + Kbe_k)$$

which can be rearranged to the algorithm

$$u_k = \frac{(1 - 0.5aT)u_{k-1} + K(1 + 0.5bT)e_k + K(0.5bT - 1)e_{k-1}}{1 + 0.5aT} \tag{9.9d}$$

Comparison with the algorithms in Example 9.5.2 again illustrates the variations possible, even without adding forms based on the rectangular rules, in algorithms that all approximate the same continuous system.

Reset windup or integral windup is a problem which can arise in both continuous and discrete systems when integral control is present and was discussed in Section 5.4. For large changes of input or large disturbances it can lead to severe transient oscillations.

To avoid the growth of integrator output that causes this, one solution in control algorithms is to clamp the integrator output to prevent values outside selected HI and LO limits:

$$\text{IF } I > \text{HI THEN } I = \text{HI} \qquad \text{IF } I < \text{LO THEN } I = \text{LO} \tag{9.10}$$

9.6 SAMPLING CHARACTERISTICS AND SIGNAL RECONSTRUCTION

As a first step toward the analysis and design of digital control systems, it is necessary to consider the effects of sampling. Sampling at intervals of T seconds is indicated in Fig. 9.5. The sampler output $f^*(t)$ equals $f(t)$ over the very short periods $\tau \ll T$ during each interval when the relay is closed, and $f^*(t)$ is zero between samples. This may be represented mathematically by multiplying $f(t)$ by a train $S(t)$ of unit impulses:

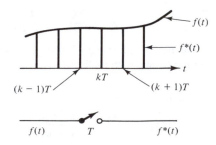

Figure 9.5 Sampling.

$$S(t) = \sum_{n=-\infty}^{\infty} \delta(t - nT) \tag{9.11}$$

Here $\delta(t)$ is a unit impulse at $t = 0$ (Laplace transform $L[\delta(t)] = 1$), and $\delta(t - nT)$ a unit impulse at $t = nT$ (delay theorem: $L[\delta(t - nT)] = 1 \cdot e^{-nTs} = e^{-nTs}$). Then, assuming that $f(t) = 0$ for $t < 0$,

$$f^*(t) = f(t)S(t) = \sum_{n=0}^{\infty} f(nT)\,\delta(t - nT)$$

$$\tag{9.12}$$

$$F^*(s) = L[f^*(t)] = \sum_{n=0}^{\infty} f(nT)e^{-nTs}$$

Let

$$\omega_s = \frac{2\pi}{T} = \text{radian sampling frequency} \tag{9.13}$$

Then for integer m:

$$F^*(s + jm\omega_s) = \sum_{n=0}^{\infty} f(nT)e^{-nT(s + jm\omega_s)}$$

$$= \sum_{n=0}^{\infty} f(nT)e^{-nTs}e^{-jnm2\pi} = \sum_{n=0}^{\infty} f(nT)e^{-nTs}$$

Hence $F^*(s)$ is periodic with radian frequency ω_s along any line parallel to the $j\omega$-axis in the s-plane:

$$F^*(s + jm\omega_s) = F^*(s) \tag{9.14}$$

Thus in Fig. 9.6 the pole–zero pattern of $F^*(s)$ in the *primary strip*, between $-0.5j\omega_s$ and $+0.5j\omega_s$, is repeated in all *complementary strips*. It must be concluded that the Laplace transform of a sampled signal in general has an infinite number of poles and zeros.

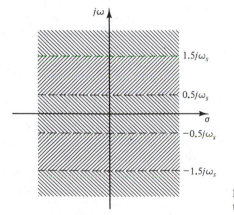

Figure 9.6 Primary and complementary strips.

*****

An alternative expression for $F^*(s)$ helps to interpret these results. The impulse train $S(t)$ of (9.11) can be written as a Fourier series $S(t) = \sum_{n=-\infty}^{\infty} C_n e^{jn\omega_s t}$, where the coefficients C_n, from the theory of Fourier series, may be expressed as

$$C_n = \frac{1}{T} \int_{-T/2}^{T/2} S(t) e^{-jn\omega_s t}\, dt = \frac{1}{T} \int_{0-}^{0+} S(t)\, dt = \frac{1}{T}$$

It follows that

$$f^*(t) = f(t)S(t) = \frac{1}{T} \sum_{n=-\infty}^{\infty} f(t) e^{jn\omega_s t}$$

and

$$F^*(s) = L[f^*(t)] = \int_0^{\infty} \frac{1}{T} \sum_{n=-\infty}^{\infty} f(t) e^{-(s-jn\omega_s)t}$$

Hence $F^*(s)$ can be expressed in terms of $F(s)$:

$$F^*(s) = \frac{1}{T} \sum_{n=-\infty}^{\infty} F(s + jn\omega_s) \tag{9.15}$$

Because of the summation from $-\infty$ to ∞, the sign of the term with n is immaterial. Of particular interest at this point is the case of frequency response, when $s = j\omega$:

$$F^*(j\omega) = \frac{1}{T} \sum_{n=-\infty}^{\infty} F\{j(\omega + n\omega_s)\} \tag{9.16}$$

This relation is illustrated in Fig. 9.7. The frequency spectrum $|F^*(j\omega)|$ of the sampler output is periodic. The *central band,* corresponding to the primary strip in Fig. 9.6, and the *sidebands* all have the same shape as the input spectrum $|F(j\omega)|$, except for the factor $1/T$.

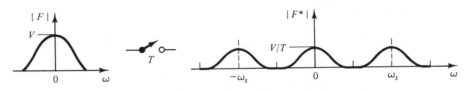

Figure 9.7 Frequency spectrum due to sampling.

The Sampling Theorem and Sampling Rates

The sidebands are due to the sampling process and must be removed to reconstruct the continuous signal from the samples. As noted earlier, this is necessary following D/A conversion of the computer output. It is also required when digital data acquisition systems are used to collect experimental data, and the frequency spectrum of the original signal is desired. The sidebands must be removed by filters. This will clearly not be possible if they overlap with the central band, as in Fig. 9.8(a). Hence follows the

> **Sampling Theorem.**   The sampling frequency ω_s must at least equal twice the value of the highest significant frequency in the signal.

Sampling rates used in practice are generally much higher and may be between 4 and 20 times the system bandwidth, depending on the desired accuracy and the load on and capacity of the computer. One rule of thumb is to choose T as one-tenth of the smallest plant time constant or the desired closed-loop time constant. Another convenient rule suggests sampling at the rate of 6 to 10 times per cycle. Thus, if the largest imaginary part of the significant system poles is 1 rad/sec, which corresponds to transient oscillations with a frequency of 1/6.28 cycle per second, or a period of 6.28 sec, $T = 1$ sec may be satisfactory.

An important purpose in Chapter 10 will be to study the effect of sampling rate on performance. This will show, for example, that for exponential transients corresponding to real poles, which largely decay in three time constants, a sampling interval somewhat below the smallest time constant may be adequate.

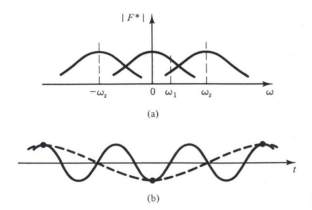

(a)

(b)   **Figure 9.8**   Aliasing or folding.

Aliasing or Folding

These are terms used to describe the consequences if the sampling theorem is not satisfied. In Fig. 9.8(a), at frequency ω_1 the central band contributes $|F(j\omega_1)|$ and the sideband $|F(j(\omega_s - \omega_1))|$. The latter is an *alias* of ω_1, and the name *folding* arises because it can also be visualized as the component of the central band an equal distance above $\omega_s/2$ being folded back inside that band. As illustrated in Fig. 9.8(b), the sampling frequency is too low to permit the two signals to be distinguished from each other.

From a somewhat different point of view, the sampling theorem is satisfied if all poles of $F(s)$ lie inside the primary strip in Fig. 9.6. It is recalled that the total transient is a superposition of components due to all poles of $F(s)$. For complex poles with imaginary part ω_d, the frequency of transient oscillations was found to be ω_d. So the highest significant frequency corresponds to the pair with largest ω_d. Suppose that one pair is located at $-a \pm 0.6j\omega_s$, outside the primary strip. Then sampling will "fold" this pair back into the primary strip, to the locations $-a \pm 0.6j\omega_s \mp j\omega_s = -a \mp 0.4j\omega_s$. So $f^*(t)$ contains components inside the primary strip which do not occur in $f(t)$.

Analog Prefilters

These are filters on the sensor outputs before sampling, and are very common in digital control systems. They were mentioned earlier, and are a direct consequence of the foregoing considerations. Analog prefilters are usually simple lag filters with time constant chosen to remove signal noise above the frequency range of interest. In continuous systems such noise is usually filtered out by low-pass characteristics of the process or the actuators. However, with sampling this noise would be folded back into the central band; that is, it would show up at low frequencies and would not be filtered out. The prefilters avoid this by limiting the bandwidth of the sampler input.

Zero-Order Hold (ZOH)

The ZOH, mentioned in Section 9.2, is a very commonly used approximation to an ideal, but physically unrealizable filter which would remove the sidebands completely and leave the central band unchanged. The ZOH constructs a staircase approximation to a continuous signal by keeping the output constant between samples, equal to the last sample value. As illustrated in Fig. 9.9, this means that for a unit impulse input $r(t) = \delta(t)$ [$R(s) = 1$] the output $c(t)$ is a box of unit height between $t = 0$ and $t = T$. This is the sum of a unit step at $t = 0$ $(1/s)$ and a negative unit step at $t = T$ $(-e^{-Ts}/s)$. So $C(s) = (1 - e^{-Ts})/s$, and the transfer function $G_h(s) = C(s)/R(s)$ is

$$G_h(s) = \frac{1 - e^{-Ts}}{s} \tag{9.17}$$

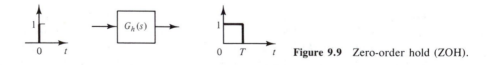

Figure 9.9 Zero-order hold (ZOH).

9.7 CONCLUSION

In this chapter a general introduction has been given to the area of digital control systems. Digital control algorithms, with emphasis on PID control, were derived by the use of finite differences. In Chapter 10 this technique will be seen in its proper context, as one of several ways in which a continuous controller can be approximated by a digital algorithm. The effects of sampling of a signal were also discussed, and will be treated in more detail in Chapter 10, which is concerned with the analysis and design of single-loop digitally controlled feedback systems.

Because of the introductory nature of this chapter, problems on the material are more appropriate at the end of the next chapter.

10

DIGITAL CONTROL SYSTEM ANALYSIS AND DESIGN

10.1 INTRODUCTION

This chapter deals with the analysis and design of single-loop digital control systems. It was found in Section 9.6 that the Laplace transform of a sampled signal in general has an infinite number of poles and zeros. As this suggests, this transform is not attractive for analysis and design.

The Z transform, introduced next, takes its place and is well adapted to represent both computer control algorithms and sampled signals.

10.2 Z TRANSFORMS AND Z TRANSFER FUNCTIONS

The Laplace transform of a sampled signal was given in (9.12):

$$F^*(s) = L[f^*(t)] = \sum_{n=0}^{\infty} f(nT)e^{-nTs} \qquad (10.1)$$

The Z transform of a sampled signal can be defined as a further transformation:

$$z = e^{Ts} \qquad F(z) = F^*(s)\bigg|_{z=e^{Ts}} \qquad (10.2)$$

$$F(z) = Z[f(t)] = \sum_{n=0}^{\infty} f(nT)z^{-n} \qquad (10.3a)$$

Only the sequence of samples $f(nT)$, $n = 0, 1, 2, \ldots$, is considered in this transform, not the response between samples. It is recalled that e^{-Ts}, and therefore

z^{-1}, means a delay of one sampling interval. $F(z)$ in (10.3a) represents a sequence of sample values, with $f(nT)$ occurring after n sampling intervals.

The number sequences f_k, $k = 0, 1, 2, \ldots$, in control algorithms are also sequences of values, and can be represented by the same transform:

$$F(z) = Z[f_k] = \sum_{k=0}^{\infty} f_k z^{-k} \tag{10.3b}$$

Table 10.2.1 shows examples derived from the definitions (10.3), using known results to write the series in closed form.

1. *Unit impulse* $\delta(t)$ *or discrete pulse* δ_k:

$$(\delta_k = 1 \text{ for } k = 0, \delta_k = 0 \text{ for } k \neq 0)$$

$$Z[\delta] = 1 + 0 \cdot z^{-1} + \cdots = 1$$

2. *Delayed unit pulses* $\delta(t - nT)$ *or* δ_{k-n}:

$$Z[\cdot] = 0 + 0 \cdot z^{-1} + \cdots + 1 \cdot z^{-n} + 0 \cdot z^{-n-1} + \cdots = z^{-n}$$

3. *Unit step* $u(t)$ *or number sequence* $f_k = 1$:

$$Z[\cdot] = 1 + 1 \cdot z^{-1} + 1 \cdot z^{-2} + \cdots = \frac{1}{1 - z^{-1}} = \frac{z}{z - 1}$$

4. *Unit ramp* t *or sequence* $f_k = kT$:

$$Z[\cdot] = 0 + Tz^{-1} + 2Tz^{-2} + \cdots = Tz^{-1}(1 + 2z^{-1} + 3z^{-2} + \cdots)$$

$$= \frac{Tz^{-1}}{(1 - z^{-1})^2} = \frac{Tz}{(z - 1)^2}$$

TABLE 10.2.1 *Z* TRANSFORM PAIRS AND THEOREMS

$f(t), f_k$	$L[f(t)]$	$Z[\cdot]$
1. $\delta(t)$, δ_k	1	1
2. $\delta(t - nT)$, δ_{k-n}	e^{-nTs}	z^{-n}
3. $u(t)$, $f_k = 1$	$\dfrac{1}{s}$	$\dfrac{1}{1 - z^{-1}} = \dfrac{z}{z - 1}$
4. t, $f_k = kT$	$\dfrac{1}{s^2}$	$\dfrac{Tz^{-1}}{(1 - z^{-1})^2} = \dfrac{Tz}{(z - 1)^2}$
5. e^{-at}	$\dfrac{1}{s + a}$	$\dfrac{1}{1 - e^{-aT}z^{-1}} = \dfrac{z}{z - e^{-aT}}$
$e^{-j\omega_0 t}$	$\dfrac{1}{s + j\omega_0}$	$\dfrac{1}{1 - e^{-j\omega_0 T}z^{-1}} \quad (a = j\omega_0)$

6. Linearity theorem: $Z[af(t) + bg(t)] = aF(z) + bG(z)$
7. Delay theorem: $Z[f(t - nT)]$ or $Z[f_{k-n}] = z^{-n}F(z)$
8. Final value theorem: $\lim_{n \to \infty} f(nT)$ or $f_n = \lim_{z \to 1} (1 - z^{-1})F(z)$

5. *Exponential decay* e^{-at} *and geometric sequences:*

$$Z[e^{-at}] = 1 + e^{-aT}z^{-1} + e^{-2aT}z^{-2} + \cdots$$

$$= 1 + (e^{aT}z)^{-1} + (e^{aT}z)^{-2} + \cdots = \frac{1}{1 - e^{-aT}z^{-1}} = \frac{z}{z - e^{-aT}}$$

For a geometric sequence b^{-k}, replace e^{-aT} by b^{-1}, and for b^k replace e^{-aT} by b, because the sequence for e^{-at} is $(e^{-aT})^k$.

6. *Linearity:* Proved from definition (10.3).

7. *Delay theorem:*

$$Z[f_{k-n}] = \sum_{k=0}^{\infty} f_{k-n}z^{-k}$$

$$= 0 + 0 \cdot z^{-1} + \cdots + 0 \cdot z^{-n+1} + f_0 z^{-n} + f_1 z^{-n-1} + \cdots$$

$$= z^{-n}[f_0 + f_1 z^{-1} + \cdots] = z^{-n}F(z)$$

since f_{k-n} is assumed to be zero for $k < n$.

The partial fraction expansion of $F(s)$ is used to find $F(z)$ for transforms not available in the table.

Example 10.2.1

$$L[\sin \omega t] = \frac{\omega}{s^2 + \omega^2} = \frac{-1/(2j)}{s + j\omega} + \frac{1/(2j)}{s - j\omega}$$

Using entries 5 and 6 in Table 10.2.1,

$$Z[\sin \omega t] = \frac{-1/(2j)}{1 - e^{-j\omega T}z^{-1}} + \frac{1/(2j)}{1 - e^{j\omega T}z^{-1}} = \frac{z^{-1}\sin \omega T}{1 - 2z^{-1}\cos \omega T + z^{-2}} \quad (10.4)$$

More examples are given below.

Z Transfer Functions

Definition. The *Z transfer function* of a system is the ratio of the *Z* transforms of its output and input sequences.

These can be used to describe the action of a control algorithm as well as the relation between sampled inputs and outputs of continuous systems.

Discrete transfer functions and digital filters. A computer control algorithm or digital filter can be represented by one or more difference equations of the form

$$y_n = \sum_{i=0}^{p} a_i x_{n-i} - \sum_{j=1}^{q} b_j y_{n-j} \quad (10.5)$$

where y_k and x_k are output and input number sequences. The filter is *nonrecursive* or *transversal* if all b_j are zero. The present output y_n then depends only on present and past inputs. Otherwise, the filter is *recursive*. Using the linearity and delay theorems, the transform of (10.5) is

$$Y(z) = \sum_{i=0}^{p} a_i z^{-i} X(z) - \sum_{j=1}^{q} b_j z^{-j} Y(z) \tag{10.6}$$

Hence, as illustrated in Fig. 10.1(a), the discrete transfer function $D(z)$ describing the filter is

$$D(z) = \frac{Y(z)}{X(z)} = \frac{\displaystyle\sum_{i=0}^{p} a_i z^{-i}}{1 + \displaystyle\sum_{j=1}^{q} b_j z^{-j}} \tag{10.7}$$

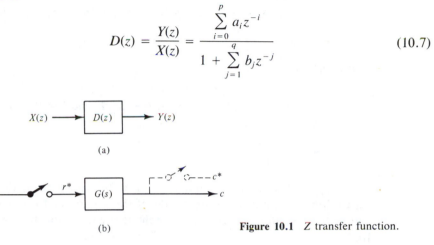

(a)

(b)   **Figure 10.1**   *Z* transfer function.

It is important to note that the algorithm corresponding to a given transfer function $D(z)$ is readily found by using these equations in the reverse order. Cross-multiplication in (10.7) gives (10.6), which, by the linearity and delay theorems, corresponds to the algorithm (10.5).

Example 10.2.2

$$D(z) = z^{-1}$$
$$Y(z) = z^{-1} X(z)$$

so $y_k = x_{k-1}$, by the delay theorem. The input number sequence is delayed by one interval.

Example 10.2.3   Control Algorithm

$$y_k = a_1 y_{k-1} - a_2 y_{k-2} + b_1 x_k + b_2 x_{k-1}$$
$$Y(z) = a_1 z^{-1} Y(z) - a_2 z^{-2} Y(z) + b_1 X(z) + b_2 z^{-1} X(z)$$
$$D(z) = \frac{Y(z)}{X(z)} = \frac{b_1 + b_2 z^{-1}}{1 - a_1 z^{-1} + a_2 z^{-2}}$$

Example 10.2.4   Integration Algorithms (9.6)

(a) $u_k = u_{k-1} + T e_{k-1}$; $U(z) = z^{-1} U(z) + T z^{-1} E(z)$;

$$D(z) = \frac{U(z)}{E(z)} = \frac{T z^{-1}}{1 - z^{-1}} \tag{10.8a}$$

(b) $u_k = u_{k-1} + T e_k$; $D(z) = \dfrac{T}{1 - z^{-1}}$ $\tag{10.8b}$

(c) $u_k = u_{k-1} + (T/2)(e_{k-1} + e_k)$ (trapezoidal rule);

$$D(z) = \frac{T}{2}\frac{1 + z^{-1}}{1 - z^{-1}} = \frac{T}{2}\frac{z + 1}{z - 1} \qquad (10.8c)$$

Example 10.2.5   PID-control Algorithms

(a) *PI algorithm* (9.7c):

$$u_k = u_{k-1} + K_p(e_k - e_{k-1}) + K_iTe_k = u_{k-1} + (K_p + K_iT)e_k - K_pe_{k-1}$$

$$D(z) = K_p + \frac{K_iT}{1 - z^{-1}} \qquad (10.9a)$$

(b) *PID algorithm* (9.8):

$$u_k = u_{k-1} + ae_k + be_{k-1} + ce_{k-2}$$

where $c \equiv K_d/T$, $a \equiv K_p + 0.5K_iT + K_d/T$, and $b \equiv 0.5K_iT - K_p - 2K_d/T$.

$$U(z) = z^{-1}U(z) + (a + bz^{-1} + cz^{-2})E(z)$$

$$D(z) = \frac{a + bz^{-1} + cz^{-2}}{1 - z^{-1}} \qquad (10.9b)$$

$D(z)$ relates number sequences. But by considering the A/D and D/A conversions on input and output side of the computer to be included, it is also a pulse transfer function relating sample sequences, and $D(z) = D^*(s)|_{z = e^{Ts}}$ as in (10.2).

Pulse transfer functions.   In Fig. 10.1(b), a sample sequence r^* is the input to a system with transfer function $G(s)$. Z transforms do not consider the response between samples, and therefore a synchronous fictitious sampler is introduced to produce c^*.

$$C(s) = G(s)R^*(s) \qquad (10.10)$$

From (9.14), $R^*(s + jn\omega_s) = R^*(s)$, so that (9.15) yields

$$C^*(s) = \frac{1}{T}\sum_{n=-\infty}^{\infty} C(s + jn\omega_s) = \frac{1}{T}\sum G(s + jn\omega_s)$$

$$\times R^*(s + jn\omega_s) = R^*(s)\left[\frac{1}{T}\sum G(s + jn\omega_s)\right]$$

The form in brackets is the ratio of output and input transforms, and also satisfies (9.15) for $G^*(s)$. Hence the key relations

$$C^*(s) = G^*(s)R^*(s) \qquad C(z) = G(z)R(z) \qquad (10.11)$$

and the definition

The *pulse transfer function* $G(z)$ is the Z transform of $G(s)$:

$$G(z) = Z[G(s)] \qquad (10.12)$$

Partial fraction expansion of $G(s)$ is used to determine the transform.

Example 10.2.6

(a) $G(s) = \dfrac{a}{s + a}$; $G(z) = \dfrac{az}{z - e^{-aT}}$ (10.13a)

(b) $G(s) = \dfrac{a}{s(s + a)}$; $G(z) = Z\left[\dfrac{1}{s} - \dfrac{1}{s + a}\right] = \dfrac{z}{z - 1} - \dfrac{z}{z - e^{-aT}}$;

$G(z) = \dfrac{z(1 - e^{-aT})}{(z - 1)(z - e^{-aT})}$ (10.13b)

(c) $G(s) = \dfrac{a}{s^2(s + a)}$; $G(z) = Z\left[\dfrac{1}{s^2} - \dfrac{1/a}{s} + \dfrac{1/a}{s + a}\right]$;

$G(z) = \dfrac{Tz^{-1}}{(1 - z^{-1})^2} - \dfrac{1/a}{1 - z^{-1}} + \dfrac{1/a}{1 - e^{-aT}z^{-1}}$ (10.13c)

where the partial fraction expansions are found in the usual way.

10.3 DYNAMIC BEHAVIOR AND THE Z-PLANE

In the *s*-plane, the correlations between dynamic behavior and the positions of system poles and zeros are important in analysis and design of continuous systems. By use of the relation $z = e^{Ts}$, these correlations can also be used to derive analogous insights in the *z-plane*.

Poles or zeros $s = -a \pm jb$ map to the positions
$z = e^{(-a \pm jb)T} = e^{-aT}e^{\pm jbT}$ (10.14)

The magnitude of z (i.e., the distance to the origin) is e^{-aT}. The angles with the positive real axis of the *z*-plane, measured positive in the counterclockwise direction, are $\pm bT$ radians. It should be observed that the *z*-plane locations depend on sampling interval T as well as the *s*-plane positions, so will move if T is changed.

Figure 10.2 provides a summary of corresponding locations in both planes:

1. *Stability:* In the *s*-plane, $a = 0$ is the boundary of the stable pole region, the imaginary axis. This axis maps into $z = e^{\pm jbT}$, a circle of unit radius about the origin as b moves along the imaginary axis, the *unit circle*. Stability requires that $a > 0$. Hence follows the basic theorem for stability of digital control systems.

   Stability theorem. For stability, all poles of the system Z transfer function must lie inside the unit circle in the *z*-plane.

2. *Time constant, T_c:* In the *s*-plane, the poles must lie to the left of a vertical at $-a = -1/T_c$ to achieve a time constant $< T_c$, so in the *z*-plane they

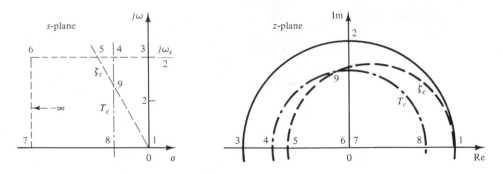

Figure 10.2 Correlations between the s- and z-planes.

must be inside a circle of radius e^{-aT}. Moving the poles closer to the origin of the z-plane increases the speed of response.

3. *Damping ratio,* ζ_c: Poles $s_{1,2} = -\zeta\omega_n \pm j\omega_n\sqrt{1-\zeta^2}$ for constant ζ lie along radial lines. In the z-plane

$$z_{1,2} = e^{-\zeta\omega_n T}e^{\pm j\omega_n\sqrt{1-\zeta^2}T}$$

As ω_n increases for constant ζ, the magnitude decreases and the phase angle increases, and z describes a logarithmic spiral. Negatively increasing ω_n gives the mirror image. Figure 10.3 shows the loci in the upper-half z-plane for several values of ζ. A specified minimum damping requires the poles to lie inside the corresponding spiral.

4. *Sampling frequency:* As the imaginary part b of the poles moves closer to the limit $\omega_s/2$ of the primary strip, the number of samples per cycle reduces, to the minimum of two. In the z-plane, this reduction occurs as the angle $e^{\pm jbT}$ of the poles moves closer to the direction of the negative real axis. This part of the real axis corresponds to the boundaries of the primary strip.

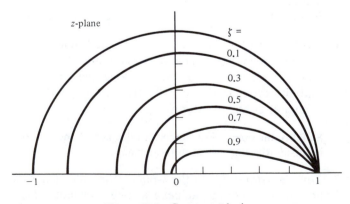

Figure 10.3 Constant ζ loci.

10.4 CLOSED-LOOP TRANSFER FUNCTIONS AND ROOT LOCI

Block Diagram Reduction

Equations (10.11), $C^*(s) = G^*(s)R^*(s)$ or $C(z) = G(z)R(z)$, which were derived from $C(s) = G(s)R^*(s)$, are the basis of block diagram reduction and are used for the examples in Fig. 10.4. In part (a), $G_1(s)$ and $G_2(s)$ are separated by a sampler and each block is represented by (10.11), but in part (b) they are not and the product $G_1(s)G_2(s)$ must be transformed. The notations $G_1G_2^*(s)$ and $G_1G_2(z)$ are used to identify such transforms. In Fig. 10.4(c) it is important to observe that a transfer function $C(z)/R(z)$ cannot be identified. This is because the input is not separated by a sampler from the first block G_1 and therefore the product $G_1(s)R(s)$ must be transformed.

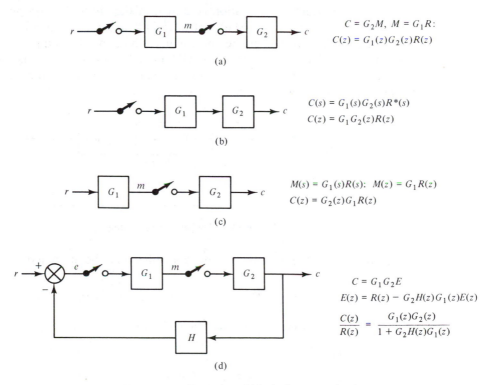

(a)

$$C = G_2M, \quad M = G_1R:$$
$$C(z) = G_1(z)G_2(z)R(z)$$

(b)

$$C(s) = G_1(s)G_2(s)R^*(s)$$
$$C(z) = G_1G_2(z)R(z)$$

(c)

$$M(s) = G_1(s)R(s): \quad M(z) = G_1R(z)$$
$$C(z) = G_2(z)G_1R(z)$$

(d)

$$C = G_1G_2E$$
$$E(z) = R(z) - G_2H(z)G_1(z)E(z)$$
$$\frac{C(z)}{R(z)} = \frac{G_1(z)G_2(z)}{1 + G_2H(z)G_1(z)}$$

Figure 10.4   Examples of block diagram reduction.

Example 10.4.1

In Fig. 10.4(a) and (b) let

$$G_1(s) = \frac{1}{s} \qquad G_2(s) = \frac{a}{s + a}$$

For Fig. 10.4(a),

$$G_1(z) = Z\left[\frac{1}{s}\right] = \frac{z}{z - 1}$$

$$G_2(z) = Z\left[\frac{a}{s + a}\right] = \frac{az}{z - e^{-aT}}, \text{ using (10.13a)}$$

$$G_1(z)G_2(z) = \frac{az^2}{(z - 1)(z - e^{-aT})}$$

For Fig. 10.4(b),

$$G_1G_2(z) = Z\left[\frac{a}{s(s + a)}\right] = \frac{z(1 - e^{-aT})}{(z - 1)(z - e^{-aT})}, \text{ using (10.13b)}$$

Note that

$$G_1(z)G_2(z) \neq G_1G_2(z)$$

A very important configuration is shown in Fig. 10.5. The output sampler is modeled in the error path. The controller D is assumed to incorporate A/D and D/A conversion as well as the control algorithm, so relates input and output pulse sequences. The zero-order hold (ZOH) of (9.17) changes the pulse sequence $u*$ into a staircase continuous input to the plant. $G(s)$ is assumed to include input power amplification and actuator dynamics.

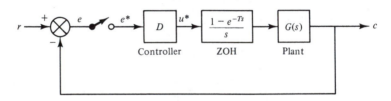

Figure 10.5 Digital control system.

As discussed below (10.9b), the controller can be modeled by $D(z) = D*(s)|_{z = e^{Ts}}$.

$$U(z) = D(z)E(z) \qquad E(z) = R(z) - C(z)$$

$$C(s) = \frac{1}{s}(1 - e^{-Ts})G(s)U*(s)$$

so, as in (10.11),

$$C(z) = G_1(z)U(z)$$

where, since $e^{-Ts} = z^{-1}$,

$$G_1(z) = Z\left[\frac{1}{s}(1 - e^{-Ts})G(s)\right] = (1 - z^{-1})Z\left[\frac{G(s)}{s}\right] \tag{10.15}$$

Combining these equations yields

$$C(z) = G_1(z)D(z)[R(z) - C(z)]$$

or

$$\frac{C(z)}{R(z)} = \frac{G_1(z)D(z)}{1 + G_1(z)D(z)} \tag{10.16}$$

Note that the equation for this configuration is of the same form as for continuous systems, with a loop gain function $G_1(z)D(z)$.

Example 10.4.2

$$G(s) = \frac{a}{s + a}$$

Using (10.15) and (10.13b) yields

$$G_1(z) = (1 - z^{-1})Z\left[\frac{a}{s(s + a)}\right] = \frac{1 - e^{-aT}}{z - e^{-aT}} \tag{10.17}$$

Then, if the PI algorithm of (10.9a) is used, the loop gain function in (10.16) is

$$G_1(z)D(z) = \frac{(1 - e^{-aT})[(K_p + K_iT)z - K_p]}{(z - 1)(z - e^{-aT})}$$

Example 10.4.3

$$G(s) = \frac{a}{s + a} \qquad D(z) = K$$

With this pure gain control, using (10.17), the loop gain function becomes

$$G_1(z)D(z) = \frac{K(1 - e^{-aT})}{z - e^{-aT}} \tag{10.18}$$

and the closed-loop transfer function is

$$\frac{C(z)}{R(z)} = \frac{K(1 - e^{-aT})}{z - e^{-aT} + K(1 - e^{-aT})} \tag{10.19}$$

For $aT = 1$, these equations become ($e^{-1} = 0.3678$)

$$G_1(z)D(z) = \frac{0.6322K}{z - 0.3678}$$

$$\frac{C(z)}{R(z)} = \frac{0.6322K}{z - 0.3678 + 0.6322K} \tag{10.20}$$

The system pole is -1 for $K = 1.3678/0.6322 = 2.164$. In contrast to the continuous system, which is always stable, sampling effects cause instability for $K > 2.164$.

Root Loci in the z-Plane

The system characteristic equation for Fig. 10.5 is, from (10.16),

$$1 + G_1(z)D(z) = 0 \tag{10.21}$$

This equation is of the same form as that for continuous systems in the s-plane. Therefore, root loci can be constructed to determine closed-loop pole positions in the z-plane by exactly the same rules as were used in the s-plane.

Example 10.4.4   Example 10.4.3 for $aT = 1$

Figure 10.6(a) shows the loci constructed from the loop gain function in (10.20).

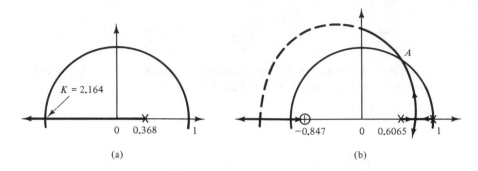

(a)   (b)

Figure 10.6   Root locus Examples 10.4.4 and 10.4.5.

The graphical application of the magnitude condition yields the root locus gain $0.6322K = 1.3678$, or $K = 2.164$ as in Example 10.4.3, where the locus crosses the unit circle. This is the stability limit.

Example 10.4.5

$$G(s) = \frac{0.5}{s(s + 0.5)} \qquad D(z) = K \qquad T = 1$$

Using (10.13c) gives

$$G_1(z) = (1 - z^{-1})Z\left[\frac{0.5}{s^2(s + 0.5)}\right]$$

$$= (1 - z^{-1})\left[\frac{z}{(z - 1)^2} - \frac{2z}{z - 1} + \frac{2z}{z - 0.6065}\right]$$

or

$$G_1(z) = 0.213\frac{z + 0.8474}{(z - 1)(z - 0.6065)} \tag{10.22}$$

$$G_1(z)D(z) = \frac{0.213K(z + 0.8474)}{(z - 1)(z - 0.6065)}$$

Figure 10.6(b) shows a partial root locus sketch. Application of the magnitude condition to the unit circle crossing A yields the stability limit $K = 2.18$. It is again important to observe that the corresponding continuous design is stable for any K. As reflected in the relation $z = e^{sT}$ and discussed further subsequently, a system with sampling generally becomes unstable as K and/or T is increased.

10.5 TRANSIENT RESPONSE OF DIGITAL CONTROL SYSTEMS

Equation (10.16) gives for the transform of the output of Fig. 10.5,

$$C(z) = \frac{G_1(z)D(z)R(z)}{1 + G_1(z)D(z)} \tag{10.23}$$

Inverse Z transformation of $C(z)$ is required to determine the response $c(nT)$ at the sampling instants. Several techniques for this are available:

1. *Long division:* Divide the numerator of $C(z)$ by its denominator to obtain a series expansion in terms of z^{-1}.

2. *Partial fraction expansion:* Probably the most commonly used method.

3. *Difference equation method:* This method leads to a recurrence relation and is very suitable for computer solution. It can also be recommended without use of the computer as often providing the least time-consuming solution for the examples and problems in this chapter.

Example 10.5.1

$$C(z) = \frac{z(z + 0.4)}{(z - 1)(z - 0.3)(z - 0.8)}$$

(a) *Long division method:*

$$C(z) = \frac{z^2 + 0.4z}{z^3 - 2.1z^2 + 1.34z - 0.24} = z^{-1} + 2.5z^{-2} + 3.91z^{-3} + 5.101z^{-4} + \cdots$$

Hence, from Table 10.2.1 or definition (10.3), the sequence of sample values $c_n \equiv c(nT)$ is

$$c_0 = 0, \quad c_1 = 1, \quad c_2 = 2.5, \quad c_3 = 3.91, \quad c_4 = 5.101, \quad \ldots$$

The method is simple and often convenient, for example when method 2 would include complex pairs of poles. But finding c_n for large n is laborious.

(b) *Partial fraction expansion method:*

$$C(z) = z \left[\frac{z + 0.4}{(z - 1)(z - 0.3)(z - 0.8)} \right]$$

$$= z \left[\frac{A}{z - 1} + \frac{B}{z - 0.3} + \frac{C}{z - 0.8} \right]$$

$$A = \frac{z + 0.4}{(z - 0.3)(z - 0.8)} \bigg|_{z=1} = 10 \qquad B = 2 \qquad C = -12$$

Hence, from Table 10.2.1,

$$c_n = 10 + 2(0.3)^n - 12(0.8)^n$$

$$(e^{-anT} = (e^{-aT})^n \qquad \text{where } e^{-aT} \text{ is 0.3 or 0.8})$$

Note that the expansion is made for $C(z)/z$ instead of $C(z)$. Then the terms of $C(z)$ have the forms available in Table 10.2.1. Otherwise, the technique is analogous to that for Laplace transforms. The advantage of the technique is that it gives a closed-form solution, from which c_n can be found directly for any value of n.

(c) *Difference equation method:*

$$\frac{C(z)}{1} = \frac{z^{-1} + 0.4z^{-2}}{1 - 2.1z^{-1} + 1.34z^{-2} - 0.24z^{-3}}$$

Cross-multiplication yields

$$(1 - 2.1z^{-1} + 1.34z^{-2} - 0.24z^{-3})C(z) = z^{-1} + 0.4z^{-2}$$

Using Table 10.2.1, the inverse can be written as

$$c_n = 2.1c_{n-1} - 1.34c_{n-2} + 0.24c_{n-3} + \delta_{n-1} + 0.4\delta_{n-2}$$

Solutions of this recurrence relation, assuming $c_i = 0$ for $i < 0$, are

$$c_0 = 0, \quad c_1 = 0 - 0 + 0 + 1 + 0 = 1, \quad c_2 = 2.1 - 0 + 0 + 0 + 0.4 = 2.5,$$

$$c_3 = 2.1 \times 2.5 - 1.34 \times 1 + 0 + 0 + 0 = 3.91, \quad \ldots$$

These methods will now be used to examine the transient response for the configuration of Fig. 10.5, and the effect of the choice of sampling interval T.

Example 10.5.2

$$G(s) = \frac{a}{s + a} \qquad D(z) = K$$

Unit step input: $R(z) = z/(z - 1)$. From (10.19) of Example 10.4.3,

$$C(z) = z\frac{K(1 - e^{-aT})}{(z - 1)[z - e^{-aT} + K(1 - e^{-aT})]}$$

$$= \frac{K}{K + 1}\left[\frac{z}{z - 1} - \frac{z}{z - e^{-aT} + K(1 - e^{-aT})}\right]$$

by partial fraction expansion, so that

$$c_n = \frac{K}{K + 1}(1 - [e^{-aT} - K(1 - e^{-aT})]^n) \tag{10.24}$$

The unit step response of the continuous system implementation is

$$c(t) = \frac{K}{K + 1}(1 - e^{-(K+1)at})$$

The steady-state value, which is the same for both, could have been found directly from the final value theorem in Table 10.2.1:

$$\lim_{n \to \infty} c_n = \lim_{z \to 1} (1 - z^{-1})C(z) = \frac{K}{K + 1} \tag{10.25}$$

Equation (10.24) gives only the sample values. The modified Z transform and submultiple sampling methods, not discussed here, are available to examine the response between samples.

Figure 10.7 shows the responses as calculated directly from (10.24) for three values of aT and for the values of gain K indicated. The time scales of the three response graphs are the same, to facilitate comparison of response speeds. It is evident that in contrast to the continuous system, the response becomes severely oscillatory as K is increased for constant aT or if T is increased for constant K. For $K = 2$, only the response for $aT = 0.25$ approximates that of the continuous system. The latter has a closed-loop time constant of

$$\frac{1}{(K + 1)a} = \frac{1}{3a}$$

so that at $1/a$ its transient will have largely decayed. For $aT = 0.25$, $T = 1/(4a)$ is somewhat smaller than this time constant, and by Section 9.6 can be satisfactory.

Figure 10.7 shows also the positions of the closed-loop system pole $z = e^{-aT} - K(1 - e^{-aT})$ for all cases, as well as the values of K for which the

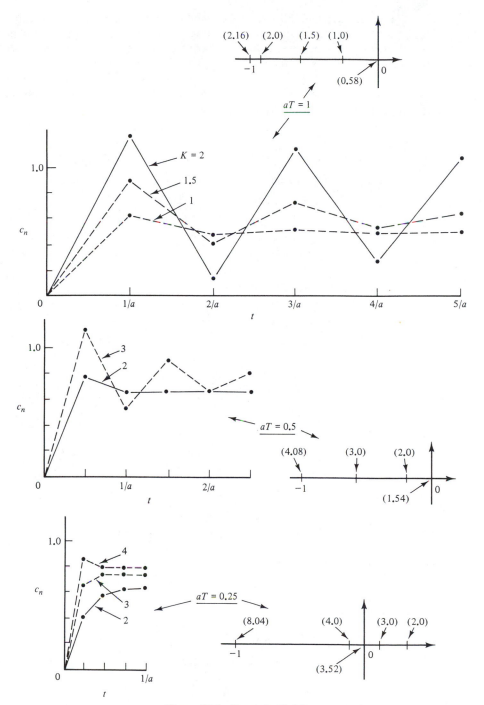

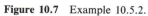

Figure 10.7   Example 10.5.2.

pole crosses the unit circle, the stability limit. Furthermore, the values of K are given for which the pole lies at the origin. Equation (10.24) shows, and interpolation in Fig. 10.7 confirms, that the response then reaches steady state, at the sampling instants, in one sampling interval. This type of response will be discussed later.

    It should also be observed that as indicated in Fig. 10.2, poles on the negative real axis correspond to an oscillatory response with only two samples per cycle.

Example 10.5.3   Transient Response for Example 10.4.5.

    For $D(z) = K = 1$, $G_1(z)D(z)$ can be written as

$$G_1(z)D(z) = \frac{0.213z + 0.1805}{z^2 - 1.6065z + 0.6065}$$

So for a unit step $R(z) = z/(z - 1)$, (10.16) yields

$$C(z) = \frac{0.213z^2 + 0.1805z}{z^3 - 2.3935z^2 + 2.1805z - 0.787} = 0.213z^{-1} + 0.690z^{-2}$$
$$+ 1.188z^{-3} + 1.505z^{-4} + 1.557z^{-5} + 1.378z^{-6} + \cdots$$

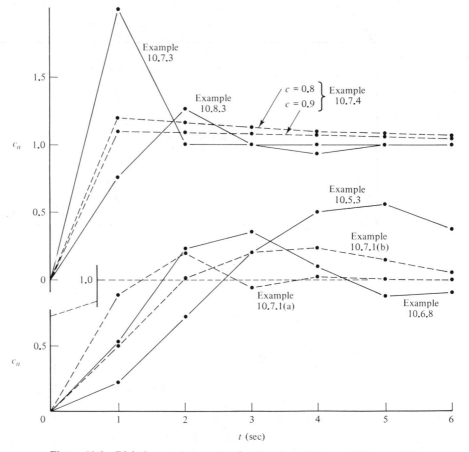

Figure 10.8   Digital control examples for the plant $G(s) = 0.5/[s(s + 0.5)]$.

by long division. So the output sample values are

$$c_0 = 0, \quad c_1 = 0.213, \quad c_2 = 0.690, \quad c_3 = 1.188,$$

$$c_4 = 1.505, \quad c_5 = 1.557, \quad c_6 = 1.378, \quad \dots$$

This sample response is plotted on Fig. 10.8, together with those for a number of alternative designs for the same plant, all for sampling interval $T = 1$ sec, discussed in subsequent examples.

The maximum overshoot is about 56%, compared to about 32% for the corresponding continuous system. The latter is found from Fig. 3.13 for the damping ratio $\zeta = 0.35$ implied by the system characteristic equation $s^2 + 0.5s + 0.5 = 0$. The characteristic equation of the present design is $z^2 - 1.394z + 0.787 = 0$, and its roots $z_{1,2} = 0.697 \pm 0.549j$ correspond to a damping ratio of about 0.2 according to the loci for constant ζ in Fig. 10.3. For a smaller sampling interval the agreement would be better.

10.6 DIGITAL FILTERS BY CONTINUOUS SYSTEM DESIGN

One common approach to the design of digital control systems is to select a compensator $D(s)$ for the equivalent continuous system and approximate it by a digital filter $D(z)$. Implementation as a control algorithm then follows immediately. For example, if

$$D(z) = \frac{U(z)}{E(z)} = \frac{K(1 + az^{-1})}{1 + bz^{-1}}$$

cross-multiplication gives

$$U(z) + bz^{-1}U(z) = KE(z) + Kaz^{-1}E(z)$$

By inversion, this corresponds to the algorithm

$$u_k = -bu_{k-1} + Ke_k + Kae_{k-1}$$

But design verification of both stability and response (e.g., by root loci) is always necessary for the chosen sampling interval.

Several techniques for translating $D(s)$ to $D(z)$ are available. These will now be discussed and illustrated by application to PID control $D(s) = K_p + (K_i/s) + K_d s$ or to the common form of compensation,

$$D(s) = \frac{U(s)}{E(s)} = \frac{K(s + b)}{s + a} \tag{10.26}$$

assumed to have been designed previously.

Finite-difference approximation.   This approach was introduced in Chapter 9 and has already been used a number of times. Equations (9.4) and (9.7) give PID control algorithms derived using backward differences, forward differences, and the trapezoidal rule. In (10.9) some of these are represented by discrete transfer functions $D(z)$. In (9.5) and (9.9), $D(s)$ of (10.26) is represented in finite-difference forms based on backward differences, forward differences, and the trapezoidal rule.

These algorithms can immediately be written as discrete transfer functions. For example, the algorithm (9.5c), rearranged to the form

$$(1 + aT)u_k - u_{k-1} = K(1 + bT)e_k - Ke_{k-1}$$

yields on transformation

$$D(z) = \frac{U(z)}{E(z)} = K\frac{1 + bT - z^{-1}}{1 + aT - z^{-1}} \tag{10.27}$$

From a more fundamental point of view, (10.8) give $D(z)$ as obtained by different algorithms for the integrator

$$D(s) = \frac{1}{s} \tag{10.28a}$$

$$D(z) = \frac{T}{z - 1} \qquad \text{forward rectangular rule} \tag{10.28b}$$

$$D(z) = \frac{Tz}{z - 1} \qquad \text{backward rectangular rule} \tag{10.28c}$$

$$D(z) = \frac{(T/2)(z + 1)}{z - 1} \qquad \text{trapezoidal rule} \tag{10.28d}$$

The inverses of these give the corresponding approximations to a derivative, $D(s) = s$.

One approach to approximating $D(s)$ by $D(z)$ is to replace s by one of these forms. The most common of these is that based on the trapezoidal rule:

$$s = \frac{2}{T}\frac{1 - z^{-1}}{1 + z^{-1}} = \frac{2}{T}\frac{z - 1}{z + 1} \tag{10.29}$$

This is known as the *bilinear transformation,* and is also often referred to as *Tustin's method:*

$$D(z) = D(s)\Big|_{s = (2/T)(z-1)/(z+1)} \tag{10.30}$$

A correction of this result based on frequency response considerations is discussed in the last part of Section 10.8.

Example 10.6.1   PID Algorithms

(a) *PI control $D(s) = K_p + K_i/s$:* With the backward rectangular rule,

$$D(z) = K_p + \frac{K_iTz}{z - 1} \tag{10.31a}$$

This is the same result as derived in (10.9a) by first formulating the algorithm.

(b) *PID control $D(s) = K_p + (K_i/s) + K_ds$:* Using the Tustin approximation for the integral and the backward rectangular approximation for the derivative, this gives

$$D(z) = K_p + \frac{K_iT}{2}\frac{z + 1}{z - 1} + K_d\frac{z - 1}{Tz} \tag{10.31b}$$

which may be verified to be identical to (10.9b).

Example 10.6.2 $D(s)$ **of (10.26)**

Substituting (10.29) gives the following approximation by Tustin's method:

$$D(z) = K\frac{(bT + 2)z + (bT - 2)}{(aT + 2)z + (aT - 2)} \tag{10.32}$$

Example 10.6.3

$$\frac{Y(s)}{X(s)} = \frac{1}{s^2 + as + b}$$

Cross-multiplying gives

$$(s^2 + as + b)Y = X$$

of which the Tustin approximation is

$$\left(\frac{4}{T^2}\frac{1 - 2z^{-1} + z^{-2}}{1 + 2z^{-1} + z^{-2}} + \frac{2a}{T}\frac{1 - z^{-1}}{1 + z^{-1}} + b\right)Y(z) = X(z)$$

This can be arranged as a discrete transfer function $D(z) = Y(z)/X(z)$, or written as a recursive algorithm for y_k.

Impulse-invariant method, or Z-transform method.   In this method, $D(z)$ is chosen to be the Z transform of $D(s)$, so that the unit pulse response sequence will equal the sampled unit impulse response of $D(s)$. However, it should be remembered that the ZOH in Fig. 10.5 produces only a staircase approximation to the continuously variable plant input of a continuous system, so that the closed-loop response may be a poor approximation.

The actual form used is often as follows:

$$D(z) = cZ[D(s)] \tag{10.33}$$

Here c is a constant chosen so that the gain of $D(z)$ will match that of $D(s)$ at a selected frequency, usually $s = 0$. By Fig. 10.2, this corresponds to $z = 1$, so that c is chosen to make

$$D(z)\bigg|_{z=1} = D(s)\bigg|_{s=0} \tag{10.34}$$

Example 10.6.4 $D(s)$ **of (10.26)**

$$Z\left[K\frac{s + b}{s + a}\right] = KZ\left[1 + \frac{b - a}{s + a}\right] = K\left[1 + \frac{b - a}{1 - e^{-aT}z^{-1}}\right]$$

$$= K\frac{1 + b - a - e^{-aT}z^{-1}}{1 - e^{-aT}z^{-1}}$$

To satisfy (10.34) requires $D(z)|_{z=1} = Kb/a$, so that

$$D(z) = K\frac{b}{a}\frac{1 - e^{-aT}}{1 + b - a - e^{-aT}}\frac{1 + b - a - e^{-aT}z^{-1}}{1 - e^{-aT}z^{-1}} \tag{10.35}$$

Zero-order-hold equivalence.   In this method $D(z)$ is considered to be generated by preceding $D(s)$ by a ZOH and following it by a sampler, as indicated in Fig. 10.9. From this, as in (10.15),

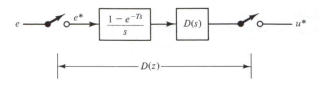

Figure 10.9 Hold equivalence.

$$D(z) = (1 - z^{-1})Z\left[\frac{D(s)}{s}\right] \tag{10.36}$$

Example 10.6.5 *D(s)* of **(10.26)**

$$
\begin{aligned}
D(z) &= K(1 - z^{-1})Z\left[\frac{b/a}{s} + \frac{1 - b/a}{s + a}\right] \\
&= K(1 - z^{-1})\left(\frac{b/a}{1 - z^{-1}} + \frac{1 - b/a}{1 - e^{-aT}z^{-1}}\right) \\
&= K\frac{1 - [1 + (b/a)(e^{-aT} - 1)]z^{-1}}{1 - e^{-aT}z^{-1}}
\end{aligned}
\tag{10.37}
$$

Pole–zero matching.   As the name implies, poles and zeros s_i of $D(s)$ are mapped as poles and zeros of $D(z)$ according to $z_i = e^{s_iT}$, and the gain of $D(z)$ is set to satisfy, say, (10.34).

If $D(s)$ has more poles than zeros, it tends to zero as frequency $\omega \to \infty$. This is simulated by making $D^*(s)$ zero at the limits of the primary strip, which by Fig. 10.2 corresponds to making $D(z)$ zero for $z = -1$. In $D(z)$, therefore, the powers of numerator and denominator are made the same by adding factors $(z + 1)$ in the numerator.

Example 10.6.6 $D(s) = a/(s + a)$

$$D(z) = \frac{0.5(1 - e^{-aT})(z + 1)}{1 - e^{-aT}z^{-1}} \tag{10.38}$$

Example 10.6.7 *D(s)* of **(10.26)**

$$D(z) = K\frac{b}{a}\frac{1 - e^{-aT}}{1 - e^{-bT}}\frac{1 - e^{-bT}z^{-1}}{1 - e^{-aT}z^{-1}} \tag{10.39}$$

These methods are used extensively. A design example follows:

Example 10.6.8 Examples 10.4.5 and 10.5.3 Continued

$$G(s) = \frac{0.5}{s(s + 0.5)}$$

As in (10.22), with $T = 1$,

$$G_1(z) = \frac{0.213(z + 0.847)}{(z - 1)(z - 0.6065)} \tag{10.40}$$

The step response results of Example 5.3 in Fig. 10.8 for $D(s) = D(z) = K = 1$

correspond to $\zeta = 0.35$ for the continuous design. Thus, assuming that accuracy requirements do not permit gain reduction, even the continuous system needs dynamic compensation. Bode plot design leads to the choice

$$D(s) = \frac{6(s + 0.5)}{s + 3} \tag{10.41}$$

The closed-loop damping ratio is then found to be $\zeta = 0.866$. Using pole–zero matching yields

$$D(z) = \frac{1 - e^{-3}}{1 - e^{-0.5}} \frac{z - e^{-0.5}}{z - e^{-3}} = 2.415 \frac{z - 0.6065}{z - 0.0498} \tag{10.42}$$

Hence

$$G_1(z)D(z) = \frac{0.514(z + 0.847)}{(z - 1)(z - 0.0498)}$$

Note that if $D(s)$ cancels a pole of $G(s)$, then $D(z)$ obtained by pole–zero matching does so in $G_1(z)$.

Figure 10.10 shows partial loci for K times $D(z)$ of (10.42), and those in Fig. 10.6(b) for $D(z) = K$ for comparison. The root locations indicated for $K = 1$ are in this case also readily found from the system characteristic equation

$$1 + G_1(z)D(z) = z^2 - 0.536z + 0.486 = 0$$

Using the constant ζ loci in Fig. 10.3, the location corresponds to a damping ratio of only about 0.3, much below the 0.866 of the continuous design. The characteristic equation of the continuous design is $1 + G(s)D(s) = s^2 + 3s + 3 = 0$, so that $\omega_n = \sqrt{3} = 1.732$ and $\zeta = 0.866$. Therefore, the system time constant is $T_c = 1/(\zeta\omega_n) = 0.67$, and the period for one cycle of the transient oscillations is $2\pi/(\omega_n\sqrt{1 - \zeta^2}) = 7.25$ sec. Thus $T = 1$ meets the sampling rate recommendations of Section 9.6 in terms of the frequency of the transient, but not in terms of the time constant, and would have to be reduced for better agreement. Use of a lag network instead of the lead $D(s)$ in (10.41) to improve stability would slow the response and improve the agreement for $T = 1$.

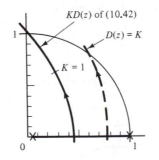

Figure 10.10   Example 10.6.8.

The closed-loop transfer function is

$$\frac{C(z)}{R(z)} = \frac{0.514z + 0.4354}{z^2 - 0.536z + 0.486}$$

and calculating the unit step sample response by the difference equation method, which is faster and easier than long division without a computer, yields

$$c_0 = 0, \quad c_1 = 0.514, \quad c_2 = 1.225, \quad c_3 = 1.356,$$

$$c_4 = 1.081, \quad c_5 = 0.870, \quad c_6 = 0.890, \quad \dots$$

This is plotted in Fig. 10.8. The overshoot is about 40%, compared to about 0% for the continuous design, as the difference in values of ζ suggests.

10.7 DIRECT DESIGN OF DIGITAL FILTERS

One alternative to approximating a continuous design is direct root locus design in the z-plane.

Example 10.7.1   Example 10.6.8 Continued

Digital filters allow simple poles, and zeros, to be placed on the negative as well as the positive real axis in the z-plane. Suppose that a pole of $D(z)$ is chosen at -0.6.

(a) If the zero of $D(z)$ is still chosen to cancel the pole of $G_1(z)$, then, for $D(z)|_{z=1} = 1$,

$$D(z) = \frac{1 + 0.6}{1 - 0.6065} \frac{z - 0.6065}{z + 0.6} = 4.066 \frac{z - 0.6065}{z + 0.6}$$

$$G_1(z)D(z) = \frac{0.866(z + 0.847)}{((z - 1)(z + 0.6))}$$

Figure 10.11 shows partial loci [curve (a)] for K times this loop gain function with the root locations for $K = 1$ identified. The loci of Fig. 10.10 are shown for comparison. The constant ζ loci in Fig. 10.3 indicate an improvement of ζ from 0.3 to almost 0.5. The poles are also closer to the origin, so the time constant is smaller. But the angle of the poles is undesirably close to the direction of the negative real axis, suggesting a low sampling rate and a poor representation of the continuous output by its samples.

The closed-loop transfer function is

$$\frac{C(z)}{R(z)} = \frac{0.866z + 0.7335}{z^2 + 0.466z + 0.1335}$$

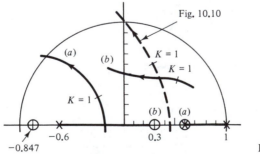

Figure 10.11   Example 10.7.1.

and the difference equation method yields the unit step response sample sequence,

$$c_0 = 0, \quad c_1 = 0.866, \quad c_2 = 1.196, \quad c_3 = 0.927,$$

$$c_4 = 1.008, \quad c_5 = 1.006, \quad c_6 = 0.996, \quad \ldots$$

The plot in Fig. 10.8 confirms the smaller overshoot, about 20%, and smaller settling time, and also the small number of samples per cycle of the transient, suggesting that the samples may be a poor representation of the response.

(b) To force the breakaway point of the complex branches to occur between 0.6065 and 1, the zero of $D(z)$ is moved from 0.6065 to 0.3:

$$D(z) = \frac{1 + 0.6 \; z - 0.3}{1 - 0.3 \; z + 0.6} = 2.286 \frac{z - 0.3}{z + 0.6}$$

Partial loci [curve (b)] are shown in Fig. 10.11. The zero at 0.3 pulls the branches to the left. The roots for $K = 1$ correspond to a damping ratio of almost 0.5, but the settling time for the samples will be larger than for curve (a).

The loop gain function is

$$G_1(z)D(z) = \frac{0.4869(z - 0.3)(z + 0.847)}{(z - 1)(z + 0.6)(z - 0.6065)}$$

and the closed-loop transfer function

$$\frac{C(z)}{R(z)} = \frac{0.4869(z^2 + 0.547z - 0.2541)}{z^3 - 0.5196z^2 - 0.09107z + 0.2402}$$

The difference equation method yields the unit step response sample sequence

$$c_0 = 0, \quad c_1 = 0.487, \quad c_2 = 1.006, \quad c_3 = 1.197,$$

$$c_4 = 1.226, \quad c_5 = 1.134, \quad c_6 = 1.043, \quad \ldots$$

which is plotted in Fig. 10.8. The overshoot is about 25% and the settling time similar to Example 10.6.8, as Fig. 10.11 suggests. The smaller angle of the poles with the positive real axis is reflected by a greater number of samples per cycle than for curve (a).

As illustrated by this example, root locus techniques apply directly to digital control systems. Polar plots and Nyquist diagrams can also be used. For Bode plots, as discussed later, a transformation is necessary. It may be noted, incidentally, that plant transport lag factors $e^{-T_{ds}}$ cause no difficulty, since the transform $z^{-T_d/T}$ represents this delay.

Deadbeat System Design

As an alternative to the use of such continuous system techniques, a direct design approach is discussed, in which $D(z)$ is calculated to make $C(z)/R(z)$ equal to a desired function $T(z)$:

$$T(z) = \frac{C(z)}{R(z)} = \frac{G_1(z)D(z)}{1 + G_1(z)D(z)} \tag{10.43a}$$

$$E(z) = R(z) - C(z) = R(z)[1 - T(z)] \tag{10.43b}$$

Solving (10.43a) for $D(z)$ gives

$$D(z) = \frac{1}{G_1(z)} \frac{T(z)}{1 - T(z)} \qquad (10.44)$$

This approach can be considered because digital control implementation is subject to relatively few constraints.

But, as one would expect, the choice of $T(z)$ to achieve desirable characteristics is constrained by characteristics of $G_1(z)$:

1. Zeros and poles of $G_1(z)$ on or outside the unit circle must be included as zeros of $T(z)$ and $1 - T(z)$, respectively. Their cancellation by $D(z)$ in $G_1(z)D(z)$ would, as in the s-plane, result in unstable closed-loop poles.
2. Zero steady-state error for an input

$$R(z) = \frac{A(z)}{(1 - z^{-1})^m} \qquad (10.45)$$

imposes the necessary condition

$$1 - T(z) = (1 - z^{-1})^m F(z) \qquad T(z) = 1 - (1 - z^{-1})^m F(z) \qquad (10.46)$$

where constraints on the choice of $F(z)$ are discussed below. This follows from (10.43b) and the final value theorem:

$$e_{ss} = \lim_{z \to 1} (1 - z^{-1})E(z) = \lim_{z \to 1} (1 - z^{-1}) \frac{A(z)[1 - T(z)]}{(1 - z^{-1})^m}$$

For a unit step, $A(z) = 1$, $m = 1$, and for a unit ramp $A(z) = Tz^{-1}$, $m = 2$.

3. If $T(z) = t_k z^{-k} + t_{k+1} z^{-k-1} + \cdots$, and if dividing numerator and denominator of $G_1(z)$ gives $G_1(z) = g_n z^{-n} + g_{n+1} z^{-n-1} + \cdots$, then from (10.44),

$$D(z) = \frac{t_k z^{-k} + \cdots}{(g_n z^{-n} + \cdots)(1 - t_k z^{-k} - \cdots)}$$

$$= d_{k-n} z^{-(k-n)} + d_{k-n+1} z^{-(k-n+1)} + \cdots$$

So $k \geq n$ is required for $D(z)$ to be realizable. Therefore, $F(z)$ must be such that the lowest power of z^{-1} in $T(z)$ is at least as high as in $G_1(z)$.

4. Finite settling time requires that $T(z)$ in (10.46) have a finite number of terms. For example, if $T(z) = z^{-1}$, the output sequence equals the input sequence with delay T. If the highest power of z^{-1} in $T(z)$ is p, settling requires p sampling intervals.

   A *deadbeat* design is one that settles to zero error at the sampling instants with p minimum. In (10.46), if $F(z) = 1$, $T(z) = z^{-1}$ for step inputs ($m = 1$) and $T(z) = 2z^{-1} - z^{-2}$ for ramp inputs ($m = 2$), so that the deadbeat response requires one and two sampling intervals, respectively, for $F(z) = 1$. The corresponding compensators are

$$D(z) = \frac{1}{G_1(z)} \frac{z^{-1}}{1 - z^{-1}} \ (m = 1) \qquad D(z) = \frac{1}{G_1(z)} \frac{2z^{-1} - z^{-2}}{(1 - z^{-1})^2} \qquad (10.47)$$

But conditions (1) and (3) may impose a different choice of $F(z)$ than $F(z) = 1$.

Example 10.7.2

$G_1(z)$ has two more poles than zeros, all inside the unit circle. By condition 3, $T(z)$ can at best be z^{-2}. Let

$$T(z) = \alpha_2 z^{-2} + \alpha_3 z^{-3} + \cdots \qquad F(z) = \beta_0 + \beta_1 z^{-1} + \beta_2 z^{-2} + \cdots$$

For zero steady-state error after a step, (10.46) then gives the condition ($m = 1$)

$$1 - (1 - z^{-1})(\beta_0 + \beta_1 z^{-1} + \beta_2 z^{-2} + \cdots) = \alpha_2 z^{-2} + \alpha_3 z^{-3} + \cdots$$

Equating coefficients shows that this equation can be satisfied for $\beta_0 = \beta_1 = \alpha_2 = 1$, $\alpha_3 = \cdots = \beta_2 = \cdots = 0$. Thus $T(z) = z^{-2}$, $F(z) = 1 + z^{-1}$, $D(z) = [1/G_1(z)]z^{-2}/(1 - z^{-2})$.

Example 10.7.3   Examples 10.4.5, 10.5.3, 10.6.8, and 10.7.1

$$G_1(z) = \frac{0.213(z + 0.847)}{(z - 1)(z - 0.6065)} \tag{10.48}$$

From (10.46) for $F(z) = 1$ and $m = 2$, $1 - T(z) = (1 - z^{-1})^2$ and $T(z) = 2z^{-1} - z^{-2}$. Note that $1 - T(z)$ also satisfies condition 1. Since $T(z) = C(z)/R(z)$,

$$C(z) = 2z^{-1}R(z) - z^{-2}R(z) \qquad c_k = 2r_{k-1} - r_{k-2}$$

For $r_k = kT$, $c_0 = 0$, $c_1 = 0$, $c_2 = 2T$, $c_3 = 3T$, ... For $r_k = 1$, $c_0 = 0$, $c_1 = 2$, $c_2 = c_3 = \cdots = 1$. So both have zero error after $2T$, but a severe overshoot exists for a step input, for which the sample sequence is plotted in Fig. 10.8.

Example 10.7.3 demonstrates a common characteristic of deadbeat design. Also, control activity is strong, to achieve zero error in minimum time, and as a result the output tends to oscillate between sampling instants. Furthermore, the method depends on cancellations of plant poles and zeros by $D(z)$. Exact cancellation is not possible in practice, and the response is very sensitive to such errors. A *staleness weighting factor* is commonly introduced to obtain a more satisfactory response. It accepts a lower speed of response in return for more damping and better behavior for inputs other than the design input.

Considering the ramp input case, (10.46) is replaced by

$$1 - T_s(z) = \frac{1 - T(z)}{1 - cz^{-1}} = \frac{(1 - z^{-1})^2 F(z)}{1 - cz^{-1}} \tag{10.49}$$

Letting

$$F(z) = \beta_0 + \beta_1 z^{-1} \qquad T_s(z) = \frac{\alpha_1 z^{-1} + \alpha_2 z^{-2}}{1 - cz^{-1}} \tag{10.50}$$

and equating coefficients, it is found that $\beta_0 = 1$ and that the equation can be satisfied with $\beta_1 = 0$ by choosing $\alpha_1 = 2 - c$, $\alpha_2 = -1$:

$$T_s(z) = \frac{C(z)}{R(z)} = \frac{(2 - c)z^{-1} - z^{-2}}{1 - cz^{-1}} \tag{10.51}$$

Cross-multiplication and inversion yields

$$c_k = cc_{k-1} + (2 - c)r_{k-1} - r_{k-2} \tag{10.52}$$

Ramp input $r_k = kT$: $c_0 = 0$, $c_1 = 0$, $c_2 = (2 - c)T$, $c_3 = [c(2 - c) + (2 - c)2 - 1]T = (3 - c^2)T$, $c_4 = (4 - c^3)T$, ... Step input $r_k = 1$: $c_0 = 0$, $c_1 = 2 - c$, $c_2 = c(2 - c) + (2 - c)1 - 1 = 1 + c - c^2$, $c_3 = 1 + c^2 - c^3$, $c_4 = 1 + c^3 - c^4$, ... Both approach zero error, faster for smaller c $(0 < c < 1)$. But to limit the step response overshoot in $c_1 = 2 - c$, c must be sufficiently large.

Example 10.7.4   Example 10.7.3 with a Staleness Weighting Factor Included

From the response equations for a step input above, Fig. 10.8 shows the response sequences for values $c = 0.9$ and $c = 0.8$ of the staleness weighting factor. These sequences are

$$c = 0.9: \quad c_0 = 0, \quad c_1 = 1.1, \quad c_2 = 1.09, \quad c_3 = 1.081,$$

$$c_4 = 1.073, \quad c_5 = 1.066, \quad c_6 = 1.059, \quad ...$$

$$c = 0.8: \quad c_0 = 0, \quad c_1 = 1.2, \quad c_2 = 1.16, \quad c_3 = 1.128,$$

$$c_4 = 1.102, \quad c_5 = 1.082, \quad c_6 = 1.066, \quad ...$$

Even allowing for considerable oscillations between samples, the overshoot is strongly reduced, but approach to equilibrium is slow (i.e., settling times are large). Using (10.50), the compensator is

$$D(z) = \frac{1}{G_1(z)} \frac{T_s(z)}{1 - T_s(z)} = \frac{(z - 0.6065)[(2 - c)z - 1]}{0.213(z + 0.847)(z - 1)} \tag{10.53}$$

10.8 FREQUENCY RESPONSE
AND THE W AND BILINEAR FORMS

To enable the use of Bode plot techniques, the *w transform* is applied, defined by

$$w = \frac{2}{T} \frac{z - 1}{z + 1} \qquad z = \frac{1 + wT/2}{1 - wT/2} \tag{10.54}$$

For frequency response, $s = j\omega$ and $z = e^{j\omega T}$, so that

$$w = \frac{2}{T} \frac{e^{j\omega T} - 1}{e^{j\omega T} + 1} = j\frac{2}{T} \frac{(e^{j\omega T/2} - e^{-j\omega T/2})/(2j)}{(e^{j\omega T/2} + e^{-j\omega T/2})/2} = j\frac{2}{T} \frac{\sin(\omega T/2)}{\cos(\omega T/2)}$$

So for frequency response w is imaginary:

$$w = jv \qquad v = \frac{2}{T} \tan \frac{\omega T}{2} \tag{10.55}$$

As ω increases from 0 to the limit $\omega_s/2 = \pi/T$ of the primary strip, and z moves along the top half of the unit circle, the new frequency variable v increases from 0 to $+\infty$. When $\omega = \pi/T$, $v = \infty$, and v is a periodic function of ω, with period $\omega_s = 2\pi/T$. Note that $v \approx \omega$ when $\omega \ll 2/T$.

Example 10.8.1   The *w* Transform $G_1(w)$ of (10.48)

$$G_1(z) = \frac{0.213(z + 0.847)}{(z - 1)(z - 0.6065)} \tag{10.56}$$

with $T = 1$. Substituting (10.54) for z gives

$$G_1(w) = \frac{(1 + w/24.144)(1 - w/2)}{w(1 + w/0.49)} \tag{10.57}$$

For frequency response, $w = jv$ is substituted, just as $s = j\omega$ in the Laplace domain. This frequency response function $G_1(jv)$ can be plotted on Bode diagrams with the frequency variable v as will be done in Example 10.8.3, and used for analysis and design in the normal manner. The plots can also be transferred to the Nichols chart. A compensator $D(w)$ can be designed by the usual techniques, and transformed to $D(z)$ by substituting (10.54) for w. $D(z)$ is then implemented as an algorithm.

Example 10.8.2

$$G(s) = \frac{1}{s} \qquad D(z) = K$$

$$G_1(z) = (1 - z^{-1})Z\left[\frac{1}{s^2}\right] = \frac{Tz^{-1}}{1 - z^{-1}} = \frac{T}{z - 1}$$

Substituting (10.54) yields

$$G_1(w) = \frac{1 - wT/2}{w} \qquad D(w) = K$$

Figure 10.12 shows the Bode plot of the loop gain function

$$G_1(jv)D(jv) = \frac{K(1 - jvT/2)}{jv}$$

The negative sign reflects a nonminimum-phase zero, and the phase angle of this factor decreases from $0°$ to $-90°$ as v increases from 0 to ∞. If the crossover frequency $v_c < 2/T$, it must equal K. If $K > 2/T$, the entire plot is above the 0 dB axis; in effect, $v_c \to \infty$. By (10.55) this means that $\omega_c T/2 = \pi/2$, so $\omega_c = \pi/T = \omega_s/2$, the limit of the primary strip, and hence too high. If a phase margin of $55°$ is desired, the numerator may contribute $-35°$ at v_c, so

$$\arctan \frac{v_c}{2/T} = 35° \qquad \text{or} \qquad v_c = \frac{1.4}{T}$$

Substituting this into (10.55) gives the actual crossover frequency $\omega_c = 1.22/T$, and $K = v_c$.

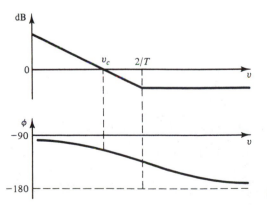

Figure 10.12 Example 10.8.2.

Example 10.8.3   Examples 10.4.5, 10.5.3, 10.6.8, 10.7.1 and 10.8.1

Equation (10.57) gives the w transform of $G_1(z)$:

$$G_1(w) = \frac{(1 + w/24.144)(1 - w/2)}{w(1 + w/0.49)} \qquad (T = 1) \qquad (10.58)$$

The solid lines in Fig. 10.13 show the Bode plot of $G_1(jv)$. The nonminimum-phase zero is encircled as a reminder that the phase angle contributed by this factor is negative. By inspection, the phase margin is inadequate. As before, it is assumed that the low-frequency asymptote may not be lowered, to avoid loss of accuracy. Phase-lag compensation can be used if the associated reduction of the crossover frequency, and speed of response, is acceptable. $T = 1$ then also gives more samples per cycle. Here, phase-lead compensation is used, and Fig. 10.13 shows by dashed lines the loop gain function for

$$D(w) = \frac{1 + w/0.5}{1 + w/5} \qquad (10.59)$$

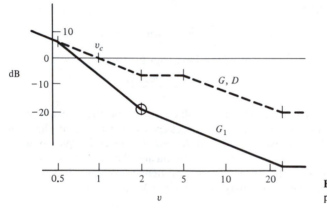

Figure 10.13   Example 10.8.3: Bode plot design.

By (10.55), the crossover frequency $v_c = 1$ corresponds to $\omega_c = 0.927$. The phase angle at $v_c = 1$ is $-125.5°$, so the phase margin $54.5°$. Using (10.54), with $T = 1$, $D(w)$ corresponds to

$$D(z) = \frac{5z - 3}{1.4z + 0.6} = \frac{U(z)}{E(z)}$$

which is equivalent to $1.4U(z) + 0.6z^{-1}U(z) = 5E(z) - 3z^{-1}E(z)$, and implementable as the algorithm

$$u_k = 0.429u_{k-1} + 3.571e_k - 2.143e_{k-1} \qquad (10.60)$$

The loop gain function $G_1(z)D(z)$ is

$$G_1(z)D(z) = \frac{1.065z^2 + 0.263z - 0.5412}{1.4z^3 - 1.6491z^2 - 0.1148z + 0.3639}$$

and by (10.43a), $C(z)$ for a unit step $R(z)$ is

$$C(z) = \frac{1.065z^3 + 0.263z^2 - 0.5412z}{1.4z^4 - 1.9841z^3 + 0.7323z^2 - 0.3255z + 0.1773}$$

Long division yields the sample response in Fig. 10.8:

$$c_0 = 0, \quad c_1 = 0.761, \quad c_2 = 1.266, \quad c_3 = 1.002,$$

$$c_4 = 0.935, \quad c_5 = 0.999, \quad c_6 = 1.0$$

It is noted that the overshoot in digital control systems is in general larger than that of continuous systems for the same phase margin. From Figs. 7.16 and 7.17, it would be about 15% for 55° phase margin.

The *bilinear transformation* was introduced in (10.29) and (10.30). Comparison with (10.54) shows that this amounts to approximating *w* by *s* at all frequencies while it is valid only for $\omega \ll 2/T$. For this reason if $D(z)$ is obtained via a continuous system design $D(s)$ it is usually necessary to *prewarp* the critical frequencies of $D(s)$ to counteract the deviations at higher frequencies. A typical factor $X(j\omega) = j(\omega/\omega_0) + 1$ of $D(s)$ will equal $X(j\omega_0) = j + 1$ at its critical frequency ω_0. It is desired to make $X(z)$ equal to this for $z = e^{j\omega_0 T}$. This is achieved by prewarping $X(s)$ to $X(s) = (s/v_0) + 1$, where v_0 and ω_0 are related by (10.55). For the example (10.26), $D(s) = K(s + b)/(s + a)$:

$$D(s) = \frac{K(b/a)(1 + s/v_b)}{1 + s/v_a} \tag{10.61}$$

The bilinear transformation $s = (2/T)(z - 1)/(z + 1)$ now gives

$$D(z) = K\frac{b}{a}\frac{v_a}{v_b}\frac{(v_b T - 2) + (v_b T + 2)z}{(v_a T - 2) + (v_a T + 2)z} \tag{10.62}$$

Example 10.8.4

$$D(s) = \frac{1}{s^2 + 0.5s + 1} \qquad T = 1$$

The critical frequency $\omega_n = 1$, so from (10.55), $v_n = (2/1) \tan (1/2) = 1.092$ and the prewarped $D(s)$, with the same damping ratio, is

$$D(s) = \frac{v_n^2}{s^2 + 2\zeta v_n s + v_n^2} = \frac{1.1925}{s^2 + 0.546s + 1.1925}$$

and substituting (10.29) yields

$$D(z) = 0.190\frac{(z + 1)^2}{z^2 - 0.893z + 0.652}$$

It should be pointed out, finally, that with the bilinear transformation (10.29) the problem of aliasing or folding will not occur. For $s = j\omega$, (10.29) gives

$$z = \frac{1 + j\omega T/2}{1 - j\omega T/2}$$

As ω increases from 0 to $+\infty$, $|z| = 1$ and the phase angle of z increases from 0° to $+180°$. So the entire $j\omega$ axis, not just the primary strip, is mapped into the 2π circumference of the unit circle. This compression, although it is responsible for the warping, eliminates the folding problem.

10.9 *CONCLUSION*

The preceding two chapters were devoted to digital control systems. Chapter 9 attempted to provide some insight into digital computer control generally, including some discussion of the components commonly present in such systems and system configurations.

Chapter 10 considered the analysis and design of single-loop digital control systems in a manner analogous to the earlier work on continuous systems, and with attention to the effects of sampling on the dynamic behavior.

PROBLEMS

10.1. Obtain digital filter algorithms and the corresponding discrete transfer functions to represent the transfer function

$$\frac{U(s)}{E(s)} = \frac{a}{s + a}$$

based on numerical integration by:
(a) The backward rectangular rule.
(b) The forward rectangular rule.
(c) The trapezoidal rule.

10.2. Obtain digital filter algorithms and the corresponding discrete transfer functions to represent

$$\frac{U(s)}{E(s)} = \frac{a}{s + a}$$

based on finite difference approximations using:
(a) Backward differences.
(b) Forward differences.
(c) Central differences.
Compare the results with those of Problem 10.1.

10.3. Show that backward differences and the backward rectangular rule lead to the same computer algorithm for

$$\frac{U(s)}{E(s)} = \frac{s + 1}{5s + 1}$$

and find the discrete transfer function.

10.4. Obtain discrete transfer functions and the corresponding control algorithms for

$$D(s) = \frac{U(s)}{E(s)} = \frac{s + 1}{5s + 1}$$

(a) By the use of central difference approximations.
(b) Via partial fraction expansion and Z transformation of $D(s)$.

10.5. For $D(s)$ of Problem 10.4:
(a) Find the algorithm and corresponding discrete transfer function based on trapezoidal integration.

(b) Show that the same result is obtained by substituting for s in $D(s)$ the Tustin transformation or bilinear transformation

$$s = \frac{(2/T)(1 - z^{-1})}{1 + z^{-1}} = \frac{(2/T)(z - 1)}{z + 1}$$

10.6. Use the Tustin method, based on trapezoidal integration and consisting of replacing s in $G(s)$ by $(2/T)(1 - z^{-1})/(1 + z^{-1})$, to obtain the Z transfer function corresponding to

$$G(s) = \frac{C(s)}{U(s)} = \frac{K}{s(s + a)}$$

and from it write an algorithm for calculation of the output sequence for a given input sequence.

10.7. Obtain the pulse transfer function corresponding to

$$G(s) = \frac{K}{(s + a)(s + b)} = \frac{C(s)}{U(s)}$$

and from it a recursive algorithm for c_k.

10.8. Obtain the pulse transfer function corresponding to

$$G(s) = \frac{4}{(s + 1)(s + 4)} = \frac{C(s)}{U(s)}$$

and from it a recursive algorithm for c_k.

10.9. Use the Tustin transformation $s = (2/T)(1 - z^{-1})/(1 + z^{-1})$, based on trapezoidal integration, to find the transform $G(z)$ corresponding to $G(s)$ of Problem 10.8, and write the corresponding algorithm.

10.10. Find the Z transfer functions $G(z)$ corresponding to the following transfer functions $G(s)$.

(a) $\dfrac{K}{s + 5}$ **(b)** $\dfrac{K}{s(s + 5)}$ **(c)** $\dfrac{K}{s^2(s + 5)}$

10.11. Find the Z transfer functions $G(z)$ corresponding to the following transfer functions $G(s)$.

(a) $\dfrac{K}{(s + 1)(s + 5)}$ **(b)** $\dfrac{K(s + 3)}{(s + 1)(s + 5)}$ **(c)** $\dfrac{K}{s(s + 1)(s + 5)}$

10.12. Determine the Z transfer functions $G_1(z)$ that result if the transfer functions $G(s)$ in Problems 10.10(a), 10.10(b), and 10.11(a) are preceded by a zero-order hold.

10.13. For satisfactory dynamic modeling of a process it is often adequate to use a transport lag in combination with one or two simple lags, as given by

1. $G(s) = Ke^{-T_{ds}}/(s + 5)$
2. $G(s) = Ke^{-T_{ds}}/[(s + 1)(s + 5)]$

If T_d can be approximated by an integer number of sample intervals, $T_d = mT$, m integer, find the corresponding Z transfer functions $G_1(z)$ if $G(s)$ is preceded by a zero-order hold.

10.14. **(a)** Show that the Z transform of

$$\frac{s + a}{(s + a + jb)(s + a - jb)} = \frac{1}{2}\left(\frac{1}{s + a - jb} + \frac{1}{s + a + jb}\right)$$

equals

$$\frac{z^2 - ze^{-aT} \cos bT}{z^2 - 2ze^{-aT} \cos bT + e^{-2aT}}$$

(b) Show that the Z transform of

$$\frac{Ke^{j\phi}}{s + a - jb} + \frac{Ke^{-j\phi}}{s + a + jb} \quad \text{is} \quad 2K\frac{z^2 \cos \phi - ze^{-aT} \cos (\phi - bT)}{z^2 - 2ze^{-aT} \cos bT + e^{-2aT}}$$

10.15. Represent each of the following computer control algorithms by a discrete transfer function and determine its stability.

(a) $u_n = e_n - 5u_{n-1} - 3u_{n-2}$
(b) $u_n = 3u_{n-1} - 2u_{n-2} + e_{n-1} + e_{n-2}$
(c) $u_n = u_{n-1} - 0.5u_{n-2} + e_n$

10.16. Find the conditions, if any, for which the algorithms in Problems 10.1 and 10.2 may be unstable.

10.17. Repeat Problem 10.3 using forward differences and the forward rectangular rule. Is the algorithm stable?

10.18. Obtain a recursive algorithm and pulse transfer function for $G(s)$ of Problem 10.8 based on backward difference approximation of derivatives. Find the conditions, if any, for which the algorithm may be unstable.

10.19. Determine the stability of systems described by the following Z transfer functions by plotting the pole positions on the z-plane:

(a) $(2z + 6)/(4z^3 - z)$
(b) $(0.2z - 0.14)/(z^2 - 0.4z + 0.4)$

10.20. (a) Use the series definition of the Z transform to show that the Z transforms of the geometric sequences a^n and $(-a)^n$, where a is positive and real, are as follows:

$$a^n: \quad \frac{z}{z - a} = \frac{1}{1 - az^{-1}}$$

$$(-a)^n: \quad \frac{z}{z + a} = \frac{1}{1 + az^{-1}}$$

(b) Use this to interpret the nature of transients corresponding to poles on the positive and negative real axis and at smaller and greater distance to the origin.

10.21. (a) Determine the locus in the s-plane which corresponds to a radial line at angle ϕ radians in the z-plane relative to the positive real axis.

(b) Find the loci in the z-plane for which transient oscillations will be sampled at the rate of, respectively, two, four, six, and eight samples per cycle.

10.22. In Fig. P10.22, if $G_1(z) = (1 - z^{-1})Z[G(s)/s]$ represents the plant and zero-order hold, prove the following results on the steady-state error sequence $e_{ss} = \lim_{k \to \infty} e_k$ for the inputs specified:

1. Unit step: $e_{ss} = 1/(1 + K_p)$; $K_p = \lim_{z \to 1} G_1(z)D(z)$

2. Unit ramp: $e_{ss} = 1/K_v$; $K_v = (1/T) \lim_{z \to 1} (z - 1)G_1(z)D(z)$

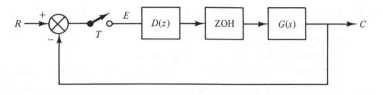

Figure P10.22

These results are analogous to those for continuous systems. K_p and K_v are often called the position and velocity error constants. What are the conditions for zero steady-state errors in (1) and (2)?

10.23. Show that the Z transfer function for the system shown in Fig. P10.23 is

$$G_1(z) = \frac{K}{a} \frac{[T + (1/a)e^{-aT} - 1/a]z + [(1/a) - (1/a)e^{-aT} - Te^{-aT}]}{(z-1)(z - e^{-aT})}$$

$$\xrightarrow{\quad} \fbox{ZOH} \to \fbox{$\dfrac{K}{s(s+a)}$} \to \qquad \textbf{Figure P10.23}$$

10.24. In Fig. P10.22 with $D(z) = 1$ and $G(s) = 1/(s + 1)$:
  (a) Find the loop gain function and the closed-loop Z transfer function.
  (b) Determine the Z transform of the output sample sequence for a unit step input and use the final value theorem to find the steady-state response.
  (c) Plot the locations of the system poles for $T = 0.1$, $T = 0.693$, $T = 1$, and $T = 2$ sec.

10.25. For the position servo model in Fig. P10.22 with $D(z) = K$, $G(s) = 1/[s(s + 1)]$, and $T = 1$:
  (a) Determine the loop gain function $G_1(z)D(z)$.
  (b) Express the closed-loop transfer function $C(z)/R(z)$.
  (c) Find the steady-state value of the output sequence, and hence the steady-state error, for unit step inputs.

10.26. Use the results of Problem 10.14(b) to find the Z transform corresponding to

$$G(s) = \frac{2}{s^2 + 2s + 2}$$

preceded by a zero-order hold.

10.27. For the temperature control system model in Fig. P10.22, with $D(z) = K$, $G(s) = 1/(s + 5)$, and $T = 0.2$:
  (a) Plot the root loci and use them to find the range of K for which the system is stable.
  (b) Calculate and plot the system pole locations for $K = 1$, $K = 2.911$, and $K = 5$.
  (c) For these values of K, find the steady-state values of the output sequence for a unit step input, and hence the steady-state errors.
  (d) Compare with the steady-state error for the corresponding continuous system, and its stability limit on K.

10.28. In Fig. P10.22 with $D(z) = K$, $G(s) = 1/(s + 2)$, and $T = 0.5$:
  (a) Express the loop gain function.
  (b) Plot the root loci.
  (c) Find the limiting value of K for stability.
  (d) Express the steady-state sample error for unit step inputs as a function of K.

10.29. In Problem 10.28, calculate and compare the values of K to achieve a system time constant $\tau = 0.25$:
  (a) By using the value of K which achieves this for the continuous system. Find the z-plane pole location corresponding to this gain and hence indicate the nature of the transient.
  (b) By finding the z-plane pole that corresponds to the s-plane pole for $\tau = 0.25$, and calculating K required to realize this z-plane pole.
  (c) Which of these alternatives is closest to the continuous system behavior?

10.30. For the system of Problem 10.25:
  (a) Sketch the loci of the z-plane poles for varying K. It can be shown that for this type of open-loop pole–zero configuration the loci include a circle centered on the zero.
  (b) Find the limiting value of K for stability.
  (c) Find K corresponding to a damping ratio $\zeta = 0.5$.
      (*Note:* The loci for constant ζ can be used, or $z = e^{Ts}$ with $T = 1$ and $s = -\zeta\omega_n \pm j\omega_n\sqrt{1 - \zeta^2} = -0.5\omega_n \pm 0.866\,j\omega_n$ can be calculated for some values of ω_n, to plot a section of the constant ζ locus where it intersects the root locus.)

10.31. For the positioning servo in Fig. P10.22, with $D(z) = K$, the motor transfer function $G(s) = 1/[s(s + 5)]$ was also considered in Problem 10.12.
  (a) Find the loop gain functions for $T = 0.1$ and $T = 1.0$.
  (b) Sketch the loci for both cases.
  (c) Find for both the limiting value of K for stability.

10.32. In the level control system model in Fig. P10.22, where $D(z) = K$ and $T = 0.2$, the plant $G(s) = 1/[(s + 1)(s + 5)]$ was also considered in Problem 10.12.
  (a) Determine the loop gain function.
  (b) Sketch the root loci.
  (c) Find the limiting value of K for stability.

10.33. In Problem 10.32:
  (a) Find K corresponding to a damping ratio $\zeta = 0.5$ (see the note for Problem 10.30).
  (b) Find the closed-loop transfer function for K found in part (a) and the steady-state error sequence.

10.34. In Problem 10.24:
  (a) From the output transform $C(z)$ in Problem 10.24(b) obtain an algorithm for the output sequence c_k.
  (b) Calculate the sample sequence c_k for the four values of T (for $T = 0.1$ up to 1 sec and for the other values up to 5 sec) and plot these sequences.

10.35. In Problem 10.27:
  (a) Calculate and plot, by partial fraction expansion, the unit step response sequence c_k, $k = 0, 1, 2, 3$, for the three values of K.

(b) Correlate the nature of the transients with the pole positions, and calculate the closed-loop time constants of the corresponding continuous designs, to compare with continuous system response.

10.36. Find the unit step response sequences for the following Z transfer functions by partial fraction expansion.

(a) $\dfrac{0.32}{(z - 0.2)(z - 0.6)}$ (b) $\dfrac{0.32z}{(z - 0.2)(z - 0.6)}$ (c) $\dfrac{0.32}{z(z - 0.2)(z - 0.6)}$

Note that (b) is (a) multiplied by z, and (c) equals (a) multiplied by z^{-1}. Verify the relations between the response sequences which are implied by this.

10.37. Find the unit step response sequence of

$$\frac{C(z)}{R(z)} = \frac{z - 0.5}{z^2 - z + 0.5}$$

by partial fraction expansion. Use graphical calculation of residues as in Section 3.3, together with the results of Problem 10.20(a) and the relations $a + jb = Re^{j\phi}$; $R = \sqrt{(a^2 + b^2)}$, $\phi = \tan^{-1}(b/a)$.

10.38. Verify the result of Problem 10.37 by:
(a) The long-division method.
(b) The difference equation method.
Calculate c_k for $k = 0, 1, 2, 3, 4, 5, 6$ by all three methods. What is the advantage of the partial fraction expansion method if solutions at higher sampling instants are of primary interest?

10.39. Determine the response of the system in Problem 10.37 to the input sequence shown in Fig. P10.39. Which technique appears most convenient for this type of problem? What is the Z transform of the sequence?

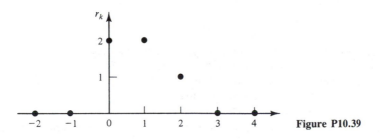

Figure P10.39

10.40. Express the solution of the following difference equations in closed form for $x_k = 1$, $k \geq 0$:
(a) $y_k - 0.7y_{k-1} = 0.3x_k$
(b) $y_k - 0.7y_{k-1} + 0.1y_{k-2} = 0.8x_k - 0.4x_{k-1}$
A closed-form solution allows y_k for a selected value of k to be calculated directly, without first calculating it for all lower k values.

10.41. For Problems 10.24 and 10.34:
(a) What is the closed-loop system time constant without sampling, and hence what is the guide for the maximum sampling interval that might be satisfactory?
(b) Compare the responses with the pole positions in Problem 24(c). The response for $T = 0.693$ is called deadbeat response.

(c) Examine the responses in terms of the s-plane to z-plane correlations of Fig. 10.2.

10.42. Determine the inverse transform f_n of

$$F(z) = \frac{z}{z^2 - 2.5z + 1}$$

(a) By long division.
(b) By partial fraction expansion.
(c) By the difference equation method.

10.43. (a) Repeat Problem 10.29 for a sampling interval $T = 0.1$ instead of $T = 0.5$.
(b) Why do the two approaches agree so much better than in Problem 10.29? With $\tau = 0.25$ desired, what value should serve as an upper limit on the sampling interval selected?
(c) Calculate the unit step response sequence for K as chosen by the equivalent poles approach. Plot c_k for $k = 0, 1, 2, 3, 4, 5$ and show c_∞.

10.44. Calculate the unit step response for the system in Problem 10.25 for $K = 1$. Plot the response c_k, $k = 0 - 7$.

10.45. In Problem 10.33:
(a) Calculate and plot the unit step response c_k, $k \leqslant 8$.
(b) Compare the approximate percentage overshoot with that predicted by Fig. 3.13 for a continuous design with $\zeta = 0.5$.
(c) Use a root locus sketch for a continuous design with $\zeta = 0.5$ to determine its resonant frequency, and hence find the approximate number of samples per cycle to suggest the reason for the result of part (b).

10.46. Determine a digital filter transfer function $D(z)$ to approximate the lead network compensator

$$D(s) = \frac{s + 1}{0.1s + 1}$$

which was designed by continuous system methods, using:
(a) Tustin's method (bilinear transformation).
(b) Impulse-invariant, or Z transform method.
(c) Zero-order-hold equivalence method.
(d) Pole–zero matching method.

10.47. Determine a digital filter transfer function $D(z)$ to approximate the phase-lag compensator

$$D(s) = \frac{0.25s + 1}{s + 1}$$

designed by continuous system methods, using:
(a) Tustin's method (bilinear transformation).
(b) Impulse-invariant, or Z transform method.
(c) Zero-order-hold equivalence method.
(d) Pole–zero matching method.

10.48. Use bilinear transformation with prewarping to find a digital approximation $D(z)$ to $D(s)$ of Problem 10.47 for (a) $T = 0.05$ and (b) $T = 0.5$.

10.49. In Fig. P10.22 with $G(s) = 1/(s + 1)$, PI control $D(s) = 3 + 8/s$ has been chosen by continuous system design, to remove steady-state error and obtain a closed-loop time constant 0.5 sec and closed-loop damping ratio 0.707.

(a) Translate $D(s)$ to $D(z)$ by the Tustin method, for $T = 0.1$.

(b) Compare the resulting closed-loop poles with those obtained by transforming the s-plane poles of the continuous design (i.e., by translating the specifications).

10.50. Repeat Problem 10.49 for $T = 0.5$. Compare T with the desired system time constant.

10.51. In Problem 10.50, with $T = 0.5$, determine $D(z)$ so that the closed-loop poles will correspond to those in the s-plane for a system time constant 0.5 sec and damping ratio 0.707. Compare K_p and K_i with those for the s-plane design ($K_p = 3$, $K_i = 8$).

10.52. In the system of Problem 10.32, PI control is to be used to eliminate the steady-state error. Assuming that $T = 0.2$ and using the Tustin model for integration, design $D(z)$ to cancel one of the plant poles and obtain a system damping ratio of about 0.5. Results and methods in Problems 10.32 and 10.33 will be useful.

10.53. In Fig. P10.22 with $G(s) = 1/[s(s + 1)]$, $D(z) = K(z - z_1)/(z - p_1)$, and $T = 0.5$:

(a) Express the loop gain function.

(b) On a relatively large scale plot, for use in several design problems, construct the upper-half-plane constant ζ locus for $\zeta = 0.5$, or copy it from Fig. 10.3.

(c) Use the result of Problem 10.22 to express the steady-state error for a unit ramp input.

10.54. In Problem 10.53, find the limiting values of K for stability, plot the root loci partially, and find K for a damping ratio of 0.5, for the following three cases:

(a) No compensation ($z_1 = p_1$).

(b) $p_1 = 0.1$ and z_1 cancels a plant pole.

(c) $p_1 = -0.6065$ and z_1 cancels a plant pole.

Also find the steady-state errors following unit ramp inputs for each case when $\zeta = 0.5$.

10.55. Calculate and plot the unit step responses for the three designs with $\zeta = 0.5$ in Problem 10.54 using the difference equation method.

10.56. (a) Use s-plane/z-plane correlations to determine the undamped natural frequencies of continuous systems which would correspond to the angular positions of the z-plane system poles for $\zeta = 0.5$ for the three designs in Problem 10.54.

(b) From these, find the numbers of samples per cycle for $T = 0.5$, and correlate these values with the step responses found in Problem 10.55 to determine for which design $T = 0.5$ may be inadequate.

(c) In a general way, correlate the order of magnitude of the settling times found in Problem 10.55 to the system pole positions.

10.57. In Problem 10.54, to obtain an intersection with the $\zeta = 0.5$ locus at a point corresponding to a higher sampling rate, modify the compensator (c) to

$$D(z) = \frac{K(z - 0.2)}{z + 0.6065}$$

(i.e., the zero no longer cancels the plant pole).

(a) Express the loop gain function as in Problem 10.53(a).

(b) Plot the loci in the general range where the intersection with the $\zeta = 0.5$ locus occurs.

(c) Find K at the intersection with this locus and express the steady-state error following unit ramp inputs.

(d) Plot the unit step response on the graph for Problem 10.55.

10.58. (a) Design a deadbeat compensator for step inputs for the system of Problem 10.53 ($T = 0.5$).

(b) Find the characteristic equation and the system pole(s) and plot the step response.

10.59. (a) Repeat Problem 10.58 for a ramp input.

(b) Calculate and plot the response sequences of the resulting system for unit step and unit ramp inputs. What is the overshoot of the step response sequence?

10.60. Introduce a staleness weighting factor into the ramp input design of Problem 10.59 to limit the overshoot of the step response sequence to 20%. Find the compensator $D(z)$ and plot the unit step response sequence.

10.61. In Fig. P10.22 with $G(s) = 1/(s + 1)$, $D(z) = K$, and $T = 0.1$:

(a) Find K for a phase margin of about 67°.

(b) Recommended sampling frequencies may be 4 to 20 times closed-loop system bandwidth. Without calculating the latter, estimate from the crossover frequency corresponding to part (a) whether $T = 0.1$ is small enough.

(c) Calculate the unit step response sequence, and find the overshoot over steady state. Compare it with the 4.5% overshoot for second-order systems when $\zeta = 0.7$ (Fig. 3.13) to comment further on part (b).

[Note that $G_1(z)$ for this system was found in Problem 10.49.]

10.62. In Fig. P10.22 with $G(s) = 1/[(s + 1)(s + 5)]$, $D(z) = K$, and $T = 0.2$, which was also considered in Problem 10.32:

(a) Determine the w transform of the loop gain function.

(b) Construct the Bode plot and find K for a phase margin of about 67°.

(c) Use the correlations in Table 7.7.1 to find an estimate of the ratio ω_s/ω_b of sampling frequency to closed-loop bandwidth, and hence an indication of whether the sampling rate is adequate.

(d) On the basis of the correlations in Fig. 7.16 and Table 7.7.1, estimate the expected closed-loop resonance peaking M_p.

10.63. Transfer the Bode plot data in Problem 10.62 to the Nichols chart to check the predictions of closed-loop resonance peaking M_p and bandwidth ω_b.

10.64. In Fig. P10.22, with $G(s) = 1/[s(s + 1)]$ and $T = 0.5$ as in Problem 10.53, and with $D(z) = K$:

(a) Find the w transform of the loop gain function.

(b) Using Bode plots, find K for a phase margin of about 67°.

(c) Use Table 7.7.1 to find an estimate of ratio ω_s/ω_b of sampling frequency to closed-loop bandwidth, and hence an indication of whether the sampling rate is satisfactory.

10.65. (a) Repeat Problem 10.64 for 53° phase margin.

(b) Estimate the resonance peaking M_p using Table 7.7.1 and Fig. 7.16, and check peaking and bandwidth estimates on the Nichols chart.

10.66. Prove the following w transforms of common compensators if the bilinear transformation is used to derive the Z transforms:

(a) PI control: $D(s) = K_p + K_i/s$; $D(w) = K_p + K_i/w$.

(b) Phase lag or phase lead:

$$D(s) = \frac{K(s + a)}{s + b} \qquad D(w) = \frac{K(w + a)}{w + b}$$

10.67. PI control must be designed for the system of Problem 10.61 to eliminate the steady-state error ($T = 0.1$).

(a) Express the loop gain function $G_1(w)D(w)$ using results in Problems 10.61 and 10.66.

(b) Using Table 7.7.1, estimate the desired crossover frequency v_c in the w domain if the phase margin is to be about 53° and the ratio of sampling frequency ω_s to closed-loop bandwidth ω_b about 10.

(c) Find the desired ratio K_i/K_p of the controller if the phase margin for v_c in part (b) is to be 53°.

(d) Construct the Bode plot, find the desired gain, and determine K_p and K_i.

(e) Find $D(z)$ from $D(w)$ and hence the control algorithm.

10.68. For the system of Problem 10.62 ($T = 0.2$), and using Problem 10.66, design PI control, that is, find the values of K_p and K_i, to achieve a phase margin of about 53° at a crossover frequency for which an estimate of the ratio of sampling frequency to closed loop bandwidth is about 10. (See Problem 10.67 for a possible sequence of solution.)

10.69. For the system of Problem 10.64 ($T = 0.5$), and using Problem 10.66, design a series compensator $D(z)$ to meet the following specifications:

1. The steady-state ramp input error sequence may not exceed 20%. (Remember that at low frequencies $v \approx \omega$.)

2. The phase margin should not be less than about 53°.

3. The ratio of sampling frequency to estimated closed-loop bandwidth should not be below about 12.

11

NONLINEAR CONTROL SYSTEMS

11.1 INTRODUCTION

As discussed in Section 1.5, most systems are nonlinear for large enough variations about the operating point, and linearization is based on the assumption that these variations are sufficiently small. But this cannot be satisfied, for example, for systems that include relays, which can switch position for very small changes. Startup and shutdown also frequently require the consideration of nonlinear effects, because of the size of the transients.

A differential equation $A\ddot{x} + B\dot{x} + Cx = f(t)$ is nonlinear if one or more of A, B, or C is a function of the dependent variable x or its derivatives. For example, $a\ddot{x} + b\dot{x}^2 + cx^3 = 0$ could represent a spring–mass–damper system of which the damping coefficient $B = b\dot{x}$ depends on $\dot{x}$ and the spring constant $C = cx^2$ on x. Note that A, B, or C of a linear system may be functions of the independent variable t. Such linear time-varying parameter systems have their own, very considerable problems, which are not discussed in this book.

The principle of superposition does not apply to nonlinear systems. Thus, if input x_1 yields output y_1 and x_2 yields y_2, it is no longer true that for an input $(c_1x_1 + c_2x_2)$ the output will be $(c_1y_1 + c_2y_2)$. This has serious consequences. In fact, the analysis and design techniques discussed so far, including the use of transfer functions and Laplace transforms, are no longer valid. Worse, there is no general equivalent technique to replace them. Instead, a number of techniques exist, each of limited purpose and limited applicability. An extensive literature exists on the subject. In this chapter only the well-known phase plane and describing function methods are discussed and an introduction given to the Liapunov, Popov, and circle criteria for the stability of nonlinear systems.

288

11.2 *NONLINEAR BEHAVIOR AND COMMON NONLINEARITIES*

As a minimum, it is important to be aware of the main characteristics of nonlinear behavior, if only to permit recognition if these are encountered experimentally or in system simulations:

1. The nature of the response depends on input and initial conditions. For example, a nonlinear system can change from stable to unstable, or vice versa, if, say, the size of a step input is doubled.

2. Instability shows itself frequently in the form of *limit cycles*. These are oscillations of fixed amplitude and frequency which can be sustained in the feedback loop even if the system input is zero. In linear systems an unstable transient grows theoretically to infinite amplitude, but nonlinear effects limit this growth.

3. The steady-state response to a sinusoidal input can contain harmonics and subharmonics of the input frequency.

4. The *jump phenomenon* is illustrated by the frequency response plot in Fig. 11.1. If the frequency of the input is reduced from high values, the amplitude of the response drops suddenly at the vertical tangent point C to the value at D. Whether or not such jumps occur depends on the size of the input, the degree of peaking, and the nonlinearity.

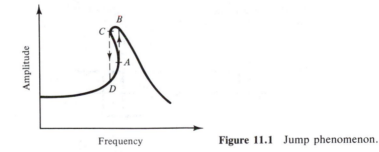

Figure 11.1   Jump phenomenon.

Figure 11.2 shows common types of nonlinearities, with x as input and y as output.

1. *Nonlinear gain:* Very common (e.g., valve flow versus pressure drop or valve opening, or force versus deflection for rubber springs).

2. *Saturation:* The output levels off to a constant limit beyond a certain value of the input. Amplifiers saturate, and valve flow cannot rise beyond pump capacity.

3. *Deadband:* An insensitive zone, for example, in instruments or relays, or due to overlap of the lands on a hydraulic control valve spool over the ports to the cylinder.

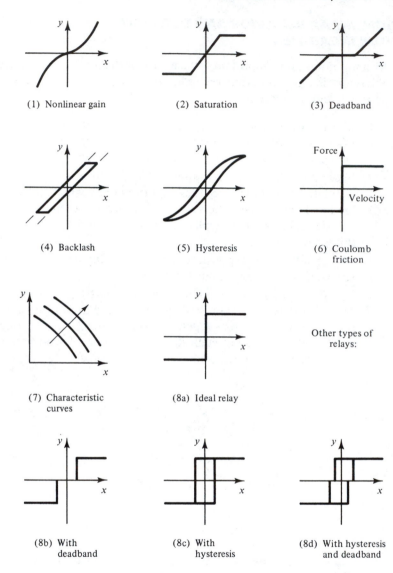

Figure 11.2 Common nonlinearities.

4. *Backlash:* Due to play in mechanical connections.

5. *Hysteresis:* In electromagnetic circuits; in materials.

6. *Coulomb friction, or dry friction:* The friction force depends only on the direction of velocity.

7. *Nonlinear characteristic curves:* The torque–speed curves of motors or the flow–pressure curves of valves.

8. *Relays, with various imperfections:* A very important class of nonlinearities.

11.3 PHASE-PLANE METHOD

The phase-plane method is a graphical method for finding the transient response of first- or second-order systems to initial conditions or simple inputs. Despite these restrictions, it is useful because of the insight it provides and because many systems approximate second-order responses. Consider the nonlinear equation

$$\ddot{x} + g(x, \dot{x})\dot{x} + h(x, \dot{x})x = 0 \qquad (11.1)$$

Substitute into this

$$\dot{x} = y \quad \ddot{x} = \dot{y} = \frac{dy}{dx}\dot{x} = y\frac{dy}{dx} \qquad (11.2)$$

Then the equation reduces to one of first order:

$$y\left(\frac{dy}{dx}\right) + g(x, y)y + h(x, y)x = 0 \qquad (11.3)$$

Rearranging this yields the *phase-plane equation*,

$$\frac{dy}{dx} = \frac{-g(x, y)y - h(x, y)x}{y} \qquad (11.4)$$

The phase plane is a plot of y versus x as indicated in Fig. 11.3. At each point (x, y), dy/dx is the slope of the *phase-plane trajectory* through that point.

   Isoclines are loci of constant trajectory slope. The *isocline equation* for $dy/dx = m$ is

$$y = \frac{-h(x, y)x}{g(x, y) + m} \qquad (11.5)$$

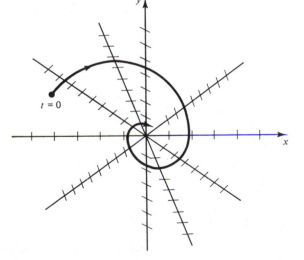

Figure 11.3 Phase plane; focus.

Example 11.3.1

$$\ddot{x} + 2\zeta\omega_n\dot{x} + \omega_n^2 x = 0 \qquad (11.6)$$

Phase-plane equation: $\dfrac{dy}{dx} = \dfrac{-(2\zeta\omega_n y + \omega_n^2 x)}{y}$

$$(11.7)$$

Isocline equation: $y = \dfrac{-\omega_n^2 x}{2\zeta\omega_n + m}$

The isoclines are straight lines through the origin, shown in Fig. 11.3 for several values of m. On each isocline, short line segments are drawn at the corresponding slope m. On $y = 0$, dy/dx is infinite, and $dy/dx = 0$ on the isocline $y = -\omega_n x/(2\zeta)$. On the y-axis, $dy/dx = -2\zeta\omega_n$. The initial condition $(x, \dot{x})(0)$ is represented by a point on the plane, and the corresponding trajectory is drawn by following the slope segments. The origin is called a *focus*.

Above the x-axis, $y = \dot{x} > 0$, so x increases. Hence the direction of motion along trajectories must be to the right above the x-axis, and to the left below it.

This graphical technique is called the *isocline method*, and is useful to sketch the nature of the phase plane portrait, as in the examples below. For numerical work, this and alternative graphical techniques have been largely replaced by computer methods.

Example 11.3.2   Overdamped System, $\zeta > 1$

Figure 11.3 represents an underdamped system, because x oscillates between positive and negative values. For the case $\zeta > 1$, consider when the slope of the isocline equals that of the trajectory (i.e., when the isocline is $y = mx$). Substituting this into the second of (11.7) yields a quadratic equation for m of which the roots are

$$m_{1,2} = \omega_n(-\zeta \pm \sqrt{\zeta^2 - 1}) \qquad (11.8)$$

For $\zeta > 1$ this gives two real slopes, both negative, as indicated in Fig. 11.4. These

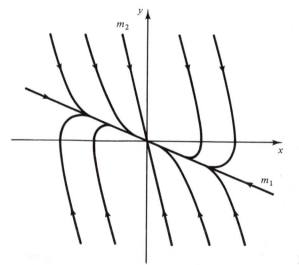

Figure 11.4   Example 11.3.2: node.

isoclines satisfy the differential equation, so are also possible trajectories. They cannot be crossed, and trajectories that approach them follow these straight lines into the origin, which is now called a *node*.

Example 11.3.3   Servo with Piecewise Linear Gain

Figure 11.5(a) shows a simple model of a motor plus load (inertia J, damping B) in a position servo of which the gain of the amplifier varies with system error $E = \theta_i - \theta_o$ as shown in Fig. 11.5(b). For a step input $\dot{E} = -\dot{\theta}_o$, $\ddot{E} = -\ddot{\theta}_o$, and the system equation $J\ddot{\theta}_o + B\dot{\theta}_o = K(E)$ is written as follows in terms of E:

$$J\ddot{E} + B\dot{E} + K(E) = 0 \tag{11.9}$$

As in (11.1) to (11.4), this yields the phase-plane equation

$$\frac{d\dot{E}}{dE} = \frac{-(B\dot{E} + K(E))}{J\dot{E}} \tag{11.10}$$

Assuming that for the lower gain K_1 the system (11.9) is overdamped, the phase-plane portrait for $-E_o \le E \le E_o$ will be of the form in Fig. 11.4, as shown in Fig. 11.5. If for gain K_2 the system is underdamped, the trajectories for $-E_o > E > E_o$ must be of the spiral form in Fig. 11.3. The foci of these spirals are the points $-E_1$ and E_1 identified in Fig. 11.5(b).

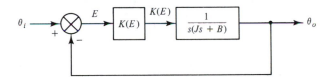

(a)

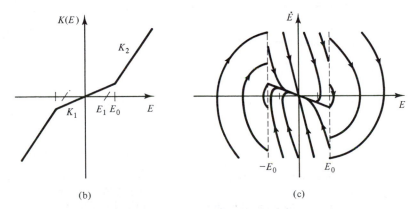

(b)

(c)

Figure 11.5   Example 11.3.3: piecewise linear gain.

Example 11.3.4   Relay Servo with a Ramp Input

Relay control of motors is a low-cost form of control which in many applications can provide adequate servo performance. In Fig. 11.6(a), the relay is assumed to

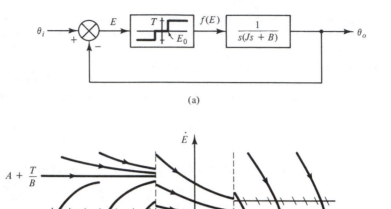

(a)

(b)

Figure 11.6   Example 11.3.4: relay servo.

have a deadband, and the input is a ramp $\theta_i = At$. With $E = \theta_i - \theta_o$, $\dot{E} = A - \dot{\theta}_o$, $\ddot{E} = -\ddot{\theta}_o$, and the system equation $J\ddot{\theta}_o + B\dot{\theta}_o = f(E)$ becomes

$$JË + BÈ + f(E) = BA \qquad f(E) = 0 \qquad \text{for} \;\; -E_0 < E < E_0,$$
$$+ T \qquad \text{for} \;\; E > E_0, \text{and} -T \qquad \text{for} \;\; E < -E_0$$

The phase-plane equation is

$$\frac{d\dot{E}}{dE} = \frac{-B\dot{E} - f(E) + BA}{J\dot{E}} = \frac{-B(\dot{E} - A)}{J\dot{E}} \qquad -E_0 < E < E_0$$

$$\frac{-B(\dot{E} - A + T/B)}{J\dot{E}} \qquad E > E_0 \qquad\qquad (11.11)$$

$$\frac{-B(\dot{E} - A - T/B)}{J\dot{E}} \qquad E < -E_0$$

It is seen that in each range $d\dot{E}/dE$ depends only on $\dot{E}$, so the isoclines are horizontal lines. For any chosen value of $\dot{E}$ the slope $d\dot{E}/dE$ can be calculated from (11.11), and line segments at this slope drawn on the isocline, as shown in Fig. 11.6(b) for an example in each range.

     In this manner the phase-plane portrait can be constructed. Note in particular the isoclines identified in Fig. 11.6(b) on which the slope $d\dot{E}/dE$ is zero, so the trajectory horizontal. These are velocity limits, on which the available torque equals that needed to overcome friction, so that none is available to change $\dot{E}$. The portrait shows that the servo cannot follow a ramp input for which $A \geqslant T/B$. The velocity limit for $E > E_0$ then lies on or above the E-axis, so that $\dot{E} \geqslant 0$ and motion cannot be in the direction of decreasing error E. Also verify that, for $A < T/B$, the final steady-state error must equal half the deadband region.

In the case of an ideal relay, the portrait is as in Fig. 11.6(b) with the central region reduced to zero width. For an ideal relay with a step input, one may verify that the pattern is that in Fig. 11.6(b) for $A = 0$. It is noted finally that relay servos are also called *on-off servos* or *bang-bang servos*.

11.4 DESCRIBING FUNCTION

The describing function technique is a frequency response method, and its main use is in stability analysis (i.e., the prediction of limit cycles). In Fig. 11.7, where G_1 and G_2 represent linear parts of the system and N a nonlinear element, the question is whether a limit cycle exists, that is, whether an oscillation can maintain itself around the loop for $R = 0$.

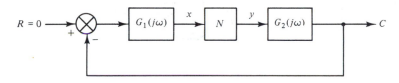

Figure 11.7   System configuration.

Limit cycles for second-order systems can also be constructed by phase-plane methods. As illustrated in Fig. 11.8, they are represented by closed curves in the phase plane. But limit cycles are distinguished from other possible closed curves in that the phase-plane trajectories tend toward or away from them asymptotically. A stable limit cycle is one that is approached by trajectories from both sides. Even the slightest disturbance causes trajectories to depart from unstable limit cycles. As indicated in the first of Fig. 11.8, an initial condition or input outside the limit cycle leads to an unstable transient growth while transients

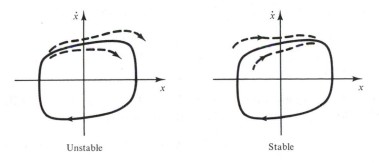

Unstable   Stable

Figure 11.8   Limit cycles.

following a sufficiently small disturbance decay to zero. Thus it is also necessary to determine the type of limit cycle.

The model for N in Fig. 11.7 used in this analysis is based on the following assumption:

The input x to the nonlinearity is sinusoidal:

$$x = A \sin \omega t \qquad (11.12)$$

There is an apparent contradiction here. In Fig. 11.9 the square-wave output of an ideal relay for a sinusoidal input is a periodic function, so that it can be represented by a *Fourier series* of the general form

$$y(t) = b_0 + \sum_{n=1}^{\infty} (a_n \sin n\omega t + b_n \cos n\omega t) \qquad (11.13)$$

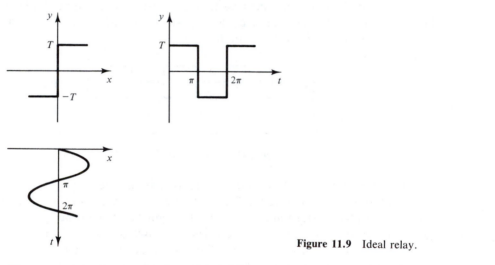

Figure 11.9   Ideal relay.

Thus y contains harmonics ($n > 1$) in addition to the fundamental Fourier component

$$y_f = a_1 \sin \omega t + b_1 \cos \omega t \qquad (11.14)$$

The harmonics would pass around the loop via G_2 and G_1 back to x, contradicting the assumption (11.12). But for most practical systems:

1. y_f is considerably larger than the harmonics.
2. G_2 acts as a low-pass filter which attenuates the harmonics much more strongly than y_f.

The combination of these effects usually justifies (11.12), and also implies that only y_f is required. Hence the definition:

The *describing function* (DF) N of a nonlinearity is the ratio $N = y_f/x$.   (11.15)

From the theory of Fourier series, the *Fourier coefficients* a_1 and b_1 for y_f in

(11.14) are

$$a_1 = \frac{\omega}{\pi} \int_0^{2\pi/\omega} y \sin \omega t \, dt \qquad b_1 = \frac{\omega}{\pi} \int_0^{2\pi/\omega} y \cos \omega t \, dt \qquad (11.16)$$

If y can be extended into an odd function of time, as in Fig. 11.9, then $b_1 = 0$ since $\cos \omega t$ is an even function of time, and the DF becomes

$$N = \frac{a_1}{A} \qquad (11.17)$$

The DF is an equivalent linear gain which depends on the amplitude A, and sometimes also the frequency ω, of the input x.

Example 11.4.1   DF of an Ideal Relay (Fig. 11.9)

Since the DF is independent of ω, $\omega = 1$ can be assumed for simplicity, so that $x = A \sin t$.

$$a_1 = \frac{1}{\pi} \int_0^{2\pi} y \sin t \, dt = \frac{2}{\pi} \int_0^\pi T \sin t \, dt = \frac{4T}{\pi} \qquad (11.18)$$

So the describing function is

$$N = \frac{4T}{\pi A} \qquad (11.19)$$

As expected, this shows that the equivalent gain decreases as the input A increases, since the output is constant.

Example 11.4.2   Relay with Deadband (Fig. 11.10)

$$a_1 = \frac{1}{\pi} \int_0^{2\pi} y \sin t \, dt = \frac{2}{\pi} \int_\alpha^{\pi-\alpha} T \sin t \, dt$$

$$= \frac{2T}{\pi} (-\cos t) \Big|_\alpha^{\pi-\alpha} = \frac{4T}{\pi} \cos \alpha \qquad (11.20)$$

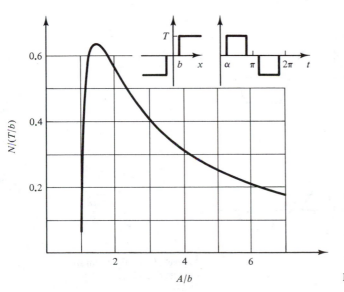

Figure 11.10 Relay with deadband.

Here α is the value of t at which x equals b, that is,

$$\alpha = \sin^{-1}\frac{b}{A} \quad \text{or} \quad A\sin\alpha = b \tag{11.21}$$

Since $\cos\alpha = \sqrt{1 - (b/A)^2}$, it follows that for $A > b$ ($N = 0$ for $A \leq b$),

$$N = \frac{4T}{\pi A}\sqrt{1 - \left(\frac{b}{A}\right)^2} \tag{11.22}$$

The plot in Fig. 11.10 is as expected, with $N = 0$ for $A < b$, N rising to a maximum close to $A = b$, where the equivalent gain is largest, and then decreasing to zero as A increases while the output is constant.

Example 11.4.3   Saturation [Fig. 11.11(a)]

$$a_1 = \frac{2}{\pi}\int_0^\pi y\sin t\, dt = \frac{4}{\pi}\int_0^{\pi/2} y\sin t\, dt$$

$$= \frac{4}{\pi}\left[\int_0^\alpha KA\sin^2 t\, dt + \int_\alpha^{\pi/2}(Kb)\sin t\, dt\right]$$

$$= \frac{4KA}{\pi}\left(\frac{\alpha}{2} - \frac{\sin 2\alpha}{4}\right) + \frac{4Kb}{\pi}\cos\alpha$$

Saturation is reached when $\sin\alpha = b/A$, so

$$\cos\alpha = \sqrt{1 - \left(\frac{b}{A}\right)^2} \qquad \sin 2\alpha = \frac{2b}{A}\sqrt{1 - \left(\frac{b}{A}\right)^2}$$

Substitution yields, for $A > b$ ($N = K$ for $A \leq b$),

$$N = \frac{2K}{\pi}\left[\sin^{-1}\frac{b}{A} + \frac{b}{A}\sqrt{1 - \left(\frac{b}{A}\right)^2}\right] \tag{11.23}$$

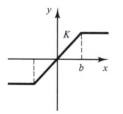
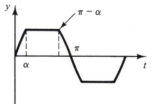

(a)

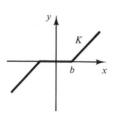
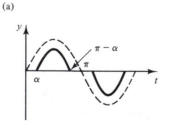

(b)

Figure 11.11   (a) Saturation; (b) Deadband.

This is plotted in Fig. 11.12. As Fig. 11.11(a) suggests, the equivalent gain decreases as A increases beyond the saturation limit.

Example 11.4.4   Deadband [Fig. 11.11(b)]

Instead of deriving N as above, one may prove and use the property that the DF of a sum of nonlinear functions equals the sum of the individual DFs. From this it may be verified that the DF equals K minus the DF (11.23) for saturation. This describing function is plotted in Fig. 11.12. For large A the equivalent gain approaches K.

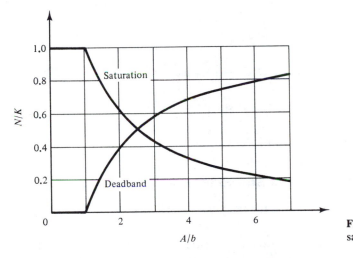

Figure 11.12 Describing functions for saturation and deadband.

Extensive tables and plots of DFs are given in books on nonlinear control systems. These include DFs that depend on frequency, and DFs where y_f and x are not in phase, as in the following example.

Example 11.4.5   Relay with Hysteresis and Deadband

In Fig. 11.13, half the width of each square pulse is

$$\alpha = 0.5\left(\pi - \sin^{-1}\frac{c}{A} - \sin^{-1}\frac{b}{A}\right)$$

$$\beta = \alpha + \sin^{-1}\frac{c}{A}$$

$$\phi = \beta - \frac{\pi}{2}$$

Hence y_f will lag x by a phase shift $-\phi$, and N is no longer real.

$$-\phi = -0.5\left(\sin^{-1}\frac{c}{A} - \sin^{-1}\frac{b}{A}\right) \tag{11.24}$$

To find the amplitude of y_f, it is noted from Example 11.4.2 that if the width of each square pulse is $(\pi - 2\alpha_1)$, then the DF is $N = (4T/\pi A)\cos\alpha_1$. If $2\alpha = \pi - 2\alpha_1$, then $\cos\alpha_1 = \cos[(\pi/2) - \alpha] = \sin\alpha$, so that the DF becomes

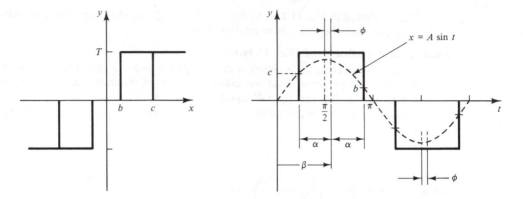

Figure 11.13   Relay with hysteresis and deadband.

$$N = \frac{4T \sin \alpha}{\pi A} e^{-j\phi} \tag{11.25}$$

where $e^{-j\phi}$ shows the phase lag of y_f relative to x.

11.5 STABILITY ANALYSIS USING DESCRIBING FUNCTIONS

Since N in Fig. 11.7 is an equivalent linear gain, and $E = R - C = -C$ for $R = 0$,

$$C = G_2 N G_1 E = -GNC \qquad C(GN + 1) = 0 \tag{11.26}$$

where

$$G = G_1 G_2 = \text{product of linear elements in loop} \tag{11.27}$$

From (11.26), the condition for a nonzero solution for C (i.e., a limit cycle) is

$$GN + 1 = 0 \qquad G = \frac{-1}{N} \tag{11.28}$$

Graphical interpretation.   Limit cycles are identified by the intersections of the polar plot of $G(j\omega)$ and a plot of $-1/N$.

Example 11.5.1   Relay-Controlled Servo (Fig. 11.14)

Physically, there must be a limit cycle since the relay must switch back and forth near $E = 0$. $-1/N = -\pi A/(4T)$, so $-1/N$ for increasing A lies along the entire negative real axis. From Chapter 7, the plot of this $G(j\omega)$ is the form shown, and therefore an intersection, and a limit cycle, will exist.

The frequency of the limit cycle is that corresponding to the intersection along the $G(j\omega)$ curve. Its amplitude A at the input to the nonlinearity can be found from the $-1/N$ plot at the intersection.

To determine the stability of limit cycles, it is recalled that a linear system

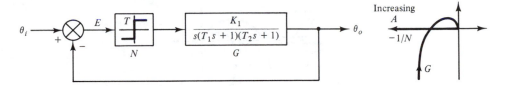

Figure 11.14 Relay-controlled servo.

with loop gain function KG is on the verge of instability when the polar plot of KG passes through the -1 point, i.e., when the polar plot of G passes through the critical point $-1/K$. Analogously, for nonlinear systems the critical point is $-1/N$, and the plot of $-1/N$ shows how in this case the point changes with conditions at the input to the nonlinearity. Use is now made of the Nyquist criterion for open-loop stable systems that the system is stable if the critical point $-1/N$ lies to the left of the polar plot of G.

Suppose that a system is in a limit cycle at amplitude A and frequency corresponding to the intersection of G and $-1/N$. Now let due to some disturbance A increase slightly, moving the critical point slightly away from the intersection along $-1/N$. If this brings it to the right of G, the system is unstable, causing A to increase, farther away from the limit cycle. If A decreased slightly from the value at the intersection, it would then be to the left of G. This indicates stability and continuing decrease of A. Thus disturbances in either direction from the limit cycle would cause trajectories to depart farther from it, showing it to be unstable.

Example 11.5.1 (continued)

From Fig. 11.14, the frequency of the limit cycle can be found from the condition that the phase angle of $G(j\omega)$ is $-180°$. G is the same as equation (7.15) of Example 7.5.2, where this frequency was found to be $\omega = 1/\sqrt{T_1 T_2}$, and $G = -K_1 T_1 T_2/(T_1 + T_2)$. The amplitude A of the limit cycle is given by $G = -1/N$, so by

$$\frac{-K_1 T_1 T_2}{T_1 + T_2} = -\frac{\pi A}{4T} \quad \text{or} \quad A = \frac{4TK_1 T_1 T_2}{\pi(T_1 + T_2)}$$

It is also readily verified that the intersection in Fig. 11.14 represents a stable limit cycle. An increase of A brings the critical point to the left of G, so causes A to decrease back to the intersection. A further decrease would cause instability and a return to the intersection.

Example 11.5.2 Relay-Controlled Servo with Deadband

In Example 11.5.1 let the ideal relay be replaced by one with deadband, of which Fig. 11.10 gives the DF. $N = 0$ for $A \leqslant b$, rises to a maximum and then decreases back to zero with increasing A. Thus $-1/N$, real and negative, changes as indicated in Fig. 11.15 with increasing A. There are two intersections, P and Q, so two limit cycles. It may be verified that P is unstable and Q stable. Thus the system is stable for an initial condition or disturbance for which A does not reach P, but goes into a limit cyle at Q if excited sufficiently severely, beyond P.

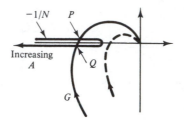

Figure 11.15   Example 11.5.2.

If G is as in Example 11.5.1 with $K_1 = 1$, $T_1 = 1$, $T_2 = 0.5$, the frequency of any limit cycle is $\omega = 1/\sqrt{T_1 T_2} = 1.414$ and its amplitude must satisfy $-1/N = G = -1/3$, so $N = 3$. Using Fig. 11.10, if $T = 1$, $b = 0.1$, then $N/(T/b) = 0.3$ and two limit cycles can exist, satisfying $A/b = 1.04$ and $A/b = 4.1$. Hence the unstable limit cycle at P has an amplitude 0.1 and the stable limit cycle at Q an amplitude of 0.41.

If the gain of G is reduced, as shown by the dashed polar plot, there is no intersection, so no limit cycle, and the system is stable.

System design should evidently aim to avoid intersections between G and $-1/N$. Phase-lead or phase-lag compensation can serve to modify G in order to achieve this.

11.6 SECOND, OR DIRECT, METHOD OF LIAPUNOV

Liapunov's method for stability analysis is in principle very general and powerful. The major drawback, which seriously limits its use in practice, is the difficulty often associated with construction of the *Liapunov function* or *V-function* required by the method.

The system dynamics must be described by a state space model, discussed in detail in Chapter 12. It is a description in terms of a set of first-order differential equations. For example, a nonlinear system might be described by a set of n first-order nonlinear differential equations

$$\dot{x}_i = f_i (x_1, x_2, \ldots, x_n, t) \qquad i = 1, \ldots, n \qquad (11.29)$$

Referring to Section 12.2 or to Appendix A, this can be written compactly in the form of a *state space model* as

$$\dot{\mathbf{x}} = \mathbf{f}(\mathbf{x}, t) \qquad (11.30a)$$

where

$$\mathbf{x} = \begin{bmatrix} x_1 \\ \cdot \\ \cdot \\ \cdot \\ x_n \end{bmatrix} \qquad \dot{\mathbf{x}} = \begin{bmatrix} \dot{x}_1 \\ \cdot \\ \cdot \\ \cdot \\ \dot{x}_n \end{bmatrix} \qquad \mathbf{f}(\mathbf{x}, t) = \begin{bmatrix} f_1 (x_1, \cdots, x_n, t) \\ \cdot \\ \cdot \\ \cdot \\ f_n (x_1, \cdots, x_n, t) \end{bmatrix} \qquad (11.30b)$$

The vector **x** is the *state vector*, and its elements are *state variables*. The origin **x** = **0** (x_1 = ... = x_n = 0) of the state space will be assumed to be an equilibrium solution, where f_i = 0, i = 1, ... , n. The phase plane in Section 11.3 is in effect a state space for second-order systems. It should be emphasized that the state-model description of a given system is not unique but depends on which variables are chosen as state variables. Section 12.2 can be consulted for a detailed discussion and for examples.

The Liapunov function, $V(x_1, ..., x_n)$, is a scalar function of the state variables. To motivate the following and to make the stability theorems plausible, let V be selected to be

$$V(\mathbf{x}) = \| \mathbf{x} \|^2 = \sum_{i=1}^{n} x_i^2 \qquad (11.31)$$

Here $\| \mathbf{x} \|$ is the Euclidian norm of **x**, the length of the vector **x**, and the distance to the origin of the state space. V is evidently positive and $V(\mathbf{0})$ = 0. Now let

$$\dot{V} = \frac{dV}{dt} = \frac{\partial V}{\partial x_1} \dot{x}_1 + \cdots + \frac{\partial V}{\partial x_n} \dot{x}_n \qquad (11.32)$$

be calculated by substituting (11.29). If $\dot{V}$ were to be found to be always negative, with $\dot{V}(\mathbf{0})$ = 0, then apparently V decreases continuously, and the state must end up in the origin of the state space, implying asymptotic stability.

It could be that $\dot{V}$ is only negative in a small enough region around the origin. Hence the following distinctions are appropriate:

1. A system is *globally asymptotically stable* if it returns to **x** = **0** after any size disturbance.
2. It is *locally asymptotically stable* if it does so after a sufficiently small disturbance.
3. It is *stable* if for a given size disturbance the solution stays inside a certain region.

To develop these concepts, the following definitions are used for the sign of V (and $\dot{V}$):

1. V is *positive (negative) definite* in a region containing **x** = **0** if it is positive (negative) everywhere except that $V(\mathbf{0})$ = 0.
2. V is *positive (negative) semidefinite* if it has uniform positive (negative) sign but is zero also at points other than the origin.
3. V is *indefinite* if both signs occur in the region.

For example, $V(x_1, x_2) = x_1^2 + x_2^2$ is positive definite, $V(x_1, x_2) = x_1 + x_2$ is indefinite, and $V(x_1, x_2, x_3) = -x_1^2 - x_2^2$ is negative semidefinite, because it is zero along the x_3-axis.

Sylvester's theorem is used to find such properties for a general quadratic form

$$Q = \sum_{i=1}^{n} \sum_{j=1}^{n} a_{ij} x_i x_j \qquad (a_{ij} = a_{ji}) \tag{11.33}$$

From Appendix A, this can also be written as

$$Q = \mathbf{x}'\mathbf{A}\mathbf{x}$$

$$\mathbf{x}' = [x_1 \quad \cdots \quad x_n]$$

$$\mathbf{A} = \begin{bmatrix} a_{11} & a_{12} & \cdots & a_{1n} \\ a_{12} & a_{22} & & \cdot \\ \cdot & & & \cdot \\ \cdot & & & \cdot \\ \cdot & & & \cdot \\ a_{1n} & \cdots\cdots & & a_{nn} \end{bmatrix} \tag{11.34}$$

Here $\mathbf{x}'$ is the transpose of $\mathbf{x}$, and $\mathbf{A}$ is a symmetrical matrix.

Sylvester's theorem. Q is positive definite if and only if all principal minors of the determinant $|\mathbf{A}|$ are larger than zero:

$$a_{11} > 0, \qquad \begin{vmatrix} a_{11} & a_{12} \\ a_{12} & a_{22} \end{vmatrix} > 0, \quad \ldots, \quad |\mathbf{A}| > 0 \tag{11.35}$$

If one or more are zero, Q is semidefinite. A matrix $\mathbf{A}$ is said to be, say, positive definite if the corresponding quadratic form is positive definite, and $-\mathbf{A}$ is then negative definite.

Example 11.6.1

$$Q = x_1^2 + 2x_2^2 + 9x_3^2 + 2x_1x_2 + 4x_1x_3 + 6x_2x_3$$

From (11.33), the coefficient of $x_i x_j$, $i \neq j$, is $a_{ij} + a_{ji} = 2a_{ij}$. Using this yields

$$a_{11} = 1, \quad a_{22} = 2, \quad a_{33} = 9, \quad a_{12} = 1, \quad a_{13} = 2, \quad a_{23} = 3$$

$$a_{11} = 1, \quad \begin{vmatrix} a_{11} & a_{12} \\ a_{12} & a_{22} \end{vmatrix} = \begin{vmatrix} 1 & 1 \\ 1 & 2 \end{vmatrix} > 0, \quad \begin{vmatrix} a_{11} & \cdot & a_{13} \\ \cdot & \cdot & \cdot \\ a_{13} & \cdot & a_{33} \end{vmatrix} = \begin{vmatrix} 1 & 1 & 2 \\ 1 & 2 & 3 \\ 2 & 3 & 9 \end{vmatrix} = 4 > 0$$

Hence Q is positive definite.

Liapunov Stability Theorem. If there exists a positive definite V, and $V \to \infty$ as $\|\mathbf{x}\| \to \infty$, the system is asymptotically stable in the region in which $\dot{V}$ is negative definite, and stable if $\dot{V}$ is negative semidefinite. The properties are global if the region extends over the entire state space.

Liapunov Instability Theorem. If there exists a V such that $\dot{V}$ is negative definite, and $V \to -\infty$ as $\|\mathbf{x}\| \to \infty$, the system is unstable in the region in which V is not positive (semi-) definite.

Remarks:

1. The reason for two theorems is that if the origin is unstable it will be impossible to find a V-function which satisfies the stability theorem. But one satisfying the instability theorem will exist and, if it can be found, will prove instability.

2. The V-function is not unique, and different choices in general will indicate different stability regions. As this implies, a system in general does not become unstable where $\dot{V}$ changes sign. The theorems give only sufficient conditions, and the predicted stability boundaries are usually quite conservative.

3. Asymptotic stability can often be proved even if $\dot{V}$ is only semidefinite: If the curve on which $\dot{V} = 0$ is found not to satisfy the system equations, it is not a trajectory, so the state cannot remain on this curve, and hence the system must be asymptotically stable.

The numerous techniques for deriving Liapunov functions, discussed in books on nonlinear control systems, often apply to certain classes of systems. As noted earlier, the difficulties here can be considerable, and experience is quite important. The quadratic form is often suitable, and will be used in the examples below to illustrate the application of the theorems.

Example 11.6.2   Nonlinear Spring–Mass–Damper System

A spring–mass–damper system for which the damping force is proportional to the third power of the velocity is described by the differential equation

$$\ddot{y} + 0.5\dot{y}^3 + y = 0$$

Let mass position y and velocity $\dot{y}$ be chosen as state variables: $x_1 = y$, $x_2 = \dot{y}$. Then $\dot{x}_1 = x_2$ and, from the differential equation, $\dot{x}_2 = \ddot{y} = -0.5\dot{y}^3 - y = -0.5x_2^3 - x_1$ are the state equations. Attempt the positive-definite V-function $V = x_1^2 + x_2^2$ ($V \to \infty$ as $\|\mathbf{x}\| \to \infty$).

$$\dot{V} = 2x_1\dot{x}_1 + 2x_2\dot{x}_2 = 2x_1x_2 - x_2^4 - 2x_1x_2 = -x_2^4$$

$\dot{V}$ is negative semidefinite, so the system is globally stable. But since the x_1-axis ($x_2 = 0$) is not a trajectory, the system is also globally asymptotically stable, by remark 3. However, usually $\dot{V}$ for this simple V-function would be indefinite, so would not prove either stability or instability.

Example 11.6.3   Aizerman's Method (Fig. 11.16)

Figure 11.16 shows a motor position servo in which loading and saturation effects on the controller/amplifier combine to produce the nonlinear characteristic $u = f(e)$ indicated.

As in Section 11.3, the system equation is $\ddot{e} + \dot{e} + f(e) = 0$. Choosing the state variables $x_1 = e$, $x_2 = \dot{e}$, the state equations are

$$\dot{x}_1 = x_2 \qquad \dot{x}_2 = -x_2 - f(x_1)$$

Let $f(x_1)/x_1 = 1$ be a linear approximation to the nonlinearity. A V-function will

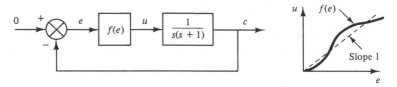

Figure 11.16 Example 11.6.3.

first be sought for the linearized system

$$\dot{x}_1 = x_2 \qquad \dot{x}_2 = -x_2 - x_1$$

Let $V = a_{11}x_1^2 + 2a_{12}x_1x_2 + a_{22}x_2^2$. Then

$$\dot{V} = -2a_{12}x_1^2 + 2(a_{11} - a_{12} - a_{22})x_1x_2 + 2(a_{12} - a_{22})x_2^2$$

Constrain $\dot{V}$ to, say, $\dot{V} = 2x_1^2 + 2x_2^2$. This requires $a_{11} = -3$, $a_{12} = -1$, $a_{22} = -2$, so that

$$V = -3x_1^2 - 2x_1x_2 - 2x_2^2$$

Sylvester's theorem will show this to be negative definite, so the linearized system is globally asymptotically stable.

This negative-definite V-function is now used for the nonlinear system. It is found that

$$\dot{V} = 2x_1f(x_1) + 4x_2f(x_1) - 4x_1x_2 + 2x_2^2$$

$$= 2\frac{f(x_1)}{x_1}x_1^2 + 4\left(\frac{f(x_1)}{x_1} - 1\right)x_1x_2 + 2x_2^2$$

Sylvester's theorem yields the following conditions for $\dot{V}$ to be positive definite:

$$k > 0, \quad k - (k-1)^2 = -k^2 + 3k - 1 > 0 \qquad \left[\frac{f(x_1)}{x_1} \equiv k\right]$$

From the roots of $k^2 - 3k + 1 = 0$, sufficient conditions for global asymptotic stability are

$$0.38 < \frac{f(x_1)}{x_1} < 2.62$$

V-functions that provide wider bounds on the sector in which the nonlinearity may be located usually exist, and certainly do in Example 11.6.3. But to obtain this improvement other than by trial and error requires more advanced methods. It is noted also that Liapunov functions are not well adapted to nonlinear system design.

11.7 POPOV AND CIRCLE CRITERIA FOR STABILITY

The Popov criterion and the circle criterion give sufficient conditions for stability of nonlinear systems in the frequency domain. They have direct graphical interpretations, similar to that with describing functions, and are convenient for design as well as analysis. The disadvantage is that predicted stability limits are frequently overly conservative. Figure 11.17 shows the system configuration considered.

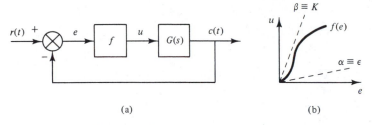

Figure 11.17   System and nonlinearity.

$G(s)$ represents the linear part and the nonlinearity f is constrained to a sector bounded by slopes α (= ε) and β (= k). The nonlinearities considered are the following:

(n1) *Single-valued and time-invariant:* a unique output u for each input e

(n2) *With hysteresis:* nonlinearities with memory, where u depends on the time history of e; for example, a relay with hysteresis, or backlash in mechanical connections

(n3) *General nonlinearities:* time-varying and perhaps including hysteresis

It will be assumed that $r(t)$ and any disturbance inputs are bounded and square integrable, that is, $\int_0^\infty r^2(t)dt < \infty$. Note that this implies that $r \rightarrow 0$ as $t \rightarrow \infty$.

Popov's Method

1. $G(s)$ is assumed to be open-loop stable.
2. The nonlinearity is assumed to satisfy

$$\varepsilon < \frac{u}{e} < K \qquad (11.36)$$

where $\varepsilon = 0$ if all poles of G are inside the left-half s-plane, and $\varepsilon > 0$, arbitrarily small, if G has poles on the imaginary axis. This is because for $f = 0$ the closed-loop poles coincide with the open-loop poles, so one or more would be on the imaginary axis and the system could not be asymptotically stable.

Use is made of a modified frequency response function $G^*(j\omega)$, defined by

$$\text{Re } G^* = \text{Re } G \qquad \text{Im } G^* = \omega \text{ Im } G \qquad \omega \geqslant 0 \qquad (11.37)$$

in the following graphical interpretation of the Popov conditions for global asymptotic stability.

Popov's theorem.   For any initial condition, the system output is bounded and tends to zero as $t \rightarrow \infty$ if the plot of $G^*(j\omega)$ lies entirely to the right of the *Popov line*, which crosses the real axis at $-1/K$ at a slope $1/q$. Here the restrictions on q and K depend on the nonlinearity:

(n1) $-\infty < q < \infty$ if $0 < K < \infty$; $0 \leqslant q < \infty$ if $K = \infty$.
(n2) $-\infty < q \leqslant 0$ and $0 < K < \infty$.
(n3) $q = 0$ and $0 < K \leqslant \infty$.

Figure 11.18 illustrates the use of these criteria. In Fig. 11.18(a) the polar plot of $G^*(\omega \geqslant 0)$ lies to the right of the Popov line shown, so the system is asymptotically stable. In Fig. 11.18(b), the intersection identified as $-1/K$ gives the maximum K for which the theorem guarantees stability, and then only for the single-valued, time-invariant nonlinearities (n1), for which $q > 0$ is admissible. For a general nonlinearity (n3) the restriction on q is more severe. As indicated in Fig. 11.18(c), with q constrained to be zero, the maximum K is now that which corresponds to the vertical tangent of G^*. If in (c) the nonlinearity were type (n1), the maximum K would be that corresponding to the intersection of G^* with the negative real axis, because the tangent at this point would be an admissible Popov line. It is important to observe that the intersections of G and G^* with the negative real axis are the same, and that this intersection therefore also gives the maximum stable gain of a linear system, by the Nyquist criterion. Hence, for nonlinearities (n1) the *Popov sector* of stability for Fig. 11.18(c), and (a) as well, is the same as the *Hurwitz sector* (i.e., the range of stable gains if the system were linear). Since the Hurwitz sector gives the largest permissible gain, an indication of the degree to which the prediction of the Popov sector could be conservative is available by inspection.

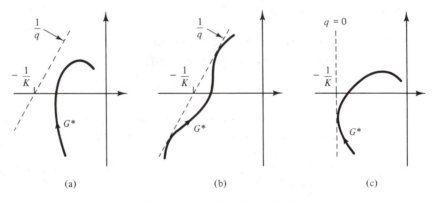

(a)   (b)   (c)

Figure 11.18 Popov's method.

The circle criterion for stability is of the same nature as the Popov theorem, with a similar form of graphical interpretation. An advantage is that it uses the Nyquist diagram or polar plot of $G(j\omega)$ and not the modified function G^*.

The Circle Criterion

Circle theorem.   A system with the nonlinearity in the sector $\alpha < u/e < \beta$ has an output that is bounded and tends to zero as $t \rightarrow \infty$ for any initial condition if:

1. $\beta > \alpha > 0$: $G(j\omega)$ does not touch or encircle the circular disk in Fig. 11.19(a).
2. $\beta > 0$, $\alpha = 0$: $G(j\omega)$ lies to the right of the vertical at $-1/\beta$ indicated in Fig. 11.19(b).
3. $\alpha < 0 < \beta$: $G(j\omega)$ lies inside the circle in Fig. 11.19(c).

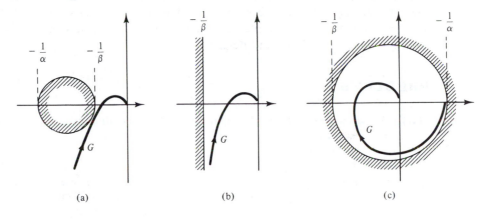

Figure 11.19 The circle criterion.

Remarks:

1. The nonlinearity may be time varying, and in cases 1 and 2 may also include memory. Indeed, case 2 corresponds directly to the Popov line for the nonlinearity (n3), where $q = 0$, since the real parts of G and G^* are the same.

2. Evidently, for case 3 G may not have poles at the origin since then $|G| \rightarrow \infty$ as $\omega \rightarrow 0$.

3. Unlike the Popov criterion, the circle criterion for case 1 can be restated to apply to open-loop unstable systems:

The Nyquist diagram of $G(j\omega)$, $-\infty < \omega < \infty$, lies outside the circle in Fig. 11.19(a) and encircles it as many times in counterclockwise direction as there are poles of $G(s)$ with positive real parts.

This extension is exactly what the Nyquist criterion would suggest. The critical point $-1/K$ for a linear system with loop gain KG can be visualized as having expanded into a circle to account for nonlinear effects.

The Popov and circle criteria do not consider the actual shape of the nonlinearity, while the describing function does. Therefore, the stability prediction resulting from DFs can be expected to be in general less conservative, and indicate stability even if G does enter the critical disk.

11.8 CONCLUSION

The behavior of nonlinear systems, types of nonlinearities, and the classical phase-plane and describing function techniques have been discussed. The second method of Liapunov was introduced, as well as the very useful Popov and circle criteria for stability analysis in the frequency domain. Numerous extensions of these frequency-domain criteria are available, and the discussion in Section 11.7 should also serve as a suitable introduction to this important area of nonlinear feedback system analysis and design.

PROBLEMS

11.1. Determine linearized models about the given operating points for the following nonlinearities:
   (a) $z = f(x, y) = x^3 + 2y^2$, $(x_0, y_0) = (1, 1)$
   (b) $y = f(x) = 3x^3 + 2x$, $x_0 = 1$

11.2. Determine a linearized gain to represent the nonlinear element $y = f(x) = x + 0.5x^2 - 0.05x^3$ for small variations about each of the following operating points.
   (a) $x_0 = 0$ **(b)** $x_0 = 1$ **(c)** $x_0 = 2$ **(d)** $x_0 = 3$

11.3. Linearize the following nonlinearities for small variations about the operating points indicated.
   (a) $z = k_1 x + k_2 x^2 y$, $(x_0, y_0) = (1, 1)$
   (b) $z = x^3 + 2x^2 y + 4xy^2 + 4y^3$, $(x_0, y_0) = (1, 1)$

11.4. Sketch the general form of the phase-plane patterns for each of the systems in Fig. P11.4.

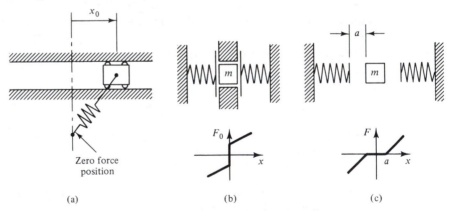

(a)   (b)   (c)

Figure P11.4   (a) Dual equilibrium; (b) preloaded springs; (c) springs and deadband.

11.5. Sketch the phase-plane portrait for step inputs for the relay servo shown in Fig. P11.5.

11.6. Repeat Problem 11.5 for ramp inputs $\theta_i = At$.

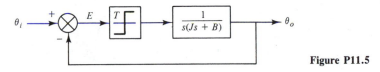

Figure P11.5

11.7. (a) Verify that Fig. P11.7 is a model for a motor position servo with Coulomb friction on the motor shaft.
   (b) Obtain the phase-plane and isocline equations for step inputs. What is the nature of the isoclines, and where do they intersect the E-axis of an $\dot{E}$-E phase plane?
   (c) Sketch the phase-plane pattern, assuming underdamped system behavior.

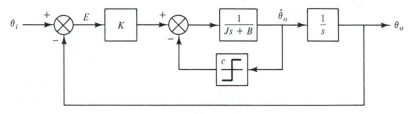

Figure P11.7

11.8. (a) Verify that Fig. P11.8 represents a motor position servo with tachometer feedback which is subject to saturation.
   (b) Obtain the phase-plane and isocline equations for step inputs. Where do the isoclines intersect the E-axis?
   (c) Sketch the phase-plane pattern, taking into account that for large $|\dot{\theta}_o|$ the tachometer feedback will be relatively less effective.

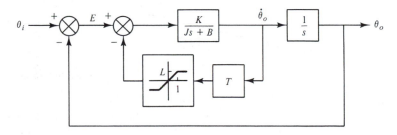

Figure P11.8

11.9. Repeat Problem 11.8 for ramp inputs.

11.10. A motor position servo with deadband in the error detector is modeled as shown in Fig. P11.10. For step inputs:
   (a) Obtain the phase-plane and isocline equations and find the isocline intersections with the E-axis.
   (b) Sketch the phase-plane pattern, assuming underdamped behavior for the value of K.
   (c) Comment on the steady-state errors.

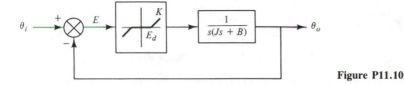

Figure P11.10

11.11. Saturation, shown in a motor position servo in Fig. P11.11, is a very common nonlinearity. Obtain the phase-plane equation and isocline equation for step inputs and sketch the phase-plane pattern.

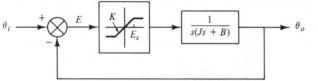

Figure P11.11

11.12. Repeat Problem 11.11 for ramp inputs. Verify the value of the steady-state error from that for small-signal operation. Compare the pattern for large error with that for a relay in Problem 11.6 and comment.

11.13. Sketch the phase-plane pattern of the system of Fig. 11.6 for step inputs.

11.14. Determine the describing function for the nonlinear element in Fig. P11.14, with input x and output y and slopes K_1 and K_2. (Use the describing functions for deadband and/or saturation.)

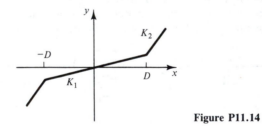

Figure P11.14

11.15. In Fig. P11.15, let $G = 10/[s(0.2s + 1)(0.05s + 1)]$. Determine the amplitude at the input of the nonlinearity and the frequency of any limit cycle that may be present for zero input.

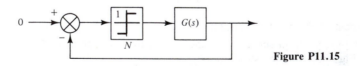

Figure P11.15

11.16. In Fig. P11.15, calculate and compare frequency and amplitude of the error signal limit cycle for
(a) $G = 10/[s(0.1s + 1)(0.01s + 1)]$
(b) $G = 10/[s(0.1s + 1)(0.02s + 1)]$

11.17. In Fig. P11.15, if $G = 5/[s(0.05s + 1)]$, will there be a limit cycle, and if so, what are its amplitude and frequency?

11.18. Let Fig. P11.15 represent a relay servo with

$$G = \frac{100}{s(0.05s + 1)(0.01s + 1)}$$

What are the amplitude and frequency of any limit cycle at the system output?

11.19. If Fig. P11.15 is a relay servo with $G = K/[s(s + 1)(s + 2)]$, find and compare amplitude and frequency of the limit cycle in the system output for $K = 1$ and $K = 6$.

11.20. In Fig. P11.20, a tachometer feedback loop has been added to a relay servo, to eliminate finite-amplitude limit cycles. Show that this is possible, and use Bode plot sketches to determine a reasonable choice for T.

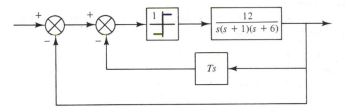

Figure P11.20

11.21. In Fig. P11.21, $G(s) = 12/[s(s + 1)(s + 5)]$ and N is a saturation element.
   (a) Determine the limit on its slope K for which no limit cycles will occur.
   (b) If $K = 5$ and the saturation limit of the input to the nonlinearity is $b = 1$, find the amplitude and frequency of any limit cycle and determine its stability.

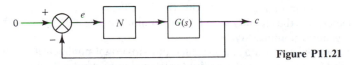

Figure P11.21

11.22. In Problem 11.21, let N be a deadband nonlinearity with a deadband from -1 to $+1$ and a slope $K = 5$.
   (a) Find amplitude and frequency of any limit cycle.
   (b) What is the nature of this limit cycle, and for what range of amplitudes at the input to the nonlinearity is the system stable, if any?

11.23. In Problem 11.21, let N be a relay with a deadband $b = 1$ and output $T = 5$. Determine amplitude and frequency of any limit cycles. Find the nature of these limit cycles, and the stability of system behavior for different amplitudes at the input to the nonlinearity.

11.24. In Fig. P11.21 with $N = f(e)$ and $G(s) = 1/[s(s + 1)]$ as in Example 11.6.3, use Liapunov functions to determine the Hurwitz sector of stability, that is, the range of linear gains of N for which the system is stable. Does the result agree with the actual size of this sector as obtained, say, from a root locus sketch?

11.25. In Problem 11.24, if $f(e)/e = K$ is a linear approximation to the nonlinearity, use Aizerman's method based on the Liapunov function for arbitrary K found in

Problem 11.24 to determine the stability sector for:

(a) $K = 2$ (b) $K = 4$ (c) $K = 10$

Compare the results also with the stable sector $0.38 < f(e)/e < 2.62$ found in Example 11.6.3 for $K = 1$.

11.26. In Fig. P11.21 with $N = f(e)$ and $G(s) = 5/[s(s + 5)]$, determine the stability sector of the nonlinearity by Aizerman's method. Use $f(e)/e = 1$ as a linear approximation.

11.27. In Fig. P11.27, N is a nonlinearity with output $f(x_1)$.

(a) Obtain a state space model with x_1 and x_2 as state variables. [*Hint:* For the last block, $-x_1/x_2 = (s - 3)/(s + 4)$, and cross-multiplication and inverse transformation gives $\dot{x}_2 = 3x_2 - \dot{x}_1 - 4x_1$, where $\dot{x}_1$ can be eliminated by using the same approach for the preceding block.]

(b) Determine the Hurwitz sector of stability (i.e., the range of linear gains in N for which the system is stable).

(c) Determine the stability sector of $f(x_1)/x_1$ by Aizerman's method using $f(x_1) = 0$ as a linear approximation to the nonlinearity.

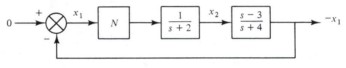

Figure P11.27

11.28. For Problem 11.27, apply the Popov stability criterion to find the Popov sector of stability and the type of nonlinearity permitted. Compare with the Hurwitz sector for an idea of to what extent the Popov sector could be conservative.

11.29. For the system of Fig. P11.21 with $G(s) = 20/[s(s + 2)(s + 4)]$:

(a) Determine the stability sector of the nonlinearity by the Popov method, assuming a general nonlinearity.

(b) Repeat part (a) for a single-valued time-invariant nonlinearity. Compare it with the Hurwitz sector of stability to determine to what extent the prediction may be conservative.

11.30. For the system of Problem 11.29, use the circle criterion to predict the stability sectors of a general nonlinearity, and compare with the Popov predictions in Problem 11.29:

(a) If the lower sector boundary approaches zero.

(b) If the lower sector boundary is $f(x, t)/x = 0.5$.

11.31. For the system of Fig. P11.21 with $G(s) = 15/[s(s + 2)(s + 6)]$:

(a) Find the Popov stability sector of N for a general nonlinearity.

(b) Repeat for a single-valued time-invariant nonlinearity.

(c) Compare both with the Hurwitz stability sector.

11.32. For the system of Problem 11.31, predict the stability sectors for a general nonlinearity by the circle criterion:

(a) If the lower sector boundary approaches zero.

(b) If the lower sector boundary is $f(x, t)/x = 0.5$.

11.33. Apply the circle criterion to the system considered in Problems 11.27 and 11.28. Compare the stability sector obtained with the Hurwitz and Popov sectors. Note that the negative gain limit of the Hurwitz sector suggests the use of the form of the circle criterion which allows a negative slope sector boundary.

11.34. For the system in Fig. P11.21 with $G(s) = (s - 1)/(s + 1)^2$:
   (a) What is the Hurwitz stability sector?
   (b) Find the Popov stability sector and the type of nonlinearity permitted.

11.35. Use the circle criterion to determine the stability sector for the system of Problem 11.34, allowing for a negative slope sector boundary, and compare with the Hurwitz and Popov sectors.

11.36. The restriction of the Popov stability sector to zero minimum slope, which is a serious disadvantage in Problems 11.28 and 11.34, where the Hurwitz sector includes negative gains, can be overcome by *pole shifting*. Show that the system of Problem 11.34, with nonlinearity $m = f(x)$ and $G(s) = -x/m$, is equivalent to one with nonlinearity m' and G' as given below, where L is a chosen constant:

$$m' = g(x) = m + Lx$$

$$G' = \frac{-x}{m'} = \frac{s - 1}{(s + 1)^2 - L(s - 1)}$$

Note that the poles of G' differ from those of G, hence the name "pole shifting," and that L can be chosen to change a sector boundary of $f(x)/x$ with negative slope to one with zero or larger slope of $g(x)/x$.

11.37. Use the result of Problem 11.36 to apply pole shifting to Problem 11.34. Choose L such that the lower (negative slope) boundary of the Hurwitz sector becomes a zero slope boundary for the nonlinearity $g(x)$. Determine the Popov stability sector for $g(x)/x$, and from it that for $f(x)/x$. Compare the result with the Hurwitz sector. What type of nonlinearity is permitted?

12

STATE SPACE ANALYSIS

12.1 INTRODUCTION

The many books largely devoted to the subject, and the continuing emphasis that it receives in the technical literature, are evidence of the importance of the state space approach. The key factor which accounts for this is that the system dynamics are described by a state space model instead of by transfer functions. A state space model is a description in terms of a set of first-order differential equations which are written compactly in a standard matrix form. This standard form has permitted the development of general computer programs, which can be used for the analysis and design of even very large systems.

Techniques for modeling and analysis are discussed in this chapter, and design is introduced in Chapter 13. Appendix A reviews the topics in matrix analysis needed.

12.2 STATE SPACE MODELS

The derivation of state space models is no different from that of transfer functions, described in Chapter 2, in that the differential equations describing the system dynamics are written first. In transfer function models these equations are transformed and variables are eliminated between them to find the relation between selected input and output variables. For state models, instead, the equations are arranged into a set of first-order differential equations in terms of selected state variables, and the outputs are expressed in these same state variables. Because the elimination of variables between equations is not an inherent part of this process, state models are often easier to obtain. A series of examples is given for illustration and to relate state models to the transfer functions used thus far.

Example 12.2.1   A Transfer Function without Zeros

$$\frac{W}{R} = \frac{5}{s^3 + 6s^2 + 9s + 3} \qquad \dddot{w} + 6\ddot{w} + 9\dot{w} + 3w = 5r$$

A state model for the system described by this transfer function or the equivalent differential equation is not unique but depends on the choice of a set of *state variables* $x_1(t)$, $x_2(t)$, and $x_3(t)$. One possible choice is the following:

$$x_1 = w \qquad x_2 = \dot{w} \qquad x_3 = \ddot{w}$$

These definitions and the differential equation yield

$$\dot{x}_1 = x_2 \qquad \dot{x}_2 = x_3 \qquad \dot{x}_3 = -3x_1 - 9x_2 - 6x_3 + 5r$$

In matrix form, and with the output w expressed also in terms of all state variables,

$$\mathbf{x} = \begin{bmatrix} x_1 \\ x_2 \\ x_3 \end{bmatrix} \qquad \dot{\mathbf{x}} = \begin{bmatrix} 0 & 1 & 0 \\ 0 & 0 & 1 \\ -3 & -9 & -6 \end{bmatrix} \mathbf{x} + \begin{bmatrix} 0 \\ 0 \\ 5 \end{bmatrix} r \qquad w = [1 \ \ 0 \ \ 0]\mathbf{x}$$

The general form of a state-space model is as follows:

$$\begin{aligned} \dot{\mathbf{x}} &= \mathbf{A}\mathbf{x} + \mathbf{B}\mathbf{u} \qquad (\textit{state equation}) \\ \mathbf{y} &= \mathbf{C}\mathbf{x} + \mathbf{D}\mathbf{u} \qquad (\textit{output equation}) \end{aligned} \tag{12.1}$$

The vector $\mathbf{x}$ of the state variables is the *state vector*. The *control vector* $\mathbf{u}$ is the scalar function r in the example, and the *output vector* $\mathbf{y}$ the scalar function w, and $\mathbf{D} = \mathbf{0}$. $\mathbf{A}$ is the *system matrix*.

Example 12.2.2   Generalization of Example 12.2.1

If the system is described by the nth-order differential equation

$$\frac{d^n w}{dt^n} + a_n \frac{d^{n-1} w}{dt^{n-1}} + \cdots + a_2 \frac{dw}{dt} + a_1 w = r$$

or the equivalent transfer function, then, with the same choice of state variables as in Example 12.2.1,

$$\mathbf{x} = \begin{bmatrix} x_1 \\ x_2 \\ \cdot \\ \cdot \\ \cdot \\ x_n \end{bmatrix} = \begin{bmatrix} w \\ \dot{w} \\ \cdot \\ \cdot \\ \cdot \\ (n-1) \\ w \end{bmatrix} \qquad \mathbf{B} = \begin{bmatrix} 0 \\ \cdot \\ \cdot \\ \cdot \\ 0 \\ 1 \end{bmatrix} \qquad \mathbf{A} = \begin{bmatrix} 0 & 1 & 0 & 0 \\ 0 & 0 & 1 & 0 \\ \cdot & & & \cdot \\ \cdot & & & \cdot \\ 0 & & & 1 \\ -a_1 & -a_2 & \cdots & -a_n \end{bmatrix} \tag{12.2}$$

$$\mathbf{C} = [1 \ \ 0 \ \ \cdots \ \ 0] \qquad \mathbf{D} = \mathbf{0}$$

This form of $\mathbf{A}$ is a *companion matrix*.

Example 12.2.3   Transfer Function with Zeros

$$\frac{W}{R} = \frac{5s^2 + 2s + 2}{s^3 + 6s^2 + 9s + 3}$$

or
$$\dddot{w} + 6\ddot{w} + 9\dot{w} + 3w = 5\ddot{r} + 2\dot{r} + 2r$$

First consider only the denominator:
$$\frac{V}{R} = \frac{1}{s^3 + 6s^2 + 9s + 3} \qquad \dddot{v} + 6\ddot{v} + 9\dot{v} + 3v = r$$

As in Example 12.2.1,
$$\mathbf{x} = \begin{bmatrix} v \\ \dot{v} \\ \ddot{v} \end{bmatrix} \qquad \dot{\mathbf{x}} = \begin{bmatrix} 0 & 1 & 0 \\ 0 & 0 & 1 \\ -3 & -9 & -6 \end{bmatrix} \mathbf{x} + \begin{bmatrix} 0 \\ 0 \\ 1 \end{bmatrix} r$$

But
$$W = (5s^2 + 2s + 2)V \qquad \text{or} \qquad w = 5\ddot{v} + 2\dot{v} + 2v = [2 \quad 2 \quad 5]\mathbf{x}$$

Hence the output equation, with $y = w$, is
$$y = \mathbf{Cx} \qquad \mathbf{C} = [2 \quad 2 \quad 5]$$

Thus the output equation represents the effect of system zeros, or derivatives of the input.

Example 12.2.4

$$\frac{Y}{U} = \frac{K(s + a)}{s + b}$$

With equal powers in numerator and denominator, the method of Example 12.2.3 can be used if the division is performed first:

$$Y = K\left(1 + \frac{a - b}{s + b}\right)U = KU + X \qquad \text{where} \quad \frac{X}{U} = \frac{K(a - b)}{s + b}$$

This yields $\dot{x} = -bx + K(a - b)u$, $y = Ku + x$.

The choices of state variables $x_1 = w$, $x_2 = \dot{w}$, $x_3 = \ddot{w}$, . . . , made thus far have the disadvantage that the variables beyond the second or third have little or no physical meaning and are difficult or impossible to measure. These features are important to identify the behavior of variables of interest and to implement a control that requires feedback from the state variables. Where possible, state variables should be chosen which are measurable and physically meaningful.

Therefore, state models should not be derived from closed-loop transfer functions, but directly from the original system equations. This, in fact, tends to simplify the modeling, because these are frequently already first-order equations, and eliminating variables between them is not necessary. Derivation from a block diagram is frequently appropriate, however, since the output signals of the blocks are often measurable physical variables. The following example illustrates a common approach for this case.

Example 12.2.5 Multivariable Control System (Fig. 12.1)

If a system block diagram can be separated into simple lag blocks and integrator blocks, a state model can be derived by identifying the output of each as a state variable. In Fig. 12.1 the output equation is seen to be

$$\mathbf{y} = \begin{bmatrix} y_1 \\ y_2 \end{bmatrix} = \begin{bmatrix} 1 & 1 & 0 & 0 \\ 0 & 0 & 1 & 1 \end{bmatrix} \mathbf{x}$$

and from equations such as $\dot{x}_1 = -x_1 + u_1$ and $u_1 = K_1(r_1 - y_1) = K_1(r_1 - x_1 - x_2)$, which can be seen by inspection of Fig. 12.1, the state equation is found to be

$$\dot{\mathbf{x}} = \begin{bmatrix} -1 - K_1 & -K_1 & 0 & 0 \\ 0 & -5 & -5K_2 & -5K_2 \\ -0.4K_1 & -0.4K_1 & -0.5 & 0 \\ 0 & 0 & -4K_2 & -2 - 4K_2 \end{bmatrix} \mathbf{x} + \begin{bmatrix} K_1 & 0 \\ 0 & 5K_2 \\ 0.4K_1 & 0 \\ 0 & 4K_2 \end{bmatrix} \mathbf{u}$$

$$\text{where } \mathbf{u} = \begin{bmatrix} r_1 \\ r_2 \end{bmatrix}$$

If a block is described by a quadratic transfer function $K/(s^2 + as + b)$, the output of the block and its derivative are frequently suitable state variables.

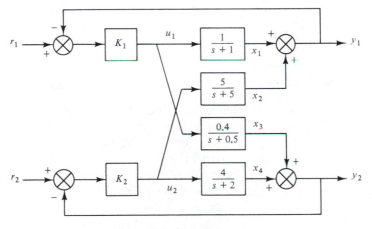

Figure 12.1   A 2 by 2 system.

The following example illustrates, for several of the systems considered in Chapter 2, the normal approach for the derivation of state models, that is, directly from the system differential equations. In each case the state variables have been chosen to be measurable and meaningful.

Example 12.2.6   Examples from Chapter 2:

(a) *Spring-mass-damper* (Example 2.2.1, Fig. 2.2):

$$m\ddot{x} + c\dot{x} + kx = f \qquad x_1 = x \qquad x_2 = \dot{x} \qquad u = f \qquad y = x_1$$

$$\dot{\mathbf{x}} = \begin{bmatrix} 0 & 1 \\ -\dfrac{k}{m} & -\dfrac{c}{m} \end{bmatrix} \mathbf{x} + \begin{bmatrix} 0 \\ \dfrac{1}{m} \end{bmatrix} u \qquad y = \begin{bmatrix} 1 & 0 \end{bmatrix} \mathbf{x}$$

(b) *RC simple lag circuit* [Example 2.3.1, Fig. 2.8(a)]:

$$e_i = iR + \frac{1}{C}\int i\,dt \qquad e_o = \frac{1}{C}\int i\,dt = y \qquad u = e_i$$

$$x = e_o = \text{capacitor voltage} \qquad u = iR + x \qquad i = C\dot{x}$$

$$\dot{x} = \frac{-x}{RC} + \frac{u}{RC} \qquad y = x$$

It is noted that the number of state variables for a system equals the number of independent energy storage elements.

(c) *Field-controlled dc motor* (Example 2.4.1, Fig. 2.9):

$$e_f = R_f i_f + L_f \dot{i}_f \qquad T_d = K_T i_f \qquad T_d = J\ddot{\theta}_o + B\dot{\theta}_o \qquad u = e_f \qquad y = \theta_o$$

$$\mathbf{x} = \begin{bmatrix} \theta_o \\ \dot{\theta}_o \\ i_f \end{bmatrix} \qquad \dot{\mathbf{x}} = \begin{bmatrix} 0 & 1 & 0 \\ 0 & \dfrac{-B}{J} & \dfrac{K_T}{J} \\ 0 & 0 & \dfrac{-R_f}{L_f} \end{bmatrix} \mathbf{x} + \begin{bmatrix} 0 \\ 0 \\ \dfrac{1}{L_f} \end{bmatrix} u$$

$$y = [1 \quad 0 \quad 0]\mathbf{x}$$

12.3 TRANSFER FUNCTION MATRICES AND STABILITY

With a state-model description of the system dynamics, the first question to be answered is how stability may be determined. To derive the stability criterion, the generalization of the concept of a transfer function is considered first, by finding the transfer function matrix that corresponds to the state model. This requires Laplace transformation of the state-model equations. The Laplace transform of a vector is the vector of the Laplace transforms of its elements, so that the transforms of $\mathbf{x}$ and $\dot{\mathbf{x}}$ are as follows:

$$\mathbf{X}(s) = L[\mathbf{x}(t)] = \begin{bmatrix} L[x_1] \\ \cdot \\ \cdot \\ L[x_n] \end{bmatrix} = \begin{bmatrix} X_1(s) \\ \cdot \\ \cdot \\ X_n(s) \end{bmatrix} \qquad L[\dot{\mathbf{x}}] = \begin{bmatrix} L[\dot{x}_1] \\ \cdot \\ \cdot \\ L[\dot{x}_n] \end{bmatrix}$$

where $L[\dot{x}_i] = sX_i(s) - x_i(0)$. Hence the transform of $\dot{\mathbf{x}} = \mathbf{Ax} + \mathbf{Bu}$ is $s\mathbf{X}(s) - \mathbf{x}_o = \mathbf{AX}(s) + \mathbf{BU}(s)$, or $(s\mathbf{I} - \mathbf{A})\mathbf{X}(s) = \mathbf{BU}(s) + \mathbf{x}_o$, where $\mathbf{I}$ is the unit matrix. Note that $s\mathbf{X}(s) = s\mathbf{IX}(s)$ and that $(s - \mathbf{A})$ is incorrect since s is a scalar and $\mathbf{A}$ is not. Then, if the system output $\mathbf{y} = \mathbf{Cx}$, so $\mathbf{Y}(s) = \mathbf{CX}(s)$:

$$\mathbf{Y}(s) = \mathbf{G}(s)\mathbf{U}(s) + \mathbf{C}(s\mathbf{I} - \mathbf{A})^{-1}\mathbf{x}_o \tag{12.3}$$

where

$$\mathbf{G}(s) = \mathbf{C}(s\mathbf{I} - \mathbf{A})^{-1}\mathbf{B} \tag{12.4}$$

Equation (12.3) shows the total response as a superposition of two separate components. The first term gives the input–output response, for $x_o = 0$, and the second the output response to initial conditions, with $u = 0$. With $C = I$ the state response $X(s)$ is obtained.

$G(s)$ of (12.4) is called the *transfer function matrix* because it relates the transforms of the input and output vectors for zero initial conditions. As illustrated in Fig. 12.2, $G(s)$ generalizes the concept of a transfer function. It is a matrix of ordinary transfer functions. For a system with r inputs and m outputs, the equation $Y = G\,U$ can be written out as

$$\begin{bmatrix} Y_1(s) \\ \cdot \\ \cdot \\ \cdot \\ Y_m(s) \end{bmatrix} = \begin{bmatrix} g_{11}(s) & \cdots & g_{1r}(s) \\ \cdot & & \cdot \\ \cdot & & \cdot \\ \cdot & & \cdot \\ g_{m1}(s) & \cdots & g_{mr}(s) \end{bmatrix} \begin{bmatrix} U_1(s) \\ \cdot \\ \cdot \\ \cdot \\ U_r(s) \end{bmatrix} \tag{12.5}$$

$$Y_i(s) = Y_{i1}(s) + \cdots + Y_{ir}(s) \qquad Y_{ij}(s) = g_{ij}(s)U_j(s)$$

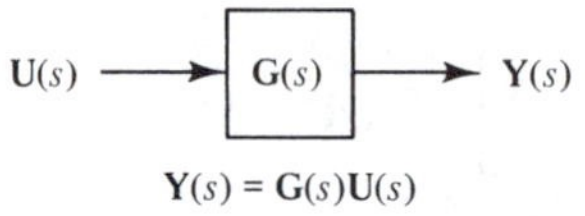

$$U(s) \longrightarrow \boxed{G(s)} \longrightarrow Y(s)$$

$$Y(s) = G(s)U(s)$$

Figure 12.2   Transfer function matrix.

The element $g_{ij}(s)$ is an ordinary transfer function which gives the part Y_{ij} of Y_i that is due to input U_j.

*****

For stability, all poles of all these transfer functions must lie in the left-half s-plane. From Appendix A, the inverse in (12.4) equals the adjoint divided by the determinant, so that the transfer function matrix becomes

$$G(s) = \frac{C\ \text{adj}\ (sI - A)B}{|sI - A|} \tag{12.6}$$

The numerator is a matrix of polynomials, and all elements of G have the same denominator. This denominator is the polynomial $|sI - A|$. Hence

Stability Theorem 12.3.1.   The system described by the state model (12.1) is stable if and only if the *eigenvalues* of the system matrix A, that is, the roots of the system

$$\text{characteristic equation } |sI - A| = 0 \tag{12.7}$$

all lie in the left-half s-plane.

The ability to exploit standard computer routines available to determine the eigenvalues of even very large matrices A is a major advantage of the state space formulation.

Example 12.3.1

$$A = \begin{bmatrix} 0 & 1 \\ -2 & -3 \end{bmatrix} \qquad sI - A = \begin{bmatrix} s & -1 \\ 2 & s+3 \end{bmatrix}$$

$|s\mathbf{I} - \mathbf{A}| = s^2 + 3s + 2$. The eigenvalues are -1 and -2, so the system is stable.

12.4 SOLUTION OF THE STATE EQUATION $\dot{\mathbf{x}} = \mathbf{Ax} + \mathbf{Bu}$

As for transfer functions, not only stability but also transient responses must be determined from the state model. Techniques for solution of the state space equation toward this end are discussed in this section.

Solution by Laplace Transforms

Based on (12.3) and (12.4), this method is often useful for smaller systems, as well as for larger ones if only certain input–output relations need to be calculated.

Example 12.4.1

$$\mathbf{A} = \begin{bmatrix} 0 & 1 \\ -2 & -3 \end{bmatrix} \qquad \mathbf{B} = \begin{bmatrix} 0 \\ 2 \end{bmatrix} \qquad \mathbf{C} = \begin{bmatrix} 1 & 0 \end{bmatrix} \qquad \mathbf{x}_0 = \begin{bmatrix} -1 \\ 0 \end{bmatrix}$$

$$s\mathbf{I} - \mathbf{A} = \begin{bmatrix} s & -1 \\ 2 & s+3 \end{bmatrix} \qquad (s\mathbf{I} - \mathbf{A})^{-1} = \frac{1}{s^2 + 3s + 2} \begin{bmatrix} s+3 & 1 \\ -2 & s \end{bmatrix}$$

since $|s\mathbf{I} - \mathbf{A}| = s^2 + 3s + 2 = (s+1)(s+2)$. Equation (12.4) gives

$$G = \mathbf{C}(s\mathbf{I} - \mathbf{A})^{-1}\mathbf{B} = \frac{[1 \quad 0]}{(s+1)(s+2)} \begin{bmatrix} s+3 & 1 \\ -2 & s \end{bmatrix} \begin{bmatrix} 0 \\ 2 \end{bmatrix} = \frac{2}{(s+1)(s+2)}$$

This is an ordinary transfer function, since the system is single input/single output, and responses to inputs can be calculated in the usual way. From (12.3), the state response to $\mathbf{x}_0$ is

$$\mathbf{X}(s) = (s\mathbf{I} - \mathbf{A})^{-1} \begin{bmatrix} -1 \\ 0 \end{bmatrix} = \begin{bmatrix} \dfrac{-2}{s+1} + \dfrac{1}{s+2} \\ \dfrac{2}{s+1} - \dfrac{2}{s+2} \end{bmatrix} \qquad \mathbf{x}(t) = \begin{bmatrix} -2e^{-t} + e^{-2t} \\ 2e^{-t} - 2e^{-2t} \end{bmatrix}$$

where partial fraction expansion was used for $\mathbf{X}(s)$.

Example 12.4.2

$$\mathbf{A} = \begin{bmatrix} -1 & 0 \\ 0 & -2 \end{bmatrix} \qquad \mathbf{B} = \begin{bmatrix} 0 \\ 1 \end{bmatrix} \qquad \mathbf{C} = \begin{bmatrix} 1 & 0 \\ 0 & 1 \end{bmatrix} \qquad \mathbf{x}_0 = \begin{bmatrix} 2 \\ -3 \end{bmatrix}$$

The output response to $\mathbf{x}_0$ and a step input $U = 1/s$ is

$$\mathbf{Y} = \mathbf{CX} = \mathbf{C}(s\mathbf{I} - \mathbf{A})^{-1}(\mathbf{B}U + \mathbf{x}_0) = \begin{bmatrix} \dfrac{1}{s+1} & 0 \\ 0 & \dfrac{1}{s+2} \end{bmatrix} \left(\begin{bmatrix} 0 \\ 1 \end{bmatrix}\dfrac{1}{s} + \begin{bmatrix} 2 \\ -3 \end{bmatrix} \right)$$

since

$$s\mathbf{I} - \mathbf{A} = \begin{bmatrix} s+1 & 0 \\ 0 & s+2 \end{bmatrix}$$

Then, using partial fractions,

$$Y = \begin{bmatrix} \dfrac{1}{s+1} & 0 \\[2ex] 0 & \dfrac{1}{s+2} \end{bmatrix} \begin{bmatrix} 2 \\[2ex] -3 + \dfrac{1}{s} \end{bmatrix} = \begin{bmatrix} \dfrac{2}{s+1} \\[2ex] \dfrac{1-3s}{s(s+2)} \end{bmatrix}$$

$$y(t) = \begin{bmatrix} 2e^{-t} \\ 0.5(1 - 7e^{-2t}) \end{bmatrix}$$

Formal Solution

The solution of the scalar equation $\dot{x} = ax + bu$, $x(0) = x_0$ is known to be $x(t)$ $= e^{at}x_0 + \int_0^t e^{a(t-\tau)}bu(\tau)\, d\tau$, where $e^{at} = 1 + at + (1/2!)a^2t^2 + (1/3!)a^3t^3 + \dots$. By analogy, for the matrix case:

$$\dot{x} = Ax \text{ (homogeneous equation): } x(t) = e^{At}x_0 \tag{12.8a}$$

$$\dot{x} = Ax + Bu: \; x(t) = e^{At}x_0 + \int_0^t e^{A(t-\tau)}Bu(\tau)\, d\tau \tag{12.8b}$$

where the matrix exponential is defined by

$$e^{At} = I + At + \left(\frac{1}{2!}\right)A^2t^2 + \left(\frac{1}{3!}\right)A^3t^3 + \dots = \sum_{k=0}^{\infty} \frac{A^k t^k}{k!} \tag{12.9}$$

1. To prove (12.8a), the solution is substituted into the differential equation, using

$$\begin{aligned}
\frac{de^{At}}{dt} &= \left(\frac{d}{dt}\right)\left(I + At + \left(\frac{1}{2!}\right)A^2t^2 + \dots\right) \\
&= A + \left(\frac{2}{2!}\right)A^2t + \left(\frac{3}{3!}\right)A^3t^2 + \dots \\
&= A\left(I + At + \left(\frac{1}{2!}\right)A^2t^2 + \dots\right) \\
&= Ae^{At} \quad \text{or} \quad e^{At}A
\end{aligned} \tag{12.10}$$

analogous to the derivative of a scalar exponential. Thus $\dot{x} = (d/dt)e^{At}x_0 = Ae^{At}x_0 = Ax$, satisfying the differential equation. Because $x(0) = e^{A.0}x_0 = (I + A.0 + \dots)x_0 = x_0$, the initial condition is also satisfied.

The matrix e^{At} relates the state at t to that at time zero, and is called the

$$\text{transition matrix } \phi(t) = e^{At} \tag{12.11}$$

Since, from (12.3), the state response to initial conditions is given by $X(s) = (sI - A)^{-1}x_0$, it also follows that

$$e^{At} = L^{-1}[sI - A)^{-1}] \tag{12.12}$$

Furthermore, $x(t_2 + t_1) = e^{A(t_2+t_1)}x_0$, but also $x(t_2 + t_1) = e^{At_2}x(t_1) = e^{At_2}e^{At_1}x_0$, so that

$$e^{A(t_1 + t_2)} = e^{At_1}e^{At_2} \qquad \phi(t_1 + t_2) = \phi(t_1)\phi(t_2) \tag{12.13}$$

analogous to scalar exponentials. With $t_2 = -t_1 = t$,

$$I = e^{-\mathbf{A}t}e^{\mathbf{A}t} \qquad (e^{\mathbf{A}t})^{-1} = e^{-\mathbf{A}t} \qquad (12.14)$$

2. To prove (12.8b), assume a solution

$$\mathbf{x}(t) = e^{\mathbf{A}t}\mathbf{f}(t) \qquad \dot{\mathbf{x}} = \mathbf{A}e^{\mathbf{A}t}\mathbf{f} + e^{\mathbf{A}t}\dot{\mathbf{f}} = \mathbf{A}\mathbf{x} + e^{\mathbf{A}t}\dot{\mathbf{f}}$$

For this to satisfy $\dot{\mathbf{x}} = \mathbf{A}\mathbf{x} + \mathbf{B}\mathbf{u}$ requires that $e^{\mathbf{A}t}\dot{\mathbf{f}} = \mathbf{B}\mathbf{u}$ or, using (12.14), $\dot{\mathbf{f}} = e^{-\mathbf{A}t}\mathbf{B}\mathbf{u}$. Integration yields $\mathbf{f}(t) = \mathbf{f}(0) + \int_0^t e^{-\mathbf{A}\tau}\mathbf{B}\mathbf{u}(\tau)\, d\tau$, and substitution then gives the solution (12.8b), since $\mathbf{x}(0) = \mathbf{f}(0)$.

Discrete-Time Solution

As illustrated in Fig. 12.3, the time axis is discretized into intervals of width T, and $u_i(t)$ is approximated by a staircase function, constant over the intervals. Using (12.8b), for $t = T$,

$$\mathbf{x}(T) = e^{\mathbf{A}T}\mathbf{x}(0) + e^{\mathbf{A}T}\int_0^T e^{-\mathbf{A}\tau}d\tau\,\mathbf{B}\mathbf{u}(0)$$

or

$$\mathbf{x}(T) = \boldsymbol{\phi}\mathbf{x}(0) + \boldsymbol{\Delta}\mathbf{u}(0)$$

where

$$\boldsymbol{\phi} = e^{\mathbf{A}T} \qquad \boldsymbol{\Delta} = e^{\mathbf{A}T}\int_0^T e^{-\mathbf{A}\tau}\, d\tau\,\mathbf{B} \qquad (12.15)$$

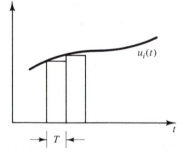

Figure 12.3 Time discretization.

This can be repeated for the second interval, with $\mathbf{x}(T)$ as initial condition, to yield $\mathbf{x}(2T) = \boldsymbol{\phi}\mathbf{x}(T) + \boldsymbol{\Delta}\mathbf{u}(T)$, and then for the succeeding intervals. For the kth interval, with $\mathbf{x}_k = \mathbf{x}(kT)$:

$$\mathbf{x}_{k+1} = \boldsymbol{\phi}\mathbf{x}_k + \boldsymbol{\Delta}\mathbf{u}_k \qquad (12.16)$$

This permits the solution to be calculated forward in time.

The series (12.9) is used to compute $\boldsymbol{\phi}$, adding terms until the next one does not change the result by more than a specified tolerance. To maintain accuracy, it turns out to be desirable to calculate $\boldsymbol{\phi}(T)$ as $[\boldsymbol{\phi}(T/n)]^n$, where n is a positive even integer chosen so that the elements of $\mathbf{A}T/n$ will be "sufficiently small."

To find Δ, (12.15) is integrated:

$$\Delta = e^{AT}(-e^{-AT}A^{-1} + A^{-1})B = (\phi - I)A^{-1}B \qquad (12.17)$$

If A is singular, $e^{A(T-\tau)}$ in (12.15) is written as a series:

$$\Delta = \int_0^T \sum_{k=0}^{\infty} \frac{A^k(T - \tau)^k}{k!} \, d\tau \, B = \sum_{k=0}^{\infty} \frac{A^k T^{k+1}}{(k + 1)!} B \qquad (12.18)$$

Approximate Discrete-Time Solution

The vector finite-difference approximation to $\dot{x}$ is of the same form as a scalar finite difference:

$$\dot{x} = \frac{1}{T}[x(t + T) - x(t)]$$

The equation $\dot{x} = Ax + Bu$ is then approximated by

$$x(t + T) = (I + AT)x(t) + TBu(t) \qquad (12.19)$$

This is of the form of (12.16) with

$$\phi = I + AT \qquad \Delta = TB \qquad (12.20)$$

Only the first terms of the previous series for ϕ and Δ are included, so that T must be chosen sufficiently small, often as one-tenth of the smallest system time constant.

Other Solutions

Standard routines such as the Runge–Kutta or predictor–corrector methods are quite satisfactory for smaller systems, but become very time consuming for, say, the large matrices in power system dynamics. The discrete-time solution or other available routines are needed for such cases.

A further method of solution is based on the use of coordinate transformations, a very important subject in its own right, and considered next.

12.5 EIGENVALUES, EIGENVECTORS, AND MODES

To simplify analysis and design for a system

$$\dot{x} = Ax + Bu \qquad y = Cx \qquad x(0) = x_0 \qquad (12.21)$$

it is often beneficial to define a new state vector z by a coordinate transformation

$$x = Tz \qquad (12.22)$$

Substituting this into (12.21) gives

$$T\dot{z} = ATz + Bu \qquad y = CTz \qquad x_0 = Tz_0$$

If T^{-1} exists, the transformed system is

$$\dot{z} = \hat{A}z + \hat{B}u \qquad y = \hat{C}z \qquad z(0) = z_0 \qquad (12.23)$$

$$\hat{A} = T^{-1}AT \qquad \hat{B} = T^{-1}B \qquad \hat{C} = CT \qquad z_0 = T^{-1}x_0$$

This definition of a new set of internal state variables should evidently not affect the eigenvalues or input–output behavior:

1.
$$|s\mathbf{I} - \hat{\mathbf{A}}| = |s\mathbf{I} - \mathbf{T}^{-1}\mathbf{A}\mathbf{T}| = |\mathbf{T}^{-1}(s\mathbf{I} - \mathbf{A})\mathbf{T}|$$
$$= |\mathbf{T}^{-1}| \, |s\mathbf{I} - \mathbf{A}| \, |\mathbf{T}| = |s\mathbf{I} - \mathbf{A}| \quad (12.24)$$

So the characteristic equation does not change.

2. The transfer function matrix of (12.4) becomes
$$\hat{\mathbf{G}} = \hat{\mathbf{C}}(s\mathbf{I} - \hat{\mathbf{A}})^{-1}\hat{\mathbf{B}} = \mathbf{C}\mathbf{T}(s\mathbf{I} - \mathbf{T}^{-1}\mathbf{A}\mathbf{T})^{-1}\mathbf{T}^{-1}\mathbf{B}$$
$$= \mathbf{C}\mathbf{T}[\mathbf{T}^{-1}(s\mathbf{I} - \mathbf{A})\mathbf{T}]^{-1}\mathbf{T}^{-1}\mathbf{B} = \mathbf{C}\mathbf{T}\mathbf{T}^{-1}(s\mathbf{I} - \mathbf{A})^{-1}\mathbf{T}\mathbf{T}^{-1}\mathbf{B}$$
$$= \mathbf{C}(s\mathbf{I} - \mathbf{A})^{-1}\mathbf{B} = \mathbf{G} \quad (12.25)$$

So **G** is also unaffected by the transformation.

$$*****$$

Of particular interest is the transformation $\mathbf{T} = \mathbf{U}$ for which $\hat{\mathbf{A}}$ is a diagonal matrix:
$$\mathbf{U}^{-1}\mathbf{A}\mathbf{U} = \Lambda = \text{diag}\,(\lambda_1, \lambda_2, \ldots, \lambda_n) \quad (12.26)$$
Since $|s\mathbf{I} - \mathbf{A}| = |s\mathbf{I} - \Lambda| = \det\{\text{diag}\,(s - \lambda_1, \ldots, s - \lambda_n)\}$
$$= (s - \lambda_1)(s - \lambda_2) \ldots\ldots (s - \lambda_n) \quad (12.27)$$
it follows that:

The elements λ_i of the diagonal matrix Λ are the eigenvalues, the roots of $|s\mathbf{I} - \mathbf{A}| = 0$.

Let

$$\mathbf{U} = [\mathbf{u}_1, \mathbf{u}_2, \ldots, \mathbf{u}_n] \,;\, \mathbf{u}_i = \begin{bmatrix} u_{1i} \\ \cdot \\ \cdot \\ \cdot \\ u_{ni} \end{bmatrix} = i\text{th column of } \mathbf{U} \quad (12.28)$$

Equation (12.26) shows that $\mathbf{A}\mathbf{U} = \mathbf{U}\Lambda$:
$$\mathbf{A}\,[\mathbf{u}_1, \mathbf{u}_2, \ldots, \mathbf{u}_n] = [\mathbf{u}_1, \mathbf{u}_2, \ldots, \mathbf{u}_n]\,\text{diag}\,(\lambda_1, \ldots, \lambda_n)$$
By equating the ith columns:
$$\mathbf{A}\mathbf{u}_i = \lambda_i\mathbf{u}_i \qquad (\lambda_i\mathbf{I} - \mathbf{A})\mathbf{u}_i = 0 \quad (12.29)$$

This is a set of homogeneous equations for $u_{1i}, \ldots, u_{ni}$, and has a nontrivial solution because the determinant $|\lambda_i\mathbf{I} - \mathbf{A}|$ of the coefficients is zero, by virtue of the fact that λ_i is an eigenvalue.

The solution $\mathbf{u}_i$ of (12.29) is the ith (right) "*eigenvector*" of **A**. Use of the "*modal matrix*" $\mathbf{U} = [\mathbf{u}_1, \ldots, \mathbf{u}_n]$ as a transformation matrix diagonalizes **A**.

Example 12.5.1

$$\mathbf{A} = \begin{bmatrix} -3 & 2 \\ -1 & 0 \end{bmatrix}; |\lambda\,\mathbf{I} - \mathbf{A}| = \lambda^2 + 3\lambda + 2.$$

The eigenvalues are $\lambda_1 = -1, \lambda_2 = -2$.

Eigenvector $\mathbf{u}_i$ is the solution of $\mathbf{A}\mathbf{u}_i = \lambda_i\mathbf{u}_i$:

$$\begin{bmatrix} -3 & 2 \\ -1 & 0 \end{bmatrix}\begin{bmatrix} u_{11} \\ u_{21} \end{bmatrix} = -1\begin{bmatrix} u_{11} \\ u_{21} \end{bmatrix}: \quad u_{11} = u_{21} \quad \mathbf{u}_1 = \begin{bmatrix} 1 \\ 1 \end{bmatrix}$$

$$\begin{bmatrix} -3 & 2 \\ -1 & 0 \end{bmatrix}\begin{bmatrix} u_{12} \\ u_{22} \end{bmatrix} = -2\begin{bmatrix} u_{12} \\ u_{22} \end{bmatrix}: \quad u_{12} = 2u_{22} \quad \mathbf{u}_2 = \begin{bmatrix} 1 \\ \frac{1}{2} \end{bmatrix}$$

Hence

$$\mathbf{U} = \begin{bmatrix} 1 & 1 \\ 1 & \frac{1}{2} \end{bmatrix}$$

The first and second rows are dependent in each case and give the same relation between u_{1i} and u_{2i}. Because $|\lambda_i\mathbf{I} - \mathbf{A}| = 0$, only the ratios of the elements of an eigenvector are fixed, and one element can always be chosen. Thus any multiples of $\mathbf{u}_1$ and $\mathbf{u}_2$ are also eigenvectors.

Numerous computer routines are available to compute both the eigenvalues and eigenvectors of large matrices.

Transition Matrix $e^{\mathbf{A}t}$

Application of the modal transformation $\mathbf{x} = \mathbf{U}\mathbf{z}$ to $\dot{\mathbf{x}} = \mathbf{A}\mathbf{x}$, $\mathbf{x}(0) = \mathbf{x}_0$ yields $\dot{\mathbf{z}} = \mathbf{\Lambda}\mathbf{z}$, $\mathbf{z}_0 = \mathbf{U}^{-1}\mathbf{x}_0$. The solution is $\mathbf{z}(t) = e^{\mathbf{\Lambda}t}\mathbf{z}_0 = e^{\mathbf{\Lambda}t}\mathbf{U}^{-1}\mathbf{x}_0$, and

$$\mathbf{x}(t) = \mathbf{U}\mathbf{z}(t) = \mathbf{U}e^{\mathbf{\Lambda}t}\mathbf{U}^{-1}\mathbf{x}_0$$

Since $\mathbf{x}(t)$ also equals $e^{\mathbf{A}t}\mathbf{x}_0$, it follows that

$$e^{\mathbf{A}t} = \mathbf{U}e^{\mathbf{\Lambda}t}\mathbf{U}^{-1} \qquad (12.30)$$

provides an additional way for calculation of the transition matrix. Here $e^{\mathbf{\Lambda}t}$ is a diagonal matrix. Using (12.12) gives

$$e^{\mathbf{\Lambda}t} = L^{-1}[(s\mathbf{I} - \mathbf{\Lambda})^{-1}] = L^{-1}[(\text{diag}\{s - \lambda_1, ..., s - \lambda_n\})^{-1}]$$

$$= L^{-1}[\text{diag}\{\frac{1}{s - \lambda_1}, ..., \frac{1}{s - \lambda_n}\}]$$

$$= \text{diag}(e^{\lambda_1 t}, e^{\lambda_2 t}, ..., e^{\lambda_n t}) \qquad (12.31)$$

Left Eigenvectors, and Eigenvector Normalization

In addition to the (right) eigenvectors above, the left eigenvectors are also used for analysis and design. The left eigenvectors $\mathbf{v}_i$ of $\mathbf{A}$ satisfy the equation and its transpose below, where the prime identifies a transpose:

$$\mathbf{v}_i'\mathbf{A} = \lambda_i\mathbf{v}_i' \qquad \mathbf{A}'\mathbf{v}_i = \lambda_i\mathbf{v}_i \qquad (12.32)$$

The second form implies that $\mathbf{v}_i$ is also the right eigenvector of $\mathbf{A}'$. Note that since $|\lambda\mathbf{I} - \mathbf{A}| = |\lambda\mathbf{I} - \mathbf{A}'|$, λ_i is also an eigenvalue of $\mathbf{A}'$.

Theorem 12.5.1

$$\mathbf{U} = [\mathbf{u}_1, ..., \mathbf{u}_n] \qquad \mathbf{V} = [\mathbf{v}_1, ..., \mathbf{v}_n] \qquad (12.33)$$

(i) $\mathbf{u}_j$ and $\mathbf{v}_i$, $j \neq i$, are orthogonal, that is, their inner product is zero: $\mathbf{v}_i'\mathbf{u}_j = 0$.

(ii) $\mathbf{u}_i$ and $\mathbf{v}_i$ can be normalized so that $\mathbf{v}_i'\mathbf{u}_i = 1$. Then

$$\mathbf{V}'\mathbf{U} = \mathbf{I}, \mathbf{V}' = \mathbf{U}^{-1}. \tag{12.34}$$

Proof. The scalar or inner product of $\mathbf{v}_i$ and $\mathbf{u}_j$ is defined by

$$\mathbf{v}_i'\mathbf{u}_j (= \mathbf{u}_j'\mathbf{v}_i) = [v_{1i} \quad v_{2i} \quad \cdots \quad v_{ni}] \begin{bmatrix} u_{1j} \\ \cdot \\ \cdot \\ \cdot \\ u_{nj} \end{bmatrix} = \sum_{k=1}^{n} v_{ki}u_{kj} \tag{12.35}$$

The vectors are orthogonal if this product is zero.

$$\mathbf{A}\mathbf{u}_j = \lambda_j\mathbf{u}_j, \text{ so } \mathbf{v}_i'\mathbf{A}\mathbf{u}_j = \lambda_j\mathbf{v}_i'\mathbf{u}_j.$$
$$\mathbf{A}'\mathbf{v}_i = \lambda_i\mathbf{v}_i \text{ or } \mathbf{v}_i'\mathbf{A} = \lambda_i\mathbf{v}_i', \text{ so } \mathbf{v}_i'\mathbf{A}\mathbf{u}_j = \lambda_i\mathbf{v}_i'\mathbf{u}_j.$$

The left sides of the two forms are the same, so subtracting both yields $(\lambda_j - \lambda_i)\mathbf{v}_i'\mathbf{u}_j = 0$. If, as will be assumed, the eigenvalues are all distinct, this proves part (i).

One element of each of $\mathbf{v}_i$ and $\mathbf{u}_i$ can be chosen, so the condition $\mathbf{v}_i'\mathbf{u}_i = 1$ of part (ii) can always be satisfied. Then

$$\mathbf{V}'\mathbf{U} = [\mathbf{v}_1 \cdots \mathbf{v}_n]'[\mathbf{u}_1 \cdots \mathbf{u}_n] = \begin{bmatrix} \mathbf{v}_1' \\ \cdot \\ \cdot \\ \cdot \\ \mathbf{v}_n' \end{bmatrix} [\mathbf{u}_1 \cdots \mathbf{u}_n] \tag{12.36}$$

$$= \begin{bmatrix} \mathbf{v}_1'\mathbf{u}_1 & \cdots & \mathbf{v}_1'\mathbf{u}_n \\ \cdot & & \\ \cdot & & \\ \cdot & & \\ \mathbf{v}_n'\mathbf{u}_1 & \cdots & \mathbf{v}_n'\mathbf{u}_n \end{bmatrix} = \mathbf{I}$$

Example 12.5.2   Example 12.5.1 Continued

$$\lambda_1 = -1 \qquad \lambda_2 = -2$$

$\mathbf{A}'\mathbf{v}_i = \lambda_i\mathbf{v}_i$, with $\mathbf{A}' = \begin{bmatrix} -3 & -1 \\ 2 & 0 \end{bmatrix}$ yields $\mathbf{v}_1 = \begin{bmatrix} -1 \\ 2 \end{bmatrix}$, $\mathbf{v}_2 = \begin{bmatrix} -1 \\ 1 \end{bmatrix}$, or any multiples. Here $\mathbf{v}_1'\mathbf{u}_1 = -1$, $\mathbf{v}_2'\mathbf{u}_2 = \frac{1}{2}$, and to make $\mathbf{v}_i'\mathbf{u}_i = 1$, $\mathbf{v}_1$ and $\mathbf{v}_2$ are rescaled to

$$\mathbf{v}_1 = \begin{bmatrix} -1 \\ 2 \end{bmatrix} \qquad \mathbf{v}_2 = \begin{bmatrix} 2 \\ -2 \end{bmatrix} \qquad (\mathbf{U}^{-1}\mathbf{A}\mathbf{U} = \mathbf{V}'\mathbf{A}\mathbf{U} = \mathbf{\Lambda})$$

Modal Decomposition of $\dot{\mathbf{x}} = \mathbf{A}\mathbf{x}$

With the transformation $\mathbf{x} = \mathbf{U}\mathbf{z}$, the equation $\dot{\mathbf{x}} = \mathbf{A}\mathbf{x}$, $\mathbf{x}(0) = \mathbf{x}_0$, becomes

$$\dot{\mathbf{z}} = \mathbf{\Lambda}\mathbf{z} \quad \text{or} \quad \dot{z}_i = \lambda_i z_i \qquad i = 1, ..., n \tag{12.37}$$

Since Λ is diagonal, the system is represented by n independent differential equations. The initial condition vector is

$$\mathbf{z}(0) = \mathbf{U}^{-1}\mathbf{x}(0) = \mathbf{V}'\mathbf{x}(0) = \begin{bmatrix} \mathbf{v}_1' \\ \cdot \\ \cdot \\ \cdot \\ \mathbf{v}_n' \end{bmatrix} \mathbf{x}(0) \quad \text{or} \quad z_i(0) = \mathbf{v}_i'\mathbf{x}_0 \qquad (12.38)$$

and the modal response to this initial condition:

$$z_i(t) = z_i(0)e^{\lambda_i t} = (\mathbf{v}_i'\mathbf{x}_0)e^{\lambda_i t} \qquad (12.39)$$

The response $\mathbf{x}$ is the sum of modal components:

$$\mathbf{x} = \mathbf{U}\mathbf{z} = [\mathbf{u}_1 \quad \ldots \quad \mathbf{u}_n]\mathbf{z} = \mathbf{u}_1 z_1 + \mathbf{u}_2 z_2 + \cdots + \mathbf{u}_n z_n$$
$$= (\mathbf{v}_1'\mathbf{x}_0)e^{\lambda_1 t}\mathbf{u}_1 + \cdots + (\mathbf{v}_n'\mathbf{x}_0)e^{\lambda_n t}\mathbf{u}_n \qquad (12.40)$$

This is the modal decomposition of $\mathbf{x}$. The ith row of $\mathbf{V}'$, or ith column of $\mathbf{V}$, is used to find the ith initial (scalar) mode size $\mathbf{v}_i'\mathbf{x}_0$ from $\mathbf{x}_0$. The (scalar) factor $e^{\lambda_i t}$ then gives the response in the mode, and $\mathbf{u}_i$ is the ith mode shape, which shows how the modal response is distributed over the elements of $\mathbf{x}$. The modal decomposition shows how the total response consists of a sum of responses in the individual modes. It also shows that if, for example, the initial condition is confined to the ith mode, that is, only $\mathbf{v}_i'\mathbf{x}_0$ is nonzero, the response will be only in the ith mode. This means that the ratios of the elements of $\mathbf{x}$ will equal those in $\mathbf{u}_i$.

Example 12.5.3   Examples 12.5.1 and 12.5.2 Continued

The modal decomposition is

$$\mathbf{x}(t) = [-1 \quad 2]\mathbf{x}_0 e^{-t} \begin{bmatrix} 1 \\ 1 \end{bmatrix} + [2 \quad -2]\mathbf{x}_0 e^{-2t} \begin{bmatrix} 1 \\ \frac{1}{2} \end{bmatrix}$$

The second mode decays faster than the first, so as t increases, the ratio of the elements of $\mathbf{x}(t)$ will increasingly approximate the first mode shape.

Complex Pairs of Eigenvalues

The case when $\mathbf{A}$ has repeated eigenvalues will not be considered. Transformation to diagonal form may then not be possible. The case of distinct complex pairs of eigenvalues is a direct generalization of that for a single pair discussed below.

If $\mathbf{A}$ has a pair of complex eigenvalues, then

$$\Lambda = \text{diag}\,(\lambda_1, \lambda_2, \ldots, \sigma + j\omega, \sigma - j\omega, \ldots, \lambda_n) \qquad (12.41)$$

and the corresponding adjacent columns of $\mathbf{U}$ are also complex conjugates. Real arithmetic is often preferable, and can be achieved by a further transformation: $\mathbf{z} = \mathbf{K}\mathbf{v}$ transforms $\dot{\mathbf{z}} = \Lambda\mathbf{z}$ to $\dot{\mathbf{v}} = \hat{\Lambda}\mathbf{v}$, where $\hat{\Lambda} = \mathbf{K}^{-1}\Lambda\mathbf{K} = \mathbf{K}^{-1}\mathbf{U}^{-1}\mathbf{A}\mathbf{U}\mathbf{K} = \mathbf{T}^{-1}\mathbf{A}\mathbf{T}$, where $\mathbf{T} = \mathbf{U}\mathbf{K}$. $\mathbf{K}$ and the resulting matrix $\hat{\Lambda}$ are

$$\mathbf{K} = \begin{bmatrix} 1 & & & & & 0 \\ & \ddots & & & & \\ & & 0.5 & -0.5_j & & \\ & & 0.5 & 0.5_j & & \\ & & & & \ddots & \\ 0 & & & & & 1 \end{bmatrix} \qquad \hat{\mathbf{\Lambda}} = \begin{bmatrix} \lambda_1 & & & & & 0 \\ & \ddots & & & & \\ & & \sigma & \omega & & \\ & & -\omega & \sigma & & \\ & & & & \ddots & \\ 0 & & & & & \lambda_n \end{bmatrix} \qquad (12.42)$$

T turns out to be real, with one column equal to the real part of the corresponding eigenvector and the next one to the imaginary part.

12.6 CONTROLLABILITY, STABILIZABILITY, OBSERVABILITY, IRREDUCIBILITY, AND MINIMALITY

The preceding sections have been concerned with methods to determine the stability and transient response of systems described by given state models. Before considering design, that is, how this behavior may be changed, it is necessary to introduce some new concepts.

Generally, there are fewer control variables than state variables. Hence the question whether it is at all possible to control all states, disregarding how this might be done.

> **Definition.** The system $\dot{\mathbf{x}} = \mathbf{A}\mathbf{x} + \mathbf{B}\mathbf{u}$, or the pair $(\mathbf{A}, \mathbf{B})$, is state *controllable* if and only if there exists a control $\mathbf{u}$ that will transfer any initial state $\mathbf{x}(0)$ to any final state $\mathbf{x}(T)$ in finite time T.

Controllability is an important concept in the theory and design of multivariable systems, and can be determined from the following theorem.

> **Theorem 12.6.1.** The pair $(\mathbf{A}, \mathbf{B})$ is state controllable if and only if the rank of the *controllability matrix*
> $$\mathbf{\Gamma} = [\mathbf{B} \quad \mathbf{A}\mathbf{B} \quad \mathbf{A}^2\mathbf{B} \quad \cdots \quad \mathbf{A}^{n-1}\mathbf{B}] \qquad (12.43)$$
> is n (i.e., it contains an $n \times n$ nonsingular matrix).

The theorem can be proved by the use of transformations from the alternative form discussed below.

Observability is a dual concept. The number of output variables measured is usually smaller than the number of state variables. Since, as is discussed later, it is desirable to use feedback from all the state variables, the question is whether it is at all possible to find all states from the measured outputs, regardless of the method used. It can be shown that this is possible, and that the state is observable, or the pair $(\mathbf{C}, \mathbf{A})$ is observable, if the following matrix has rank n:

$$\textit{Observability matrix} \quad [\mathbf{C}' \quad \mathbf{A}'\mathbf{C}' \quad \mathbf{A}'^2\mathbf{C}' \quad \cdots \quad \mathbf{A}'^{n-1}\mathbf{C}'] \qquad (12.44)$$

where $\mathbf{A}'$ and $\mathbf{C}'$ are the transposes of $\mathbf{A}$ and $\mathbf{C}$.

*****

An alternative controllability theorem, used to explain the concept of stabilizability, is based on the modal approach. With $\mathbf{x} = \mathbf{Uz}$, the state equation $\dot{\mathbf{x}} = \mathbf{Ax} + \mathbf{Bu}$ becomes, using (12.34),

$$\dot{\mathbf{z}} = \mathbf{\Lambda z} + (\mathbf{U}^{-1}\mathbf{B})\mathbf{u} = \mathbf{\Lambda z} + (\mathbf{V'B})\mathbf{u} \qquad (12.45)$$

Theorem 12.6.2.   The pair $(\mathbf{A}, \mathbf{B})$ is controllable if and only if $\mathbf{V'B}$ has no rows consisting entirely of zero elements.

The "only if" part is obvious, since if the ith row is zero, then, from (12.45), $\dot{z}_i = \lambda_i z_i$. Therefore, the ith mode is uncontrollable, since it is not affected by the control.

Now, if an uncontrollable mode is stable, it will decay in any case. This is clearly much less serious than an uncontrollable unstable mode, because in this case the system cannot be stabilized.

Theorem 12.6.3.   The pair $(\mathbf{A}, \mathbf{B})$ is *stabilizable* if it is controllable, or if it is uncontrollable but the uncontrollable modes are stable.

Thus, if a system has unstable eigenvalues and Theorem 12.6.1 shows it to be uncontrollable, the unstable modes can be checked one at a time to determine whether the system is stabilizable.

Example 12.6.1

$$\mathbf{A} = \begin{bmatrix} 0 & -1 \\ -3 & 2 \end{bmatrix} \qquad \mathbf{B} = \begin{bmatrix} 1 \\ -3 \end{bmatrix}$$

The controllability matrix (12.43) is

$$[\mathbf{B} \quad \mathbf{AB}] = \begin{bmatrix} 1 & 3 \\ -3 & -9 \end{bmatrix}$$

and has a rank of 1 since the determinant is zero (i.e., the matrix is singular). $|s\mathbf{I} - \mathbf{A}| = s^2 - 2s - 3$, and the eigenvalues are $\lambda_1 = -1$, $\lambda_2 = +3$. Therefore, to determine whether this uncontrollable system is stabilizable, the left eigenvector $\mathbf{v}_2$ is found from $\mathbf{A'v}_2 = +3\mathbf{v}_2$. It is found that $\mathbf{v}_2' = [1 \quad -1]$, so that $\mathbf{v}_2'\mathbf{B} = 4 \neq 0$. Hence the unstable mode is controllable, so the system is stabilizable.

Finally, the relationships between controllability and observability and cancellations between numerator and denominator of a transfer function are introduced. This involves the new concepts of irreducibility and minimal realizations of transfer functions.

In $\dot{\mathbf{x}} = \mathbf{Ax} + \mathbf{Bu}$, $\mathbf{y} = \mathbf{Cx}$, if $\mathbf{c}$ is the ith row of $\mathbf{C}$ and $\mathbf{b}$ the jth column of $\mathbf{B}$, then the ith output y and jth input u are related by

$$\dot{\mathbf{x}} = \mathbf{Ax} + \mathbf{b}u \qquad y = \mathbf{cx} \qquad (12.46)$$

and, by (12.6), the corresponding transfer function is

$$G(s) = \frac{\mathbf{c} \text{ adj } (s\mathbf{I} - \mathbf{A})\mathbf{b}}{|s\mathbf{I} - \mathbf{A}|} = \frac{p(s)}{q(s)} \qquad (12.47)$$

where $q(s) = |s\mathbf{I} - \mathbf{A}|$ is of order n. The state model (12.46), with n state variables, is called a *realization* of the transfer function $G(s)$.

Now, it can easily be that $p(s)$ and $q(s)$ have common factors (i.e., that they are not relatively prime or coprime). If this is the case, then (12.46) is not a *minimal realization*. A realization $\{\mathbf{A}, \mathbf{b}, \mathbf{c}\}$ of $G(s)$ is minimal if it has the smallest possible number of state variables, and it can be shown that

> A realization $\{\mathbf{A}, \mathbf{b}, \mathbf{c}\}$ of $G(s)$ is minimal if and only if $q(s) = |s\mathbf{I} - \mathbf{A}|$ and $p(s) =$ $\mathbf{c}\ adj\ (s\mathbf{I} - \mathbf{A})\mathbf{b}$ are relatively prime (i.e., have no common factors).

To relate this to the concepts of controllability and observability, the following can be shown:

1. If $q(s)$ is of order n, then an nth-order realization $\{\mathbf{A}, \mathbf{b}, \mathbf{c}\}$ of $G(s)$ will be controllable and observable if and only if $G(s) = p(s)/q(s)$ is irreducible, that is, if $p(s)$ and $q(s)$ have no common factors.
2. A realization $\{\mathbf{A}, \mathbf{b}, \mathbf{c}\}$ of $G(s)$ is minimal if and only if $\{\mathbf{A}, \mathbf{b}\}$ is controllable and $\{\mathbf{c}, \mathbf{A}\}$ observable.

Thus, if a realization $\{\mathbf{A}, \mathbf{b}, \mathbf{c}\}$ is made of a transfer function $G(s)$ which is reducible, this state model will be found to be uncontrollable or unobservable or both. And if the common factors in $p(s)$ and $q(s)$ are canceled and a state model made of the resulting transfer function, this model will hide the modes corresponding to the canceled factors.

12.7 CONCLUSION

In this chapter the modeling and analysis of systems in state space has been introduced, and several new concepts have been defined which are important for both analysis and design. This material is the subject of a very extensive literature, ranging from the practical to the highly theoretical. The great advantage of the state space approach is that the standard matrix formulation has allowed the development of standard computer programs suitable for the analysis, and design, of large systems.

PROBLEMS

12.1. Derive state space models for systems described by the following transfer functions.

(a) $\dfrac{C(s)}{R(s)} = \dfrac{6}{s^3 + 3s^2 + 5s + 1}$ (b) $\dfrac{C(s)}{R(s)} = \dfrac{6(s + 1)}{s^3 + 3s^2 + 5s + 1}$

12.2. Derive state space models for systems described by the following transfer functions.

(a) $\dfrac{C(s)}{R(s)} = \dfrac{5}{(s + 1)(s^2 + 3s + 5)}$ (b) $\dfrac{C(s)}{R(s)} = \dfrac{s + 2}{s(s^2 + 6s + 1)}$

12.3. Derive state models for the following systems.

(a) $\dfrac{C(s)}{R(s)} = \dfrac{4s + 3}{3s^2 + 5s + 4}$ (b) $\dfrac{C(s)}{R(s)} = 2\,\dfrac{s + 3}{s + 10}$ (c) $\dfrac{C(s)}{R(s)} = \dfrac{s^2 + 3s + 4}{s^2 + 7s + 9}$

12.4. Obtain three alternative state models for the system in Fig. P12.4 with $G_1 = 5/(s + 1)$, $G_2 = 1/(s + 5)$, and $G_3 = 1/(s + 10)$:

(a) One where the system matrix **A** is a companion matrix.

(b) One where **A** is a diagonal matrix. (Use partial fraction expansion of $C(s)/R(s)$ [= $(X_1/R) + (X_2/R) + (X_3/R)$].)

(c) One where the physically meaningful and measurable outputs of the simple lag blocks are chosen to be the state variables.

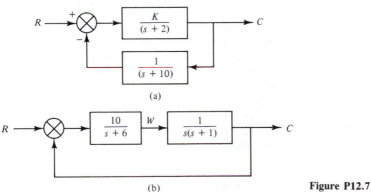

$R \longrightarrow \boxed{G_1} \longrightarrow \boxed{G_2} \longrightarrow \boxed{G_3} \longrightarrow C$

Figure P12.4

12.5. Repeat Problem 12.4 for $G_1 = 1/(s + 3)$, $G_2 = 1/(s + 1)$, and $G_3 = 1/(s + 2)$.

12.6. Obtain state models for the system in Fig. P12.4, using a meaningful selection of state variables, if $G_3 = 1/[(s + 1)(s + 2)]$ is the plant and G_1G_2 is a controller D given by:

(a) $D(s) = 1 + \dfrac{1.2}{s}$ (b) $D(s) = \dfrac{s + 3}{s + 10}$

12.7. Selecting meaningful state variables, derive state models for the systems in Fig. P12.7.

$R \longrightarrow \otimes \xrightarrow{\;+\;} \boxed{\dfrac{K}{(s + 2)}} \longrightarrow C$

$\boxed{\dfrac{1}{(s + 10)}}$

(a)

$R \longrightarrow \otimes \longrightarrow \boxed{\dfrac{10}{s + 6}} \xrightarrow{W} \boxed{\dfrac{1}{s(s + 1)}} \longrightarrow C$

(b)   **Figure P12.7**

12.8. Figure P12.8 shows a motor position servo with velocity feedback and a disturbance input D in addition to the reference input R. Obtain a state model.

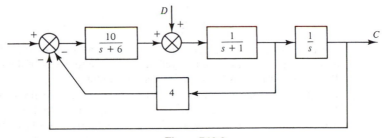

$\otimes \xrightarrow{+} \boxed{\dfrac{10}{s + 6}} \xrightarrow{+} \otimes \longrightarrow \boxed{\dfrac{1}{s + 1}} \longrightarrow \boxed{\dfrac{1}{s}} \longrightarrow C$

$\boxed{4}$

Figure P12.8

12.9. Determine state models for the systems shown in Fig. P12.9.

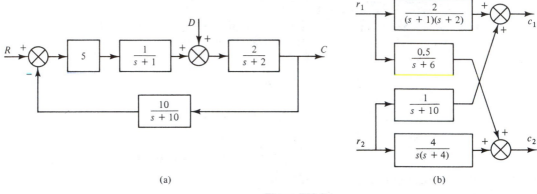

(a)   (b)

Figure P12.9

12.10. Derive a state model for the two-input/two-output feedback control system shown in Fig. P12.10.

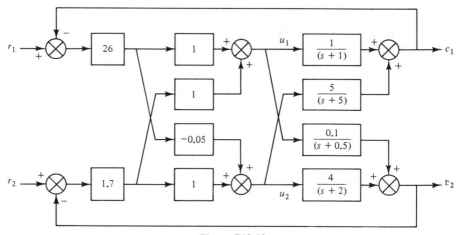

Figure P12.10

12.11. The point model equations for a nuclear reactor with allowance for six groups of delayed neutrons are

$$\dot{n} = \frac{\delta k - \beta}{\Lambda} n + \sum_{i=1}^{6} \lambda_i C_i \qquad \dot{C}_i = \frac{\beta_i}{\Lambda} n - \lambda_i C_i \qquad i = 1, \dots, 6$$

where n is the neutron density, C_i the concentration of the ith group precursor, and $\lambda_i C_i$ the growth rate of delayed neutrons due to decay of the ith group precursor. Put the equations in the form of a state model.

12.12. For the two-tank system in Fig. P12.12, q_1 and q_2 are liquid volume flow rates, h_1 and h_2 are liquid levels, where h_2 is the system output, and A_1 and A_2 are tank areas equal to $A_1 = 8$, $A_2 = 4$. The linearized models for flow through the resistors

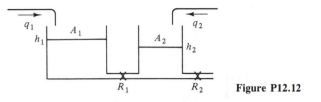

Figure P12.12

R_1 and R_2 can be taken to be $w_1 = (h_1 - h_2)/3$, $w_2 = h_2/2$. Derive a state model, selecting meaningful and measurable state variables.

12.13. Derive a state model, with $y(t)$ as output, for the representation of a train shown in Fig. P12.13.

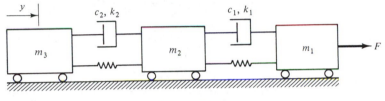

Figure P12.13

12.14. Derive a linearized state model for the system in Fig. P12.14, where h_1 and h_2 are levels, with h_2 the output, q_1 and q_2 volume flow rates, A_1 and A_2 areas, and R_1 and R_2 linearized valve resistances.

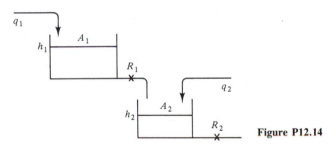

Figure P12.14

12.15. Derive a state model to represent the field-controlled dc servomotor shown in Fig. P12.15. Choose θ, $\dot{\theta}$, and i_f as state variables. θ is the output.

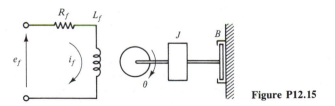

Figure P12.15

12.16. Derive a state model representation, with displacement y of mass m as output, for the mechanical system in Fig. P12.16.

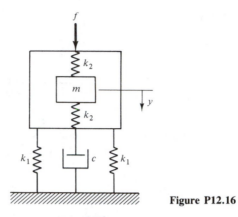

Figure P12.16

12.17. Derive two state models for the system with transfer function

$$\frac{C(s)}{R(s)} = \frac{s + 3}{s(s + 1)(s + 2)}$$

(a) One for which the system matrix is a companion matrix.
(b) One for which the system matrix is diagonal.
[Use partial fraction expansion of $C(s)/R(s)$ and define $C/R = X_1/R + X_2/R + X_3/R$.]

12.18. Determine the transfer function matrices $\mathbf{G}(s)$ corresponding to the following state models:

(a) $\dot{\mathbf{x}} = \begin{bmatrix} 0 & -1 \\ 0 & -2 \end{bmatrix} \mathbf{x} + \begin{bmatrix} 0 \\ 1 \end{bmatrix} u; \mathbf{y} = \begin{bmatrix} 1 & 0 \\ 0 & 1 \end{bmatrix} \mathbf{x}$

(b) $\dot{\mathbf{x}} = \begin{bmatrix} 0 & 1 \\ -2 & -3 \end{bmatrix} \mathbf{x} + \begin{bmatrix} 0 & 1 \\ 1 & 0 \end{bmatrix} \mathbf{u}; \mathbf{y} = \begin{bmatrix} 1 & 0 \\ 1 & -1 \end{bmatrix} \mathbf{x}$

12.19. For

$$\dot{\mathbf{x}} = \begin{bmatrix} -3/4 & -1/4 \\ -1/4 & -3/4 \end{bmatrix} \mathbf{x} + \begin{bmatrix} 0 \\ 1 \end{bmatrix} u$$

(a) Determine stability.
(b) Find the transfer function relating x_1 and u.

12.20. A single-input/single-output system is described by the state model

$$\dot{\mathbf{x}} = \begin{bmatrix} -2 & 0 & 0 \\ -1 & -4 & 0 \\ -1 & -1 & -5 \end{bmatrix} \mathbf{x} + \begin{bmatrix} 1 \\ 1 \\ 1 \end{bmatrix} u \qquad y = [0 \quad 0 \quad 1] \mathbf{x}$$

Find the corresponding transfer function $Y(s)/U(s)$. Note that instead of by inverting $s\mathbf{I} - \mathbf{A}$, this transfer function can be found also by transforming the individual state equations and eliminating variables between them.

12.21. In Problem 12.19:
(a) Evaluate the matrices ϕ and Δ in the discrete-time solution of (12.16) for $T = 1$.

(b) Use them to find $x(1)$, $x(2)$, and $x(3)$ for a unit step input, with zero initial state.

12.22. For $\dot{x} = Ax + bu$, $A = \begin{bmatrix} 0 & -1 \\ 0 & -2 \end{bmatrix}$, $b = \begin{bmatrix} 0 \\ 1 \end{bmatrix}$:

(a) Determine the transition matrix $\phi(t) = e^{At} = L^{-1}(sI - A)^{-1}$ by Laplace transforms.
(b) Verify part (a) using the series $\Sigma A^k t^k / k!$
(c) Find the state response to a unit step input.

12.23. For the system

$$\dot{x} = \begin{bmatrix} 1 & -2 \\ 2 & -3 \end{bmatrix} x$$

(a) Determine stability.
(b) Find the transition matrix $\phi = e^{At} = L^{-1}(sI - A)^{-1}$.
(c) Find the state response to initial conditions $x_1(0) = 2$, $x_2(0) = 1$.

12.24. For the system

$$\dot{x} = \begin{bmatrix} -1 & 0 \\ 1 & -2 \end{bmatrix} x$$

(a) Determine the transition matrix.
(b) Use it to evaluate matrix ϕ in the discrete-time solution of (12.16) for $T = 1$ and express the state response $x(1)$, $x(2)$, $x(3)$ for an arbitrary initial condition x_0.

12.25. Determine the output response of the system

$$\dot{x} = \begin{bmatrix} -1 & 1 \\ 0 & -2 \end{bmatrix} x + \begin{bmatrix} 0 \\ 1 \end{bmatrix} u \qquad y = \begin{bmatrix} 1 & 0 \\ 0 & 1 \end{bmatrix} x \qquad x_0 = \begin{bmatrix} +1 \\ +1 \end{bmatrix}$$

to the initial conditions and a unit step input.

12.26. Determine the response of the system

$$\dot{x} = \begin{bmatrix} 0 & 1 \\ -6 & -5 \end{bmatrix} x + \begin{bmatrix} 0 \\ 1 \end{bmatrix} u \qquad y = [1 \quad 0]x$$

to a unit step input.

12.27. For the system

$$\dot{x} = \begin{bmatrix} 0 & 1 \\ -6 & -5 \end{bmatrix} x$$

determine the eigenvalues and the eigenvectors, and use these results to find the transition matrix.

12.28. For the system

$$\dot{x} = \begin{bmatrix} 0 & 1 \\ -2 & -3 \end{bmatrix} x + \begin{bmatrix} 0 \\ 2 \end{bmatrix} u \qquad y = [3 \quad 1]x$$

(a) Determine the eigenvalues and right and left (normalized) eigenvectors.

(b) Transform the state model to diagonal form.

(c) Express the modal decomposition of the state response to initial condition

$$\mathbf{x}_0 = \begin{bmatrix} 2 \\ 3 \end{bmatrix}.$$

(d) Verify part (c) by calculation of the transition matrix.

12.29. For the system

$$\dot{\mathbf{x}} = \begin{bmatrix} 0 & -1 \\ 1 & 0 \end{bmatrix} \mathbf{x}$$

determine the transition matrix by the use of eigenvalues and eigenvectors:

(a) By the standard method, using complex arithmetic.

(b) Using the approach of (12.42), with real arithmetic.

12.30. For the system of Problem 12.24:

(a) Find the eigenvalues and normalized right and left eigenvectors.

(b) Use them to find the transition matrix.

(c) Find the modal form of the system model.

(d) Find the modal decomposition of the response to initial conditions $\mathbf{x}_0$ and use it to verify part (b).

12.31. For the system

$$\dot{\mathbf{x}} = \begin{bmatrix} -3 & 2 \\ 4 & -5 \end{bmatrix} \mathbf{x} \qquad \mathbf{x}_0 = \begin{bmatrix} 5 \\ 3 \end{bmatrix}$$

(a) Find the eigenvalues and normalized right and left eigenvectors.

(b) Express the response to $\mathbf{x}_0$ by modal decomposition.

12.32. Verify the solution in Problem 12.31(b) by calculation of the transition matrix by two different techniques.

12.33. Determine the transition matrix of each of the following systems by the use of eigenvalues and eigenvectors.

(a) $\dot{\mathbf{x}} = \begin{bmatrix} 0 & 1 \\ 0 & -2 \end{bmatrix} \mathbf{x}$ **(b)** $\dot{\mathbf{x}} = \begin{bmatrix} -1 & 1 \\ 0 & -2 \end{bmatrix} \mathbf{x}$

12.34. For a system with transfer function

$$\frac{Y(s)}{W(s)} = \frac{1}{s(s+1)(s+2)}$$

(a) Obtain a state model.

(b) Find normalized right and left eigenvectors.

(c) Determine the transition matrix.

12.35. Check the controllability of the following pairs $(\mathbf{A}, \mathbf{B})$.

(a) $\begin{bmatrix} 0 & 1 \\ -2 & -3 \end{bmatrix}, \begin{bmatrix} 0 \\ 2 \end{bmatrix}$ **(b)** $\begin{bmatrix} 1 & 1 \\ 1 & 0 \end{bmatrix}, \begin{bmatrix} 1 \\ 0 \end{bmatrix}$

(c) $\begin{bmatrix} 0 & 1 & 0 \\ 0 & 0 & 1 \\ 0 & -1 & -2 \end{bmatrix}, \begin{bmatrix} 0 \\ 1 \\ 1 \end{bmatrix}$ **(d)** $\begin{bmatrix} 1 & 0 & 0 \\ 0 & 0 & 1 \\ 1 & -2 & -1 \end{bmatrix}, \begin{bmatrix} 0 \\ 0 \\ 1 \end{bmatrix}$

12.36. Check the controllability and stabilizability of the following pairs (**A**, **B**).

(a) $\begin{bmatrix} -3 & 1 \\ -2 & 1.5 \end{bmatrix}$, $\begin{bmatrix} 0 \\ 1 \end{bmatrix}$ (b) $\begin{bmatrix} 0 & 1 \\ 2 & -1 \end{bmatrix}$, $\begin{bmatrix} 1 \\ 1 \end{bmatrix}$

12.37. Check the controllability and stabilizability of the following pairs (**A**, **B**).

(a) $\begin{bmatrix} -3 & 1 \\ -2 & 1.5 \end{bmatrix}$, $\begin{bmatrix} 1 \\ 4 \end{bmatrix}$ (b) $\begin{bmatrix} 0 & 1 & 0 \\ 0 & 0 & 1 \\ a & b & c \end{bmatrix}$, $\begin{bmatrix} 0 \\ 0 \\ 1 \end{bmatrix}$

12.38. Check the observability of the model with state equation

$$\dot{\mathbf{x}} = \begin{bmatrix} 0 & 1 & 0 \\ 0 & 0 & 1 \\ -6 & -11 & -6 \end{bmatrix} \mathbf{x} + \begin{bmatrix} 0 \\ 0 \\ 1 \end{bmatrix} u$$

for the following output equations.

(a) $y = \begin{bmatrix} 3 & 2 & 1 \end{bmatrix} \mathbf{x}$ (b) $y = \begin{bmatrix} 4 & 5 & 1 \end{bmatrix} \mathbf{x}$

12.39. Noting that the system matrix in Problem 12.38 is a companion matrix and that $s^3 + 6s^2 + 11s + 6 = (s + 1)(s + 2)(s + 3)$:

(a) Derive the system transfer function for the output equation in Problem 12.38(b). Remember that the output equation shows numerator dynamics.

(b) Is this transfer function reducible?

(c) Does the state model represent a minimal realization of this transfer function?

(d) Relate these results to an earlier statement that pole–zero cancellation can be used to improve output response but does not consider the response to initial conditions.

INTRODUCTION TO STATE SPACE DESIGN

13.1 INTRODUCTION

In this brief introduction to a very large area of activity, one goal is to provide a framework to facilitate further study. The basic pole assignment and optimal control approaches to design are introduced, as are the topics of modal control, integral control, feedforward, and constant or dynamic output feedback. A brief section on digital control systems concludes the chapter.

13.2 POLE ASSIGNMENT AND OPTIMAL CONTROL DESIGN USING STATE FEEDBACK

For system design, consider the classical feedback control configuration in Fig. 13.1. The plant is described by its state space model, and it is assumed that all state variables are available for use in feedback. For now, the system inputs are taken to be zero, with the purpose of control being to change the initial state x_0 to $x = 0$. The *state feedback* control in Fig. 13.1 is then

$$u = -Kx \tag{13.1a}$$

and the closed-loop system equation is

$$\dot{x} = Ax + Bu = (A - BK)x \tag{13.1b}$$

$(A - BK)$ is the closed-loop system matrix and must have stable eigenvalues for the response to tend to zero.

The following theorem is of fundamental importance.

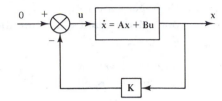

Figure 13.1   State feedback.

Feedback Control Theorem 13.2.1.   Any specified set of closed-loop eigen-values can be obtained by state feedback with **K** consisting of constant gains if and only if the pair (**A**, **B**) is controllable.

The "only if" part is obvious, and the "if" part will be proved by design of **K** via modal control in the next section. Constant gains can achieve this, while dynamic compensation was often found necessary just for adequate performance in classical single input/single output design, because there is feedback from all states instead of only from a single output.

Two major approaches to the design of **K** are the following:

1. *Pole assignment:* The closed-loop eigenvalues are placed in specified locations.
2. *Optimal control:* A specified mathematical performance criterion is minimized.

The famous optimal regulator problem is formulated and considered briefly in this section, and techniques for pole assignment are discussed in this section and the next. A section is also devoted to the multivariable generalizations of integral control and feedforward, for systems with disturbances and nonzero inputs. In Section 13.5 this framework for multivariable state space design is completed by discussion of constant and dynamic compensation for output feedback, when not all states are available for feedback.

Pole Assignment Using Companion Matrices

Consider a single-input plant described by the nth-order differential equation

$$\frac{d^n y}{dt^n} + a_n \frac{d^{n-1} y}{dt^{n-1}} + \cdots + a_2 \frac{dy}{dt} + a_1 y = c_m \frac{d^{m-1} u}{dt^{m-1}} + \cdots + c_2 \frac{du}{dt} + c_1 u \quad (13.2)$$

As in Examples 12.2.2 and 12.2.3, this can be represented by the state model

$\dot{\mathbf{x}} = \mathbf{A}\mathbf{x} + \mathbf{b}u$
$y = \mathbf{C}\mathbf{x}$
$\mathbf{C} = [c_1 \quad c_2 \quad \cdots \quad c_m \quad 0 \quad \cdots \quad 0]$

$$\mathbf{A} = \begin{bmatrix} 0 & 1 & \cdots & 0 \\ . & & & . \\ . & & & . \\ . & & & . \\ 0 & & & 1 \\ -a_1 & -a_2 & \cdots & -a_n \end{bmatrix} \qquad \mathbf{b} = \begin{bmatrix} 0 \\ . \\ . \\ . \\ 0 \\ 1 \end{bmatrix} \qquad (13.3)$$

The open-loop eigenvalues are the eigenvalues of the companion matrix $\mathbf{A}$ and, from (13.2), they are also the roots of the plant characteristic equation

$$s^n + a_n s^{n-1} + \cdots + a_2 s + a_1 = 0 \tag{13.4}$$

which has the same coefficients as the last row of $\mathbf{A}$, with opposite signs and in reverse order. Now assume a state feedback

$$u = -\mathbf{k}'\mathbf{x} = -[k_1 k_2 \cdots k_n]\mathbf{x} \tag{13.5}$$

where the prime identifies the transpose of a column vector $\mathbf{k}$. The closed-loop system matrix, in (13.6a), is then again a companion matrix, so the closed-loop eigenvalues must be the roots of the closed-loop characteristic equation (13.6b).

$$\mathbf{A} - \begin{bmatrix} 0 \\ \cdot \\ \cdot \\ \cdot \\ 0 \\ 1 \end{bmatrix} [k_1 \ \cdots \ k_n] = \begin{bmatrix} 0 & 1 & \cdots & 0 \\ \cdot & & & \cdot \\ \cdot & & & \cdot \\ \cdot & & & 1 \\ -a_1 - k_1 & -a_2 - k_2 & \cdots & -a_n - k_n \end{bmatrix} \tag{13.6a}$$

$$s^n + (a_n + k_n)s^{n-1} + \cdots + (a_2 + k_2)s + (a_1 + k_1) = 0 \tag{13.6b}$$

Let the desired closed-loop eigenvalues be $\hat{\lambda}_1, \ldots, \hat{\lambda}_n$, so the desired characteristic equation

$$(s - \hat{\lambda}_1)(s - \hat{\lambda}_2) \cdots (s - \hat{\lambda}_n) = s^n + a_{dn}s^{n-1} + \cdots + a_{d2}s + a_{d1} = 0 \tag{13.7}$$

Equation (13.6b) will be equal to this if

$$\mathbf{k}' = [a_{d1} - a_1 \quad a_{d2} - a_2 \quad \cdots \quad a_{dn} - a_n] \tag{13.8}$$

Example 13.2.1

$$\dot{\mathbf{x}} = \begin{bmatrix} 0 & 1 & 0 \\ 0 & 0 & 1 \\ -6 & -11 & -6 \end{bmatrix} \mathbf{x} + \begin{bmatrix} 0 \\ 0 \\ 3 \end{bmatrix} u$$

With $u = -\mathbf{k}'\mathbf{x}$, the elements in the last row of $\mathbf{A} - \mathbf{bk}'$ are $-6 - 3k_1$, $-11 - 3k_2$, and $-6 - 3k_3$. If $\hat{\lambda}_1 = -2$, $\hat{\lambda}_2 = -3$, and $\hat{\lambda}_3 = -4$, the desired characteristic equation is $(s + 2)(s + 3)(s + 4) = s^3 + 9s^2 + 26s + 24 = 0$, so the desired elements in the last row of $\mathbf{A} - \mathbf{bk}'$ are -24, -26, and -9. Hence

$$\mathbf{k}' = [6 \quad 5 \quad 1]$$

If, as is usual, the state model is not in companion form, computer routines for a coordinate transformation to this form are available. Transformations to particular canonical forms are also important in several techniques for multiple-input systems. A common alternative here is to reduce them in effect to single-input systems by fixing the ratios of the inputs according to $\mathbf{u} = \boldsymbol{\alpha}w$, where $\boldsymbol{\alpha}$ is a vector of constants and w the new single input. Then $\dot{\mathbf{x}} = \mathbf{Ax} + \mathbf{Bu} = \mathbf{Ax} + (\mathbf{B}\boldsymbol{\alpha})w = \mathbf{Ax} + \mathbf{b}w$, and $\boldsymbol{\alpha}$ must be chosen such that $(\mathbf{A}, \mathbf{b})$ is controllable.

Optimal Regulator Problem

The eigenvalues above could be placed far from the imaginary axis, making the speed of response arbitrarily fast. But this would require large control inputs and actuator capacities. A tank, say, can be filled arbitrarily fast if the supply flow rate and the rate at which it can be changed are large enough. However, this implies a high cost of control. An *optimal control* implies a trade-off between performance and cost of control, and this is also what determines the choice of desired eigenvalues in pole assignment techniques. In optimal control, the control is sought which minimizes the value of a *performance index J*, which is often of the standard form

$$J = \frac{1}{2} \int_0^\infty (\mathbf{x'Qx} + \mathbf{u'Ru})\, dt$$

(13.9)

$$\mathbf{Q} = \text{diag}\,(q_i) \qquad \mathbf{R} = \text{diag}\,(r_i)$$

Here $\mathbf{Q}$ and $\mathbf{R}$ are diagonal *weighting matrices* and $\mathbf{x'Qx}$ and $\mathbf{u'Ru}$ are scalar *quadratic forms* which measure, respectively, the performance and the cost of control:

$$\mathbf{x'Qx} = [x_1 \ \cdots \ x_n] \begin{bmatrix} q_1 & \cdots & 0 \\ & \cdot & \\ & \cdot & \\ & \cdot & \\ 0 & \cdots & q_n \end{bmatrix} \begin{bmatrix} x_1 \\ \cdot \\ \cdot \\ \cdot \\ x_n \end{bmatrix} = [x_1 \ \cdots \ x_n] \begin{bmatrix} q_1 x_1 \\ \cdot \\ \cdot \\ q_n x_n \end{bmatrix}$$

$$= \sum_{i=1}^n q_i x_i^2$$

(13.10a)

$$\mathbf{u'Ru} = \sum_{j=1}^r r_j u_j^2 \qquad J = \frac{1}{2} \int_0^\infty \left(\sum_i q_i x_i^2 + \sum_j r_j u_j^2 \right) dt$$

(13.10b)

Thus the optimal control minimizes a weighted sum of areas under x_i^2 and u_j^2 curves such as that shown in Fig. 13.2, with the weightings determined by the choice of the elements of $\mathbf{Q}$ and $\mathbf{R}$. Increasing the r_j relative to the q_i increases the weighting on the control and has the effect of reducing the control inputs at the expense of the response.

There are several ways, omitted here, of proving the following key result.

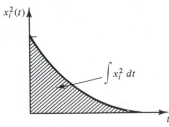

Figure 13.2 Optimal control.

Optimal Regulator Theorem 13.2.2.   The optimal control is a constant-gain state feedback

$$\mathbf{u}_{opt} = -\mathbf{Kx} \qquad \mathbf{K} = \mathbf{R}^{-1}\mathbf{B'P} \tag{13.11a}$$

where $\mathbf{P}$ is a symmetric matrix obtained by solution of the algebraic matrix *Riccati equation*

$$\mathbf{PA} + \mathbf{A'P} + \mathbf{Q} - \mathbf{PBR}^{-1}\mathbf{B'P} = 0 \tag{13.11b}$$

and, under mild restrictions, gives a stable closed-loop system.

In very simple cases the Riccati equation can be solved directly, but usually computer solution is required. A number of computer routines for this purpose are available.

It is necessary to observe that the choice of desired eigenvalues in pole assignment and of the weighting matrices in optimal control generally involves considerable trial and error until the resulting system is satisfactory from the point of view of the transient responses of states and control inputs to, say, step inputs.

13.3 MODAL CONTROL FOR POLE ASSIGNMENT USING STATE FEEDBACK

In this alternative to the pole assignment technique of Section 13.2, use is made of the modal transformation $\mathbf{x} = \mathbf{Uz}$ and the properties $\mathbf{U}^{-1} = \mathbf{V'}$, $\mathbf{v}_i'\mathbf{u}_j = 0$ $(i \neq j)$, $\mathbf{v}_i'\mathbf{u}_i = 1$, $\mathbf{Au}_i = \lambda_i\mathbf{u}_i$ corresponding to this transformation. From $\mathbf{z} = \mathbf{U}^{-1}\mathbf{x} = \mathbf{V'x}$, the ith modal variable is $z_i = \mathbf{v}_i'\mathbf{x}$, where $\mathbf{v}_i'$ is the ith row of $\mathbf{V'}$.

Consider a single-input system with state feedback

$$\dot{\mathbf{x}} = \mathbf{Ax} + \mathbf{b}w \qquad w = -\mathbf{k'x} \tag{13.12}$$

Let z_i, z_j be fed back with gains k_i, k_j to generate w according to

$$w = -k_i z_i - k_j z_j = -\mathbf{k'x} \qquad \mathbf{k'} = k_i\mathbf{v}_i' + k_j\mathbf{v}_j' \tag{13.13}$$

The closed-loop system equation is then

$$\dot{\mathbf{x}} = (\mathbf{A} - k_i\mathbf{bv}_i' - k_j\mathbf{bv}_j')\mathbf{x} \tag{13.14}$$

Using the modal properties above yields

$$(\mathbf{A} - k_i\mathbf{bv}_i' - k_j\mathbf{bv}_j')\mathbf{u}_k = \mathbf{Au}_k = \lambda_k\mathbf{u}_k \qquad k \neq i, j \tag{13.15}$$

By definition, this means that λ_k and $\mathbf{u}_k$, $k \neq i, j$, are also eigenvalues and eigenvectors of the closed loop and have not been affected by the feedback from the ith and jth modal variables.

Note that this property of $\mathbf{u}_k$ has not been shown, and is in fact not true, for $\mathbf{v}_k$. Modal control uses feedback from selected modal variables to change only these eigenvalues to more desirable locations. The new eigenvalues and eigenvectors $\hat{\lambda}_i$, $\hat{\lambda}_j$, $\hat{\mathbf{u}}_i$, $\hat{\mathbf{u}}_j$ satisfy, by definition,

$$(\mathbf{A} - k_i\mathbf{bv}_i' - k_j\mathbf{bv}_j')\hat{\mathbf{u}}_i = \hat{\lambda}_i\hat{\mathbf{u}}_i$$
$$(\mathbf{A} - k_i\mathbf{bv}_i' - k_j\mathbf{bv}_j')\hat{\mathbf{u}}_j = \hat{\lambda}_j\hat{\mathbf{u}}_j \tag{13.16}$$

Since any n-dimensional vector can be written as a series in the n modal components, let

$$\hat{\mathbf{u}}_i = \sum_{k=1}^{n} q_{ik}\mathbf{u}_k \qquad \hat{\mathbf{u}}_j = \sum_{k=1}^{n} q_{jk}\mathbf{u}_k \qquad \mathbf{b} = \sum_{k=1}^{n} p_k\mathbf{u}_k \tag{13.17}$$

The terms in the first of (13.16) are then

$$\mathbf{A}\hat{\mathbf{u}}_i = \mathbf{A}\sum q_{ik}\mathbf{u}_k = \sum q_{ik}\mathbf{A}\mathbf{u}_k = \sum \lambda_k q_{ik}\mathbf{u}_k$$
$$-k_i\mathbf{bv}_i'\hat{\mathbf{u}}_i = -k_i\mathbf{b}\sum q_{ik}\mathbf{v}_i'\mathbf{u}_k = -k_i\mathbf{b}q_{ii} = -k_i q_{ii}\sum p_k\mathbf{u}_k$$
$$-k_j\mathbf{bv}_j'\hat{\mathbf{u}}_i = -k_j\mathbf{b}\sum q_{ik}\mathbf{v}_j'\mathbf{u}_k = -k_j\mathbf{b}q_{ij} = -k_j q_{ij}\sum p_k\mathbf{u}_k$$
$$\hat{\lambda}_i\hat{\mathbf{u}}_i = \hat{\lambda}_i\sum q_{ik}\mathbf{u}_k$$

Substituting these and premultiplying by $\mathbf{v}_i'$ yields the first equation below, and by $\mathbf{v}_j'$ the second:

$$(\lambda_i - \hat{\lambda}_i - k_i p_i)q_{ii} - k_j p_i q_{ij} = 0$$

$$-k_i p_j q_{ii} + (\lambda_j - \hat{\lambda}_i - k_j p_j)q_{ij} = 0$$

Nontrivial solutions for q_{ii} and q_{ij} require that the determinant of the coefficients be zero:

$$(\lambda_i - \hat{\lambda}_i)(\lambda_j - \hat{\lambda}_i) - k_i p_i(\lambda_j - \hat{\lambda}_i) - k_j p_j(\lambda_i - \hat{\lambda}_i) = 0$$

The first of (13.18) follows by rearranging this, and the second by an analogous solution from the second of (13.16).

$$\frac{k_i p_i}{\lambda_i - \hat{\lambda}_i} + \frac{k_j p_j}{\lambda_j - \hat{\lambda}_i} = 1 \qquad \frac{k_i p_i}{\lambda_i - \hat{\lambda}_j} + \frac{k_j p_j}{\lambda_j - \hat{\lambda}_j} = 1 \tag{13.18}$$

Solving this set for k_i and k_j and using $\mathbf{v}_i'\mathbf{b} = \sum p_k\mathbf{v}_i'\mathbf{u}_k = p_i$, $\mathbf{v}_j'\mathbf{b} = p_j$, yields the following theorem.

Modal Control Theorem 13.3.1

(i) In (13.13), the gains k_i, k_j to move λ_i, λ_j to $\hat{\lambda}_i$, $\hat{\lambda}_j$ are

$$k_i = \frac{(\lambda_i - \hat{\lambda}_i)(\lambda_i - \hat{\lambda}_j)}{(\mathbf{v}_i'\mathbf{b})(\lambda_i - \lambda_j)} \qquad k_j = \frac{(\lambda_j - \hat{\lambda}_j)(\lambda_j - \hat{\lambda}_i)}{(\mathbf{v}_j'\mathbf{b})(\lambda_j - \lambda_i)} \tag{13.19}$$

(ii) To move only λ_i to $\hat{\lambda}_i$ ($\lambda_j = \hat{\lambda}_j$, $k_j = 0$):

$$k_i = \frac{\lambda_i - \hat{\lambda}_i}{\mathbf{v}_i'\mathbf{b}} \tag{13.20}$$

(iii) It can be shown that the generalized form of (13.19) for moving $\lambda_1, \ldots, \lambda_m$ to $\hat{\lambda}_1, \ldots, \hat{\lambda}_m$, using feedback from $z_1, \ldots, z_m$, is

$$k_i = \frac{\prod\limits_{j=1}^{m} (\lambda_i - \hat{\lambda}_j)}{(\mathbf{v}_i'\mathbf{b}) \prod\limits_{j=1,\neq i}^{m} (\lambda_i - \lambda_j)} \qquad i = 1, \ldots, m \qquad (13.21)$$

Example 13.3.1

$$\dot{\mathbf{x}} = \begin{bmatrix} -3/4 & -1/4 \\ -1/4 & -3/4 \end{bmatrix} \mathbf{x} + \begin{bmatrix} 0 \\ 1 \end{bmatrix} u$$

$|\lambda\mathbf{I} - \mathbf{A}| = \lambda^2 + 1.5\lambda + 0.5$ yields the open-loop eigenvalues $\lambda_1 = -1$, $\lambda_2 = -0.5$. Let it be desired to speed up the response by designing state feedback which will make the (closed-loop) system eigenvalues equal to $\hat{\lambda}_1 = -1$, $\hat{\lambda}_2 = -2$. Hence only $z_2 = -\mathbf{v}_2'\mathbf{x}$ is required since only the second eigenvalue needs to be moved.

$$\mathbf{A}'\mathbf{v}_2 = \lambda_2\mathbf{v}_2 \qquad \begin{bmatrix} -3/4 & -1/4 \\ -1/4 & -3/4 \end{bmatrix}\begin{bmatrix} v_{12} \\ v_{22} \end{bmatrix} = -0.5\begin{bmatrix} v_{12} \\ v_{22} \end{bmatrix} \qquad \mathbf{v}_2 = \begin{bmatrix} 1 \\ -1 \end{bmatrix}$$

or any multiple. Then, using (13.20),

$$k_2 = \frac{-0.5 + 2}{[1 \quad -1]\begin{bmatrix} 0 \\ 1 \end{bmatrix}} = -1.5$$

and $\mathbf{k}' = k_2\mathbf{v}_2' = [-1.5 \quad 1.5]$. Verify that -1 and -2 are the eigenvalues of $(\mathbf{A} - \mathbf{b}\mathbf{k}')$. Note that normalization of $\mathbf{v}_2$ is not necessary since it occurs in numerator and denominator of $\mathbf{k}'$.

Example 13.3.2

$$\dot{\mathbf{x}} = \begin{bmatrix} 0 & -2 \\ 1 & -2 \end{bmatrix} \mathbf{x} + \begin{bmatrix} 0 \\ 1 \end{bmatrix} u$$

$|\lambda\mathbf{I} - \mathbf{A}| = \lambda^2 + 2\lambda + 2$ yields the open-loop eigenvalues $\lambda_1 = -1 + j$, $\lambda_2 = -1 - j$. This is a complex pair, so both must be moved simultaneously. Let the desired locations for the closed-loop system be $\hat{\lambda}_1 = -4 + 4j$, $\hat{\lambda}_2 = -4 - 4j$. The equations $\mathbf{A}'\mathbf{v}_1 = (-1 + j)\mathbf{v}_1$, $\mathbf{A}'\mathbf{v}_2 = (-1 - j)\mathbf{v}_2$ yield the eigenvectors $\mathbf{v}_1' = [1 \quad -1 + j]$, $\mathbf{v}_2' = [1 \quad -1 - j]$. Then, from (13.19), the required gains are

$$k_1 = \frac{-1 + j + 4 - 4j}{-1 + j}\frac{-1 + j + 4 + 4j}{-1 + j + 1 + j} = -3\frac{3 + 5j}{2j}\frac{j}{j} = -7.5 + 4.5j$$

$$k_2 = -7.5 - 4.5j$$

Equation (13.13) now gives the feedback

$\mathbf{k}' = k_1\mathbf{v}_1' + k_2\mathbf{v}_2'$

$$= (-7.5 + 4.5j)[1 \quad -1 + j] + (-7.5 - 4.5j)[1 \quad -1 - j] = [-15 \quad 6]$$

The closed-loop system matrix

$$\mathbf{A} - \mathbf{b}\mathbf{k}' = \begin{bmatrix} 0 & -2 \\ 1 & -2 \end{bmatrix} - \begin{bmatrix} 0 \\ 1 \end{bmatrix}[-15 \quad 6] = \begin{bmatrix} 0 & -2 \\ 16 & -8 \end{bmatrix}$$

may be verified to have the desired eigenvalues.

Multiple Input Systems: $\dot{\mathbf{x}} = \mathbf{Ax} + \mathbf{Bu}$

For a plant with r inputs, $\mathbf{B}$ consists of r columns and may be written as $\mathbf{B} = [\mathbf{b}_1 \quad \mathbf{b}_2 \quad \cdots \quad \mathbf{b}_r]$. The state equation and the state feedback control may then be written as

$$\dot{\mathbf{x}} = \mathbf{Ax} + \mathbf{Bu} = \mathbf{Ax} + \mathbf{b}_1 u_1 + \cdots + \mathbf{b}_r u_r$$

$$\mathbf{u} = -\mathbf{Kx} \qquad u_i = -\mathbf{k}_i'\mathbf{x} \qquad i = 1, \ldots, r \tag{13.22}$$

where u_i is the ith element of $\mathbf{u}$ and $\mathbf{k}_i'$ the ith row of $\mathbf{K}$. From Theorems 13.2.1 and 12.6.2, input u_i can be used to place the eigenvalues for which $\mathbf{v}_j'\mathbf{b}_i \neq 0$. A mode is usually controllable from more than one input, so the design is not unique. The alternatives have the same closed-loop poles, but the transient responses may differ widely. This can be interpreted in terms of different zero locations, and more advanced techniques seek to optimize these locations.

An approach to modal control which is sometimes convenient, and can be used also for single-input systems, is to move (real) eigenvalues one at a time, as illustrated in the following example of a system with two inputs.

Example 13.3.3   Two-Tank Level Control System

Figure 13.3 shows a process consisting of two interconnected tanks. R_0, R_1, and R_2 are linearized valve resistance values and h_1, h_2 and A_1, A_2 represent tank levels and surface areas, respectively. The flow rates q_1 and q_2 into the tanks are controlled by the signals u_1 and u_2 via valves and actuators, and are modeled by the linearized relations

$$q_1 = K_1 u_1 \qquad q_2 = K_2 u_2$$

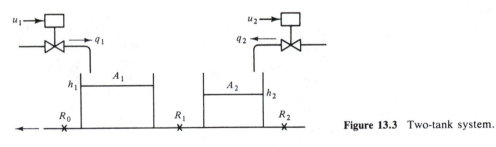

Figure 13.3 Two-tank system.

From the results in Chapter 2, the differential equations for the levels in the tanks are as follows:

$$\dot{h}_1 = -\frac{1}{A_1}\left(\frac{1}{R_1} + \frac{1}{R_0}\right)h_1 + \frac{1}{A_1 R_1}h_2 + \frac{1}{A_1}q_1$$

$$\dot{h}_2 = \frac{1}{A_2 R_1}h_1 - \frac{1}{A_2}\left(\frac{1}{R_1} + \frac{1}{R_2}\right)h_2 + \frac{1}{A_2}q_2$$

Let $A_1 = 1$, $A_2 = 0.5$, $R_0 = 1$, $R_1 = 0.5$, $R_2 = 2$, $K_1 = 1$, and $K_2 = 0.5$. Then, eliminating q_1 and q_2, the equations become

$$\dot{h}_1 = -3h_1 + 2h_2 + u_1 \qquad \dot{h}_2 = 4h_1 - 5h_2 + u_2$$

These correspond to the following state model:

$$\mathbf{x} = \begin{bmatrix} x_1 \\ x_2 \end{bmatrix} = \begin{bmatrix} h_1 \\ h_2 \end{bmatrix} \qquad \mathbf{u} = \begin{bmatrix} u_1 \\ u_2 \end{bmatrix} \qquad \dot{\mathbf{x}} = \begin{bmatrix} -3 & 2 \\ 4 & -5 \end{bmatrix} \mathbf{x} + \begin{bmatrix} 1 & 0 \\ 0 & 1 \end{bmatrix} \mathbf{u}$$

$|\lambda \mathbf{I} - \mathbf{A}| = \lambda^2 + 8\lambda + 7$ yields the open-loop eigenvalues $\lambda_1 = -1$, $\lambda_2 = -7$. The eigenvectors $\mathbf{v}_i$ are found from $\mathbf{A}'\mathbf{v}_1 = -1\mathbf{v}_1$, $\mathbf{A}'\mathbf{v}_2 = -7\mathbf{v}_2$ to be $\mathbf{v}_1' = [2 \quad 1]$, $\mathbf{v}_2' = [1 \quad -1]$. Since $\mathbf{v}_i'\mathbf{b}_j \neq 0$ for all i, j, either input can be used to place either or both eigenvalues. Let u_1 be used to move only λ_1 to a desired position $\hat{\lambda}_1 = -4$ for the closed-loop system. Using (13.20), the required control is $u_1 = -\mathbf{k}_1'\mathbf{x}$, with

$$\mathbf{k}_1' = k_1\mathbf{v}_1' = \frac{-1 + 4}{[2 \quad 1]\begin{bmatrix} 1 \\ 0 \end{bmatrix}} [2 \quad 1] = [3 \quad 1.5]$$

The new system is $\dot{\mathbf{x}} = \mathbf{A}\mathbf{x} + \mathbf{b}_1 u_1 + \mathbf{b}_2 u_2 = \hat{\mathbf{A}}\mathbf{x} + \mathbf{b}_2 u_2$, where

$$\hat{\mathbf{A}} = \mathbf{A} - \mathbf{b}_1\mathbf{k}_1' = \begin{bmatrix} -3 & 2 \\ 4 & -5 \end{bmatrix} - \begin{bmatrix} 1 \\ 0 \end{bmatrix}[3 \quad 1.5] = \begin{bmatrix} -6 & 0.5 \\ 4 & -5 \end{bmatrix}$$

This has the eigenvalues $\hat{\lambda}_1 = -4$, $\lambda_2 = -7$, as expected. Input u_2 is chosen next to move λ_2 to a desired position $\hat{\lambda}_2 = -9$. The eigenvector $\hat{\mathbf{v}}_2$ of $\hat{\mathbf{A}}$ is found from

$$\hat{\mathbf{A}}'\hat{\mathbf{v}}_2 = -7\hat{\mathbf{v}}_2 \colon \hat{\mathbf{v}}_2' = [-4 \quad 1]$$

[Note that the vector $\mathbf{v}_2$ changes when λ_1 is moved; only u_2 is not affected, as mentioned below (13.15).] Using (13.20), the required control is $u_2 = -\mathbf{k}_2'\mathbf{x}$, with

$$\mathbf{k}_2' = k_2\mathbf{v}_2' = \frac{-7 + 9}{[-4 \quad 1]\begin{bmatrix} 0 \\ 1 \end{bmatrix}} [-4 \quad 1] = [-8 \quad 2]$$

The total feedback $\mathbf{u} = -\mathbf{K}\mathbf{x}$ and the closed-loop matrix, which may be verified to have the eigenvalues -7 and -9, are given by

$$\mathbf{K} = \begin{bmatrix} \mathbf{k}_1' \\ \mathbf{k}_2' \end{bmatrix} = \begin{bmatrix} 3 & 1.5 \\ -8 & 2 \end{bmatrix} \qquad \mathbf{A} - \mathbf{B}\mathbf{K} = \begin{bmatrix} -3 & 2 \\ 4 & -5 \end{bmatrix} - \begin{bmatrix} 3 & 1.5 \\ -8 & 2 \end{bmatrix}$$

$$= \begin{bmatrix} -6 & 0.5 \\ 12 & -7 \end{bmatrix}$$

13.4 MULTIVARIABLE INTEGRAL CONTROL AND FEEDFORWARD

Thus far, only the response to an initial condition $\mathbf{x}(0)$ has been considered, with $\mathbf{x} = \mathbf{0}$ as the desired state. Reference inputs $\mathbf{y}_{\text{ref}}$ (i.e., the desired output $\mathbf{y}$) or disturbance inputs $\mathbf{w}$ were implied to be transient only, with zero steady-state values. In practice, however, this is generally not the case. Now, in single-variable design it was found that if y_{ref} and w have constant steady-state values, integral control can provide a stable design (i.e., $\dot{\mathbf{x}} \to \mathbf{0}$ as $t \to \infty$) with zero steady-state error (i.e., $y \to y_{\text{ref}}$ as $t \to \infty$). The generalization of integral control to multivariable systems is of great practical importance and will be achieved by solution of the following design problem:

> Let $\mathbf{y}_{ref}$ and $\mathbf{w}$ be vectors of constant reference and disturbance inputs of the system described by
>
> $$\dot{\mathbf{x}} = \mathbf{Ax} + \mathbf{Bu} + \mathbf{Ew} \qquad \mathbf{y} = \mathbf{Cx} + \mathbf{Du} + \mathbf{Fw} \qquad (13.23)$$
>
> where $\mathbf{x}$, $\mathbf{u}$, $\mathbf{w}$, $\mathbf{y}$, and $\mathbf{y}_{ref}$ are vectors of dimensions n, r, q, m, and m. Design $\mathbf{u}$ such that
>
> $$\dot{\mathbf{x}} \longrightarrow \mathbf{0} \text{ as } t \longrightarrow \infty \qquad \text{for stability.}$$
> $$\mathbf{y} \longrightarrow \mathbf{y}_{ref} \text{ as } t \longrightarrow \infty \qquad \text{for zero steady-state errors} \qquad (13.24)$$

Note that the state space model had to be augmented to allow for the presence of disturbance inputs. The last condition in (13.24) means that the steady-state value of each output can be set independently by manipulation of the corresponding reference input. This is often required, even if limited interaction during transients may be quite acceptable. Such a system is said to be *statically* or *asymptotically decoupled* or *noninteracting* since each reference input affects only the corresponding output for $t \to \infty$.

To solve this design problem, first define m new state variables:

$$\mathbf{p} = \int_0^t (\mathbf{y} - \mathbf{y}_{ref}) \, dt \qquad (13.25)$$

Then (13.23) become

$$\dot{\mathbf{x}} = \mathbf{Ax} + \mathbf{Bu} + \mathbf{Ew}$$
$$\dot{\mathbf{p}} = \mathbf{y} - \mathbf{y}_{ref} = \mathbf{Cx} + \mathbf{Du} + \mathbf{Fw} - \mathbf{y}_{ref} \qquad (13.26)$$

or, in matrix form,

$$(\dot{\mathbf{z}} =) \begin{bmatrix} \dot{\mathbf{x}} \\ \dot{\mathbf{p}} \end{bmatrix} = \begin{bmatrix} \mathbf{A} & \mathbf{0} \\ \mathbf{C} & \mathbf{0} \end{bmatrix} \begin{bmatrix} \mathbf{x} \\ \mathbf{p} \end{bmatrix} + \begin{bmatrix} \mathbf{B} \\ \mathbf{D} \end{bmatrix} \mathbf{u} + \begin{bmatrix} \mathbf{E} & \mathbf{0} \\ \mathbf{F} & -\mathbf{I} \end{bmatrix} \begin{bmatrix} \mathbf{w} \\ \mathbf{y}_{ref} \end{bmatrix} \qquad (13.27)$$

In the steady state $\dot{\mathbf{x}} = \mathbf{0}$, $\dot{\mathbf{p}} = \mathbf{0}$, since $\mathbf{w}$ and $\mathbf{y}_{ref}$ are constant, and the vectors satisfy

$$\begin{bmatrix} \mathbf{E} & \mathbf{0} \\ \mathbf{F} & -\mathbf{I} \end{bmatrix} \begin{bmatrix} \mathbf{w} \\ \mathbf{y}_{ref} \end{bmatrix} = -\begin{bmatrix} \mathbf{A} & \mathbf{0} \\ \mathbf{C} & \mathbf{0} \end{bmatrix} \begin{bmatrix} \mathbf{x}_s \\ \mathbf{p}_s \end{bmatrix} - \begin{bmatrix} \mathbf{B} \\ \mathbf{D} \end{bmatrix} \mathbf{u}_s \qquad (13.28)$$

Substituting this into (13.27), and defining the deviations from this steady state by

$$\mathbf{z} = \begin{bmatrix} \mathbf{z}_1 \\ \mathbf{z}_2 \end{bmatrix} = \begin{bmatrix} \mathbf{x} - \mathbf{x}_s \\ \mathbf{p} - \mathbf{p}_s \end{bmatrix} \quad (\dot{\mathbf{z}} = \begin{bmatrix} \dot{\mathbf{x}} \\ \dot{\mathbf{p}} \end{bmatrix}) \qquad \mathbf{v} = \mathbf{u} - \mathbf{u}_s \qquad (13.29)$$

yields

$$\dot{\mathbf{z}} = \hat{\mathbf{A}}\mathbf{z} + \hat{\mathbf{B}}\mathbf{v} \qquad \hat{\mathbf{A}} = \begin{bmatrix} \mathbf{A} & \mathbf{0} \\ \mathbf{C} & \mathbf{0} \end{bmatrix} \qquad \hat{\mathbf{B}} = \begin{bmatrix} \mathbf{B} \\ \mathbf{D} \end{bmatrix} \qquad (13.30)$$

The problem is now in the standard form of the preceding sections, with $z = 0$ as the desired state. Therefore, pole assignment or optimal control techniques can be used to design a constant-gain state feedback matrix K from z:

$$v = -Kz = -K_1 z_1 - K_2 z_2$$
$$u - u_s = -K_1(x - x_s) - K_2(p - p_s) \tag{13.31}$$

Here K has been partitioned appropriately into $K = [K_1 \quad K_2]$. The steady-state terms must balance so, using (13.25), the control u equals

$$u = -K_1 x - K_2 \int_0^t (y - y_{ref}) \, dt \tag{13.32}$$

This consists of proportional state feedback and integral control of output error, and represents a multivariable generalization of PI control.

Figure 13.4 illustrates the feedback configuration. In the steady state, $\dot{p} = y - y_{ref} = 0$, so the errors are zero. The control is called *robust* because this remains true for any plant parameter variations as long as the system remains stable.

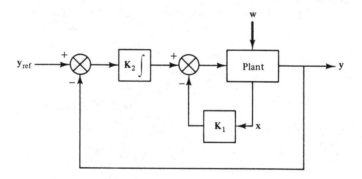

Figure 13.4   Multivariable PI control.

The controllability of $(\hat{A}, \hat{B})$ is examined by the following theorem.

Theorem 13.4.1.   The pair $(\hat{A}, \hat{B})$ is controllable if and only if (A, B) is controllable and

$$G = \begin{bmatrix} A & B \\ C & D \end{bmatrix}$$

has full rank $(n + m)$, equal to the number of rows. Since the number of columns is $(n + r)$, $r \geq m$ is necessary (i.e., there must be at least as many plant inputs as there are outputs to be regulated to zero errors).

Example 13.4.1   Two-Tank Process of Example 13.3.3

In Fig. 13.3, let it be desired to control the level in the second tank with zero steady-state error for reference inputs and disturbances with constant steady-state values. Thus the system output is taken to be $y = [0 \quad 1]x$. Assume that the second input is absent ($u_2 = 0$), and that a disturbance flow rate w enters the first tank via a returns line from elsewhere in the process. This adds a term w/A_1 to the equation for $\dot{h}_1$ in Example 13.3.3, which equals w for the numerical values in that example.

The pertinent matrices are then as follows:

$$\mathbf{A} = \begin{bmatrix} -3 & 2 \\ 4 & -5 \end{bmatrix} \quad \mathbf{B} = \begin{bmatrix} 1 \\ 0 \end{bmatrix} \quad \mathbf{E} = \begin{bmatrix} 1 \\ 0 \end{bmatrix} \quad \begin{matrix} \mathbf{C} = [0 \quad 1] \\ \mathbf{D} = \mathbf{F} = 0 \end{matrix}$$

The matrices for integral control design are

$$\mathbf{G} = \begin{bmatrix} -3 & 2 & 1 \\ 4 & -5 & 0 \\ 0 & 1 & 0 \end{bmatrix} \quad \hat{\mathbf{A}} = \begin{bmatrix} -3 & 2 & 0 \\ 4 & -5 & 0 \\ 0 & 1 & 0 \end{bmatrix}$$

$$\hat{\mathbf{B}} = \begin{bmatrix} 1 \\ 0 \\ 0 \end{bmatrix} \quad \lambda\mathbf{I} - \hat{\mathbf{A}} = \begin{bmatrix} \lambda + 3 & -2 & 0 \\ -4 & \lambda + 5 & 0 \\ 0 & -1 & \lambda \end{bmatrix}$$

The pair $(\mathbf{A}, \mathbf{B})$ may be verified to be controllable and $|\mathbf{G}| = 4$, so $\mathbf{G}$ has full rank $n + m = 3$, and the controllability theorem 13.4.1 is satisfied. The eigenvalues of $\hat{\mathbf{A}}$ are $\lambda_1 = 0$, $\lambda_2 = -1$, $\lambda_3 = -7$. These are those of $\mathbf{A}$ as in Example 13.3.3 and an open-loop eigenvalue at the origin due to the integral control. Let it be sufficient to move this eigenvalue to $\hat{\lambda}_1 = -2$, leaving the others unchanged. Hence only the eigenvector $\mathbf{v}_1$ is needed, and is found from $\hat{\mathbf{A}}'\mathbf{v}_1 = 0 \cdot \mathbf{v}_1$ to be $\mathbf{v}_1' = [1 \quad 0.75 \quad 1.75]$. Then in $\mathbf{v} = -\mathbf{k}'\mathbf{z}$, $\mathbf{k}' = [(0 + 2)/(\mathbf{v}_1'\hat{\mathbf{B}})]\mathbf{v}_1' = [2 \quad 1.5 \quad 3.5]$. So in (13.31), $\mathbf{K}_1 = [2 \quad 1.5]$, $K_2 = 3.5$, and (13.32) gives

$$u = -2x_1 - 1.5x_2 - 3.5 \int_0^t (y - y_{\text{ref}}) \, dt$$

To calculate the response of the level $h_2 = x_2 = z_2$ in the second tank with this control to unit step inputs of w and y_{ref}, the control $u = -[2 \quad 1.5 \quad 3.5]z$ is substituted into the original state-model equations to obtain

$$\dot{\mathbf{z}} = \begin{bmatrix} -3 & 2 & 0 \\ 4 & -5 & 0 \\ 0 & 1 & 0 \end{bmatrix} \mathbf{z} - \begin{bmatrix} 1 \\ 0 \\ 0 \end{bmatrix} [2 \quad 1.5 \quad 3.5]\mathbf{z} + \begin{bmatrix} 1 & 0 \\ 0 & 0 \\ 0 & -1 \end{bmatrix} \begin{bmatrix} w \\ y_{\text{ref}} \end{bmatrix}$$

$$= \begin{bmatrix} -5 & 0.5 & -3.5 \\ 4 & -5 & 0 \\ 0 & 1 & 0 \end{bmatrix} \mathbf{z} + \begin{bmatrix} 1 & 0 \\ 0 & 0 \\ 0 & -1 \end{bmatrix} \begin{bmatrix} w \\ y_{\text{ref}} \end{bmatrix} = \overline{\mathbf{A}}\mathbf{z} + \overline{\mathbf{B}}\overline{\mathbf{r}}$$

Since only z_2 is needed, the output matrix can be taken to be $\overline{\mathbf{C}} = [0 \quad 1 \quad 0]$, and the transfer function matrix from $\overline{\mathbf{r}}$ to z_2 is

$$\mathbf{G}(s) = \overline{\mathbf{C}}(s\mathbf{I} - \overline{\mathbf{A}})^{-1}\overline{\mathbf{B}}$$

Because of the form of $\overline{\mathbf{C}}$, only the second row of the inverse

$$(s\mathbf{I} - \overline{\mathbf{A}})^{-1} = \begin{bmatrix} s+5 & -0.5 & 3.5 \\ -4 & s+5 & 0 \\ 0 & -1 & s \end{bmatrix}^{-1}$$

is required, and is readily found to be $[4s \quad s(s+5) \quad -14]/\Delta$, where $\Delta = (s +$

1)$(s + 2)(s + 7)$ is known from the assigned poles. Then the transform $Z_2(s)$ of $z_2(t)$ is

$$Z_2(s) = \left[\frac{4s}{(s + 1)(s + 2)(s + 7)} \quad \frac{14}{(s + 1)(s + 2)(s + 7)} \right] \left[\begin{array}{c} W(s) \\ Y_{ref}(s) \end{array} \right]$$

Responses to unit steps in w and y_{ref} in turn can now be calculated by partial fraction expansion, and are shown in Fig. 13.5. As expected, h_2 has zero steady-state error in response to y_{ref} and a transient deviation in response to w.

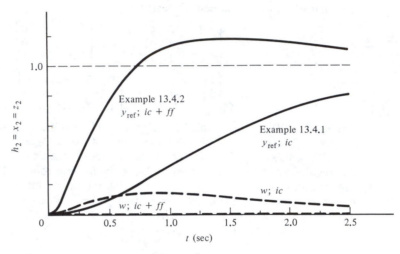

Figure 13.5   Examples 13.4.1 and 13.4.2: integral control and feedforward.

Feedforward Control for Measurable Disturbances

While with integral control the static errors can be made zero, the errors during the transients may be large. Feedforward control, discussed in Section 8.9, is an important technique in practice to reduce the effect of disturbances if these are measurable.

Equations (13.26) can be written as follows:

$$\left[\begin{array}{c} \dot{x} \\ y - y_{ref} \end{array} \right] = G \left[\begin{array}{c} x \\ u \end{array} \right] + H \left[\begin{array}{c} w \\ y_{ref} \end{array} \right] \qquad G = \left[\begin{array}{cc} A & B \\ C & D \end{array} \right] \qquad H = \left[\begin{array}{cc} E & 0 \\ F & -I \end{array} \right] \quad (13.33)$$

If the steady-state solution for which $\dot{x} \to 0$ and $y \to y_{ref}$ is x_s, u_s, then

$$G \left[\begin{array}{c} x_s \\ u_s \end{array} \right] = -H \left[\begin{array}{c} w \\ y_{ref} \end{array} \right] \qquad \left[\begin{array}{c} x_s \\ u_s \end{array} \right] = -G^{-1}H \left[\begin{array}{c} w \\ y_{ref} \end{array} \right] \quad (13.34)$$

if G is square and nonsingular. (For rectangular matrices pseudoinverses can be used.) Substituting the first form into (13.33) yields the following in terms of the deviations from steady state:

$$\hat{x} = x - x_s \quad (\dot{\hat{x}} = \dot{x}) \qquad \begin{bmatrix} \dot{\hat{x}} \\ y - y_{ref} \end{bmatrix} = G \begin{bmatrix} \hat{x} \\ \hat{u} \end{bmatrix}$$
$$\hat{u} = u - u_s$$

Hence, by substituting for **G** from (13.33):

$$\dot{\hat{x}} = A\hat{x} + B\hat{u} \qquad y - y_{ref} = C\hat{x} + D\hat{u} \tag{13.35}$$

The problem is now in standard form, with desired state $\hat{x} = 0$. Hence previous optimal control or pole assignment techniques can be used to design a state feedback:

$$\hat{u} = -K\hat{x} \qquad u - u_s = -K(x - x_s) \qquad u = -Kx + [K \quad I]\begin{bmatrix} x_s \\ u_s \end{bmatrix} \tag{13.36}$$

Substituting (13.34) now gives the control:

$$u = -Kx - [K \quad I]G^{-1}H\begin{bmatrix} w \\ y_{ref} \end{bmatrix} \tag{13.37}$$

This is a proportional state feedback plus feedforward control from y_{ref} and **w**, which can be implemented if **w** is measurable. Note that feedforward may also be used without the use of feedback, by putting **K** = **0**, if feedback is not required. But feedforward control is not robust; that is, zero steady-state errors are not maintained with plant parameter variations. Therefore, integral control is usually also included.

Example 13.4.2   Example 13.4.1 with Feedforward

Using (13.33) and results in Example 13.4.1:

$$G^{-1} = \frac{1}{4}\begin{bmatrix} 0 & 1 & 5 \\ 0 & 0 & 4 \\ 4 & 3 & 7 \end{bmatrix} \qquad H = \begin{bmatrix} 1 & 0 \\ 0 & 0 \\ 0 & -1 \end{bmatrix} \qquad G^{-1}H = \frac{1}{4}\begin{bmatrix} 0 & -5 \\ 0 & -4 \\ 4 & -7 \end{bmatrix}$$

With the same assigned poles, $K = K_1 = [2 \quad 1.5]$.

$$[K \quad I]G^{-1}H = \frac{1}{4}[4 \quad (-10 - 6 - 7)] = [1 \quad -5.75]$$

Using this in the feedforward term in (13.37), the control of Example 13.4.1 is augmented to that illustrated in Fig. 13.6:

$$u = -2x_1 - 1.5x_2 - 3.5\int_0^t (y - y_{ref})\, dt - w + 5.75 y_{ref}$$

To calculate responses corresponding to those in Example 13.4.1 for this control, the term $\hat{B}[-1 \quad 5.75]\bar{r}$ must be added to the right side of the equation for $\dot{z}$ in the preceding example and $\bar{B}\bar{r}$ becomes

$$\bar{B}\bar{r} = \left\{ \begin{bmatrix} 1 & 0 \\ 0 & 0 \\ 0 & -1 \end{bmatrix} + \begin{bmatrix} 1 \\ 0 \\ 0 \end{bmatrix}[-1 \quad 5.75] \right\}\bar{r} = \begin{bmatrix} 0 & 5.75 \\ 0 & 0 \\ 0 & -1 \end{bmatrix}\bar{r}$$

The first column contains only zeros, so feedforward control has eliminated the

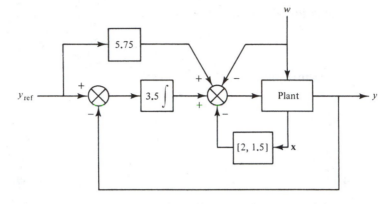

Figure 13.6 Feedforward control plus PI control.

effect of the disturbance input. The transfer function from $Y_{ref}(s)$ to $Z_2(s)$ is found to be

$$\frac{Z_2(s)}{Y_{ref}(s)} = \frac{14 + 23s}{(s + 1)(s + 2)(s + 7)}$$

from which partial fraction expansion yields the unit step response shown in Fig. 13.5. Feedforward of the reference input is responsible for an overshoot of about 17.5%, but also for a major improvement of the speed of response.

13.5 *OUTPUT FEEDBACK DESIGN*

Constant-gain state feedback $\mathbf{u} = -\mathbf{Kx}$ can provide specified eigenvalues or an optimal and stable design. But it assumes that all states can be measured, whereas in practice often only the output $\mathbf{y}$ can be considered to be available for feedback. This is *output feedback.* One solution to the system design problem in this case, and like the others in this section the subject of an extensive literature, is to use a *dynamic observer,* implemented as a computer algorithm. For observable systems, the observer determines an estimate $\hat{\mathbf{x}}$ of the state $\mathbf{x}$ from the measured output $\mathbf{y}$. The Kalman filter gives an optimal estimate if $\mathbf{y}$ is contaminated by noise. Constant-gain feedback from $\hat{\mathbf{x}}$ is then again used for optimal control or pole assignment.

Frequently, however, this would be considered too complex, and a simpler design sought by the use of output feedback which is constant or of low dynamic order. For the system

$$\dot{\mathbf{x}} = \mathbf{Ax} + \mathbf{Bu} \qquad \mathbf{y} = \mathbf{Cx} \tag{13.38}$$

these lead to the following equations:

1. *Constant-gain output feedback:*

$$\mathbf{u} = -\mathbf{Ky} = -\mathbf{KCx} \qquad \dot{\mathbf{x}} = (\mathbf{A} - \mathbf{BKC})\mathbf{x} \tag{13.39}$$

2. *Dynamic output feedback:* The output **y** contributes directly to **u** and also indirectly via a dynamic compensator:

$$\mathbf{u} = \mathbf{Hz} + \mathbf{Ny} \qquad \dot{\mathbf{z}} = \mathbf{Fz} + \mathbf{Gy} \qquad (13.40)$$

The vectors **x**, **y**, **u**, and **z** have dimensions n, m, r, and p. These equations can be combined as follows, where the definitions of **v**, $\hat{\mathbf{A}}$, and so on, are implied by the equations.

$$\begin{bmatrix} \dot{\mathbf{x}} \\ \dot{\mathbf{z}} \end{bmatrix} = \begin{bmatrix} \mathbf{A} & \mathbf{0} \\ \mathbf{0} & \mathbf{0} \end{bmatrix}\begin{bmatrix} \mathbf{x} \\ \mathbf{z} \end{bmatrix} + \begin{bmatrix} \mathbf{B} & \mathbf{0} \\ \mathbf{0} & \mathbf{I} \end{bmatrix}\begin{bmatrix} \mathbf{u} \\ \dot{\mathbf{z}} \end{bmatrix} : \dot{\mathbf{v}} = \hat{\mathbf{A}}\mathbf{v} + \hat{\mathbf{B}}\hat{\mathbf{u}} \qquad (13.41a)$$

$$\begin{bmatrix} \mathbf{u} \\ \dot{\mathbf{z}} \end{bmatrix} = \begin{bmatrix} \mathbf{N} & \mathbf{H} \\ \mathbf{G} & \mathbf{F} \end{bmatrix}\begin{bmatrix} \mathbf{y} \\ \mathbf{z} \end{bmatrix} \qquad \begin{bmatrix} \mathbf{y} \\ \mathbf{z} \end{bmatrix} = \begin{bmatrix} \mathbf{C} & \mathbf{0} \\ \mathbf{0} & \mathbf{I} \end{bmatrix}\begin{bmatrix} \mathbf{x} \\ \mathbf{z} \end{bmatrix} : \hat{\mathbf{y}} = \hat{\mathbf{C}}\mathbf{v} \quad \hat{\mathbf{u}} = \mathbf{P}\hat{\mathbf{y}} \qquad (13.41b)$$

Substitution of (13.41b) into (13.41a) yields the closed-loop equation

$$\dot{\mathbf{v}} = (\hat{\mathbf{A}} + \hat{\mathbf{B}}\mathbf{P}\hat{\mathbf{C}})\mathbf{v} \qquad (13.41c)$$

Comparison with (13.38) and (13.39) shows that the problem has been reformulated into one with constant-gain feedback. The unknown matrices **N**, **H**, **G**, and **F** are combined in the single matrix **P**, which is designed for the $(n + p)$th-order system. So techniques for the design of constant-gain output feedback apply also to the design of dynamic output feedback compensation.

*****

Both pole assignment and optimal control design procedures are available, but not all eigenvalues can be placed in specified locations. Using techniques by means of which a permissible number are placed, the remainder could be anywhere, including the right-half plane. In fact, a stable design may not be possible below a certain dynamic order of feedback.

Standard computer routines for numerical parameter optimization, such as Rosenbrock's method, are often used to find the best parameter values in **K**, or in **P**. Starting with constant-gain feedback, the assumed order of compensation can be increased a step at a time until the results are satisfactory. As an example of design problem formulation, it can be shown that a suboptimal constant-gain output feedback problem may be formulated as follows:

The average value of $J = \frac{1}{2}\int_0^\infty (\mathbf{x}'\mathbf{Qx} + \mathbf{u}'\mathbf{Ru}) \, dt$ for a random initial condition $\mathbf{x}_0$ is minimized by choosing **K** to minimize trace (**V**), where **V** is the symmetric solution of the *Liapunov equation*

$$(\mathbf{A} - \mathbf{BKC})'\mathbf{V} + \mathbf{V}(\mathbf{A} - \mathbf{BKC}) = -\mathbf{Q} - \mathbf{C}'\mathbf{K}'\mathbf{RKC} \qquad (13.42)$$

This equation is solved at each stage of the *hillclimbing* procedure to find the optimal **K**. A condition for the procedure to converge is that the closed-loop system matrix $(\mathbf{A} - \mathbf{BKC})$ have stable eigenvalues. To achieve this, a preliminary optimization may be used which is also useful as an output feedback pole assignment technique. The eigenvalue with largest real part is moved as far to the left as possible by

$$\text{minimizing max Re } \{\lambda_i(\mathbf{A} - \mathbf{BKC}), \quad i = 1, \ldots, n\} \tag{13.43}$$

Alternatively, a measure of the distances of the eigenvalues to selected desired locations can be minimized. Modal control has also been applied, and techniques are available to control selected dominant modes, using measurement of the outputs to approximate the corresponding modal variables.

13.6 STATE SPACE METHODS FOR DIGITAL SIMULATION AND CONTROL

This chapter is concluded with an introduction into the use of state space methods in digital simulation and control. The brief discussion of this important area will demonstrate a strong analogy with the methods for continuous systems.

The discrete-time and approximate discrete-time solutions of $\dot{\mathbf{x}} = \mathbf{A}\mathbf{x} + \mathbf{B}\mathbf{u}$, $\mathbf{y} = \mathbf{C}\mathbf{x}$ in Section 12.4 have in effect also provided digital simulations of this state model:

$$\mathbf{x}_{k+1} = \mathbf{P}\mathbf{x}_k + \mathbf{Q}\mathbf{u}_k \qquad \mathbf{y}_k = \mathbf{C}\mathbf{x}_k \tag{13.44}$$

Here $\mathbf{P} = \boldsymbol{\phi}$ and $\mathbf{Q} = \boldsymbol{\Delta}$, where $\boldsymbol{\phi}$ and $\boldsymbol{\Delta}$ are given in (12.15) in terms of the time interval T and the matrices $\mathbf{A}$ and $\mathbf{B}$. Such discrete state models may also describe single-variable or multivariable control algorithms. For example, for a higher-order difference equation

$$x_{k+n} + a_1 x_{k+n-1} + \cdots + a_n x_k = b u_k$$

a possible state vector and associated discrete state model are as follows:

$$\mathbf{x}_k = \begin{bmatrix} x_k \\ x_{k+1} \\ \cdot \\ \cdot \\ \cdot \\ x_{k+n-1} \end{bmatrix} \qquad \mathbf{x}_{k+1} = \begin{bmatrix} 0 & 1 & 0 & \cdots & & 0 \\ 0 & 0 & 1 & & & \\ \cdot & & & & & \\ \cdot & & & & \cdot & \\ \cdot & & & & & \\ & & & & & 1 \\ -a_n & \cdot & & \cdot & \cdots & -a_1 \end{bmatrix} \mathbf{x}_k + \begin{bmatrix} 0 \\ \cdot \\ \cdot \\ \cdot \\ 0 \\ b \end{bmatrix} u_k \tag{13.45}$$

Note that $\mathbf{P}$ is a companion matrix in this case.

*****

A Z transfer function matrix $\mathbf{G}(z)$ can be derived from (13.44) as for the continuous case. From the definition of the Z transform, $Z[\mathbf{x}_{k+1}] = \Sigma_{k=0}^{\infty} \mathbf{x}_{k+1} z^{-k}$. Let $n = k + 1$ to obtain

$$Z[\mathbf{x}_{k+1}] = \sum_{n=1}^{\infty} \mathbf{x}_n z^{-n+1} = z \sum_{n=1}^{\infty} \mathbf{x}_n z^{-n} = z \left(\sum_{n=0}^{\infty} \mathbf{x}_n z^{-n} - \mathbf{x}_0 \right)$$

Using this, the transforms of (13.44) are

$$z\mathbf{X}(z) - z\mathbf{x}_0 = \mathbf{P}\mathbf{X}(z) + \mathbf{Q}\mathbf{U}(z) \qquad \mathbf{Y}(z) = \mathbf{C}\mathbf{X}(z)$$

Thus, analogous to (12.3) and (12.4),

$$\mathbf{Y}(z) = \mathbf{G}(z)\mathbf{U}(z) + \mathbf{C}(z\mathbf{I} - \mathbf{P})^{-1}z\mathbf{x}_0 \tag{13.46}$$

where

$$\mathbf{G}(z) = \mathbf{C}(z\mathbf{I} - \mathbf{P})^{-1}\mathbf{Q} \tag{13.47}$$

is the Z transfer function matrix and gives the input–output response for zero initial conditions, while the second term in (13.46) gives the response to initial conditions. For $\mathbf{C} = \mathbf{I}$, the state response is obtained.

$\mathbf{G}(z)$ consists of ordinary Z transfer functions and represents a stable system if all poles of all these transfer functions lie inside the unit circle of the z-plane. But all have the same poles, which are the roots of the characteristic equation

$$|z\mathbf{I} - \mathbf{P}| = 0 \tag{13.48}$$

Hence follows the

Stability Theorem 13.6.1.   The system (13.44) is stable if and only if all eigenvalues [i.e., all roots of (13.48)] lie inside the unit circle.

It is noted that, analogous to the Routh–Hurwitz criterion, the Jury theorem is available to determine whether any eigenvalues lie outside the unit circle.

*****

Transient responses are also calculated by techniques equivalent to those for continuous systems. Solution of the discrete state model equation (13.44) gives

$$\mathbf{x}_1 = \mathbf{P}\mathbf{x}_0 + \mathbf{Q}\mathbf{u}_0, \quad \mathbf{x}_2 = \mathbf{P}\mathbf{x}_1 + \mathbf{Q}\mathbf{u}_1 = \mathbf{P}^2\mathbf{x}_0 + \mathbf{P}\mathbf{Q}\mathbf{u}_0 + \mathbf{Q}\mathbf{u}_1, \quad \dots$$

$$\mathbf{x}_k = \mathbf{P}^k\mathbf{x}_0 + \sum_{i=0}^{k-1}\mathbf{P}^{k-1-i}\mathbf{Q}\mathbf{u}_i \tag{13.49}$$

The first term represents the response to initial conditions, and comparison with (13.46) yields the following analog of (12.12):

$$\mathbf{P}^k = Z^{-1}[(z\mathbf{I} - \mathbf{P})^{-1}z] \tag{13.50}$$

The techniques for Z transform inversion can be used as well, in the manner of the use of Laplace transforms in Section 12.4. Solutions based on transformation of $\mathbf{P}$ to diagonal form are also applicable. Design techniques based on pole assignment and optimal control are used for discrete-time systems by procedures which in essence are also quite analogous to those for continuous systems.

13.7 CONCLUSION

Chapters 12 and 13 have provided an introduction to analysis and design in state space, the subject of many books and innumerable technical papers. These techniques provide the only feasible approach for large systems studies, and are used extensively for modeling, analysis, and design of such systems. Examples are electrical power systems, electrical power plants, chemical processes and refineries, aerospace systems, transportation systems, and multivariable process control generally.

The great advantage is that the standard matrix formulation has allowed the development of standard computer programs suitable for the analysis and design of large systems. The choice of desired locations in pole assignment and of the weighting matrices in optimal control, however, frequently requires considerable trial-and-error adjustment to obtain a satisfactory transient response. Extensive work has been done to improve this, for example by considering the system zeros in addition to the poles, and on the problem of output feedback both with and without dynamic compensation.

PROBLEMS

13.1. For a system with plant transfer function $G(s) = 2/(s^2 + 4s + 5)$, design a state feedback control $u = -\mathbf{k'x}$ to place the closed-loop eigenvalues at $-3 \pm 2j$.

13.2. For the plant

$$\dot{\mathbf{x}} = \begin{bmatrix} 0 & 1 \\ -2 & -3 \end{bmatrix} \mathbf{x} + \begin{bmatrix} 0 \\ 2 \end{bmatrix} u$$

design a state feedback control to place the closed-loop eigenvalues at $-2 \pm 2j$.

13.3. A single-input/single-output plant has the transfer function

$$G(s) = \frac{4}{s(0.2s + 1)(0.05s + 1)}$$

  (a) Obtain a state model in companion form.
  (b) Design a state feedback that will place the closed-loop poles at -1 and $-2 \pm 4j$.

13.4. Noting that $s^3 + 9s^2 + 23s + 15 = (s + 1)(s + 3)(s + 5)$, determine a state feedback control for the system

$$\dot{\mathbf{x}} = \begin{bmatrix} 0 & 1 & 0 \\ 0 & 0 & 1 \\ -15 & -23 & -9 \end{bmatrix} \mathbf{x} + \begin{bmatrix} 0 \\ 0 \\ 4 \end{bmatrix} u$$

which will place the closed-loop poles at -3, -4, and -5.

13.5. Verify the solution of Problem 13.2 by modal control design.

13.6. For a system with plant transfer function $G(s) = 2/(s^2 + 3s + 2)$, use modal control to design a state feedback which places the closed-loop eigenvalues at $-2 \pm 2j$.

13.7. Verify the solution of Problem 13.4 by modal control design.

13.8. For the Example 13.2.1, perform the design also by modal control, if it is given that the roots of $s^3 + 6s^2 + 11s + 6 = 0$ are -1, -2, and -3. Move only one pole to get the desired result.

13.9. Repeat Problem 13.8 by moving all three poles to get the desired result.

13.10. Figure P13.10 shows a plant model of an aircraft roll control system. Using modal control and state variables as indicated, design state feedback to place the closed-loop eigenvalues at -5, -5, and -50.

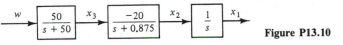

Figure P13.10

13.11. The two-tank hydraulic system in Fig. P12.12 is represented by the state model

$$\dot{\mathbf{x}} = \begin{bmatrix} -1 & 1 \\ 1 & -2 \end{bmatrix} \mathbf{x} + \begin{bmatrix} 0.1 & 0 \\ 0 & 0.1 \end{bmatrix} \mathbf{w}$$

Use modal control to design state feedback that will place the closed-loop eigenvalues at -10. Use w_1 to assign the smallest open-loop eigenvalue and w_2 for the largest.

13.12. In the notations of equations (13.23), let $A = 0$, $B = 1$, $C = 1$, $D = 0$, $E = 1$, and $F = 0$. Determine whether the conditions for integral control hold, and design state feedback plus integral control to place the poles of the closed-loop system at $-p$.

13.13. For the system

$$\dot{\mathbf{x}} = \begin{bmatrix} -1 & 0 \\ 0 & -2 \end{bmatrix} \mathbf{x} + \begin{bmatrix} 1 \\ 1 \end{bmatrix} u + \begin{bmatrix} 1 & 0 \\ 0 & 1 \end{bmatrix} \begin{bmatrix} w_1 \\ w_2 \end{bmatrix} \qquad y = [1 \quad 0]\mathbf{x}$$

where w_1 and w_2 are constant disturbances, with a constant reference input y_{ref}, check whether the conditions for integral control are satisfied, and design a control system with integral control and state feedback which places the closed-loop eigenvalues at -1, -2, and -3.

13.14. Repeat Problem 13.13 for desired eigenvalues -10, -10, and -10.

13.15. For the system

$$\dot{\mathbf{x}} = \begin{bmatrix} -3 & 1 \\ 3 & -5 \end{bmatrix} \mathbf{x} + \begin{bmatrix} 1 \\ 0 \end{bmatrix} u$$

subject to constant disturbances, design state feedback plus integral control so that no closed-loop eigenvalues will be to the right of -1, and the steady-state error between x_2 and a constant reference input will be zero.

13.16. For the system

$$\dot{\mathbf{x}} = \begin{bmatrix} -3 & 1 \\ 3 & -5 \end{bmatrix} \mathbf{x} + \begin{bmatrix} 1 \\ 0 \end{bmatrix} u + \begin{bmatrix} 1 \\ 2 \end{bmatrix} d$$

where d is a constant measurable disturbance, design feedforward control to minimize the effect of d on the error between x_2 and a constant reference input y_{ref}.

13.17. The plant in a dc motor speed control system is described by the state model

$$\dot{\mathbf{x}} = \begin{bmatrix} 0 & 1 \\ -2 & -3 \end{bmatrix} \mathbf{x} + \begin{bmatrix} 0 \\ 2 \end{bmatrix} u + \begin{bmatrix} -1 \\ 0 \end{bmatrix} w \qquad y = [1 \quad 0]\mathbf{x}$$

where x_1 is the motor speed, x_2 the armature current, u the armature voltage, and w the load torque. Check whether the conditions for integral control are satisfied, and design a state feedback plus integral control so that the closed-loop poles will be located at -1, -1, and -2.

13.18. Assuming that in Problem 13.17 the load torque w is measured, design feedforward control to reduce transient speed errors, and express the total control, consisting of this feedforward and the state feedback and integral control of Problem 13.17.

13.19. The linearized model of a distillation column is given by

$$\dot{\mathbf{x}} = \begin{bmatrix} -0.8 & 0.1 \\ -0.1 & -1.2 \end{bmatrix} \mathbf{x} + \begin{bmatrix} -1 & -0.7 \\ 0.2 & 0.3 \end{bmatrix} \mathbf{u} + \begin{bmatrix} 2 \\ 0.5 \end{bmatrix} d \qquad \mathbf{y} = \mathbf{Ix}$$

where the time unit is in hours. Design a feedforward plus state feedback controller to minimize the effect of the constant disturbance d on the error between $\mathbf{y}$ and the constant $\mathbf{y}_{ref}$ and to achieve a dominant time constant not exceeding 2 hours.

13.20. Suppose that in Example 13.3.1 only the output $y = [1 \quad 0]\mathbf{x}$ (i.e., the first state variable) is available for use in feedback control. Evaluate the use of constant-gain output feedback $u = -ky$ to improve on the speed of response of the plant with respect to both settling time and rise time. Select a suitable value for k.

13.21. A system has the discrete transfer function

$$\frac{C(z)}{R(z)} = \frac{2z + 6}{4z^3 - z}$$

Derive an equivalent difference state model.

13.22. Determine the stability of a system described by the difference state model $\mathbf{x}_{k+1} = \mathbf{Px}_k + \mathbf{Qu}_k$ if

$$\mathbf{P} = \begin{bmatrix} 0.4 & 0.2 \\ 0.3 & -0.1 \end{bmatrix}$$

13.23. Derive a difference state model to represent the algorithm $x_{k+2} = 3x_{k+1} - 2x_k + u_k$ and use it to determine the stability of this algorithm.

13.24. Derive a difference equation state model to represent the control algorithm

$$x_k = -4x_{k-1} - 6x_{k-2} + 2u_{k-1} + u_{k-2}$$

and use it to determine its stability.

13.25. Derive a difference state model to represent the algorithm $x_k = -5x_{k-1} - 3x_{k-2} + u_k$ and use it to determine its stability.

13.26. Determine the vector difference state model for digital simulation of the continuous system

$$\dot{\mathbf{x}} = \begin{bmatrix} 0 & 1 \\ -2 & -3 \end{bmatrix} \mathbf{x} + \begin{bmatrix} 0 \\ 1 \end{bmatrix} u$$

if the computation interval $T = 1$. Also determine the stability of the simulation.

13.27. Referring to Problem 12.21, determine a difference state model for digital simulation of the continuous system

$$\dot{\mathbf{x}} = \begin{bmatrix} -3/4 & -1/4 \\ -1/4 & -3/4 \end{bmatrix} \mathbf{x} + \begin{bmatrix} 0 \\ 1 \end{bmatrix} u$$

if the computation interval $T = 1$, and check the stability of the simulation.

13.28. Derive a digital simulation, in the form of a difference state model, for the system $G(s) = K/[(s + a)(s + b)] = C(s)/U(s)$ by finding first an equivalent continuous state model.

13.29. Determine a closed-form solution of the difference equation $x_{n+2} - 3x_{n+1} + 2x_n = u_n$ for $u_n = 0$ and the initial conditions $x_0 = 1$, $x_1 = 0.5$ by the use of a difference state model and Z transforms.

13.30. For the difference state model used in Problem 13.29, design state feedback so that the system eigenvalues will be located at the positions $(0.5 \pm 0.2j)$ in the z-plane.

MULTIVARIABLE SYSTEMS IN THE FREQUENCY DOMAIN

14.1 INTRODUCTION

Frequency-domain techniques provide an alternative to the state space approach, and often seek to generalize and exploit the tools and insights which have made classical control techniques so successful. The system dynamics are described by transfer function matrices, which are well adapted for the design of partially or completely decoupled or noninteracting control systems. To allow system outputs to be adjusted independently, it is often desirable that, at least in the steady state, manipulation of input i affect only the corresponding output i. This translates to the natural requirement of a diagonal steady-state transfer function matrix. The very successful Nyquist array techniques will be considered, among others, to introduce this important area of design.

14.2 SYSTEM CONFIGURATION AND EQUATIONS

In this section the system equations will be derived which form the basis for the formulation of the basic system stability theorem in Section 14.3.

In Fig. 14.1, $\mathbf{G}$, $\mathbf{K}$ and $\mathbf{H}$ are transfer function matrices, as defined for the discussion of stability in state space, and $\mathbf{r}$, $\mathbf{c}$, $\mathbf{e}$, and $\mathbf{u}$ are vectors of transforms of inputs, outputs, errors, and actuating signals.

$$\mathbf{G} = \{g_{ij}(s)\} \qquad \mathbf{c} = \{c_i(s)\} \tag{14.1}$$

$$c_i = \sum_{j=1}^{r} c_{ij} \qquad c_{ij} = g_{ij}u_j \tag{14.2}$$

$$\mathbf{H} = \text{diag}\,\{h_i\} \tag{14.3}$$

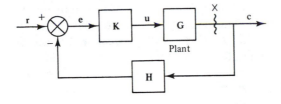

Figure 14.1 System configuration.

As discussed later, h_i can be taken to be real and constant. The system equations $\mathbf{c} = \mathbf{Gu}$, $\mathbf{u} = \mathbf{Ke}$ and $\mathbf{e} = \mathbf{r} - \mathbf{Hc}$ yield $\mathbf{c} = \mathbf{GK}(\mathbf{r} - \mathbf{Hc})$, so that

$$\mathbf{c} = (\mathbf{I} + \mathbf{GKH})^{-1}\mathbf{GKr} \tag{14.4}$$

Let

$$\mathbf{Q} = \mathbf{GK} \tag{14.5}$$

$$\mathbf{T} = \mathbf{GKH} = \mathbf{QH} \qquad \textit{loop gain matrix} \tag{14.6}$$

$$\mathbf{F} = \mathbf{I} + \mathbf{T} = \mathbf{I} + \mathbf{QH} = (\mathbf{Q} + \mathbf{H}^{-1})\mathbf{H} \qquad \textit{return difference matrix} \tag{14.7}$$

Then the closed-loop system is described by

$$\mathbf{c} = \mathbf{Rr} \tag{14.8}$$

$$\mathbf{R} = \mathbf{F}^{-1}\mathbf{Q} \qquad \textit{closed-loop transfer matrix}$$

In terms of inverse matrices, since the inverse of a product equals the product of the inverses in reverse order,

$$\hat{\mathbf{R}} = \hat{\mathbf{Q}}\mathbf{F} \qquad (\text{notation: } \hat{\mathbf{Q}} \equiv \mathbf{Q}^{-1}; \hat{\mathbf{R}} \equiv \mathbf{R}^{-1}) \tag{14.9}$$

Alternatively, since $\mathbf{F} = \mathbf{I} + \mathbf{QH}$:

$$\hat{\mathbf{R}} = \hat{\mathbf{Q}} + \mathbf{H} \tag{14.10}$$

*****

For *single-input/single-output* (SISO) *systems*, $R = Q/(1 + QH) = Q/F$. The numerator of F is the closed-loop characteristic polynomial (clcp) because its roots are the closed-loop system poles. Its denominator is the open-loop characteristic polynomial (olcp) because the roots equal those of QH, the open-loop poles. Hence the relations

$$\frac{\text{clcp}}{\text{olcp}} = F = \frac{Q}{R} = \frac{\hat{R}}{\hat{Q}} \qquad (F = 1 + T; \hat{R} = \hat{Q} + H) \tag{14.11}$$

For *multiple-input/multiple-output* (MIMO) *systems*, each transfer function in (14.11) is replaced by the determinant of a matrix:

$$\frac{\text{clcp}}{\text{olcp}} = \det \mathbf{F} = \det (\mathbf{I} + \mathbf{T}) = \det [(\mathbf{Q} + \mathbf{H}^{-1})\mathbf{H}] \tag{14.12a}$$

$$= \frac{\det \mathbf{Q}}{\det \mathbf{R}} \tag{14.12b}$$

$$= \frac{\det \hat{\mathbf{R}}}{\det \hat{\mathbf{Q}}} = \frac{\det (\hat{\mathbf{Q}} + \mathbf{H})}{\det \hat{\mathbf{Q}}} \tag{14.12c}$$

Here (14.12b) and (14.12c) follow from (14.12a) and the system equations, because the determinant of a product of matrices equals the product of their determinants. In (14.12a), the denominator of det $\mathbf{F}$ is the olcp because it is the product of the denominators of the elements of $\mathbf{T}$. To show that the numerator of det $\mathbf{F}$ is the clcp, in Fig. 14.1 let the loops be broken at $\times$ and a vector signal $\mathbf{a}(s)$ injected to the right. The "returned" signal at the left is $-\mathbf{Ta}$, and the difference between injected and returned signals is

$$\mathbf{a} - (-\mathbf{Ta}) = (\mathbf{I} + \mathbf{T})\mathbf{a} = \mathbf{Fa}$$

Hence the name of $\mathbf{F}$ in (14.7). Closing the break forces this difference to be zero:

$$\mathbf{Fa} = (\mathbf{I} + \mathbf{T})\mathbf{a} = (\mathbf{I} + \mathbf{QH})\mathbf{a} = \mathbf{0} \tag{14.13}$$

For SISO systems the roots of $1 + QH = 0$ are the closed-loop poles because for these values of s (14.13) can have a nontrivial (nonzero) solution for a. In the MIMO case, the condition for a nontrivial solution is known to be that the determinant of the coefficient matrix be zero:

$$\det \mathbf{F} = 0 \tag{14.14}$$

So the numerator of det $\mathbf{F}$ is the clcp.

The basic stability theorem in the next section as well as the particular theorems discussed subsequently are all based on one or other of the equations (14.12).

14.3 BASIC STABILITY THEOREM

Equations (14.11) for SISO systems and (14.12) for MIMO systems can be written as follows:

$$\frac{\text{clcp}}{\text{olcp}} = X = K\frac{(s + z_1)(s + z_2) \cdots (s + z_n)}{(s + p_1)(s + p_2) \cdots (s + p_m)} \tag{14.15}$$

The olcp is usually available in a factored form so that $-p_1, \ldots, -p_m$ can be assumed to be known. But the closed-loop system eigenvalues $-z_1, \ldots, -z_n$ are not, and must for stability all lie in the left-half s-plane. To determine this, a plot of X is made on a complex plane as s travels once clockwise around the Nyquist contour D in Fig. 14.2. D consists of the imaginary axis and a semicircle of radius $R \to \infty$, and in effect encloses the entire right-half s-plane. Poles on the imaginary axis are excluded by semicircular indentations.

As discussed in Section 7.3, each pole $-p_i$ inside D contributes one counterclockwise encirclement of the origin in the plot of X as s travels once clockwise around D, and each zero $-z_i$ inside D a clockwise encirclement. Hence the

Principle of the argument: If

$$p_c, p_0 = \text{number of roots of clcp, olcp inside } D \tag{14.16}$$
then the plot of X will encircle its origin $(p_c - p_0)$ times clockwise.

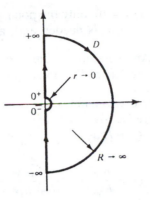

Figure 14.2 Nyquist contour D.

Hence follows, since $p_c = 0$ is required for stability:

> **Theorem 14.3.1.**   If a plot of X as s travels once clockwise around D encircles the origin N_X times clockwise, the system is stable if and only if
>
> $$N_X = -p_0 \quad \text{(negative = counterclockwise)} \tag{14.17a}$$

If X is a ratio, of transfer functions in (14.11) or of determinants in (14.12), then N_X is the difference of clockwise encirclements by numerator and denominator. Thus, if N_F, N_R, N_Q, $\hat{N}_Q$, and $\hat{N}_R$ are the clockwise origin encirclements of, respectively, det $\mathbf{F}$, det $\mathbf{R}$, det $\mathbf{Q}$, det $\hat{\mathbf{Q}}$, and det $\hat{\mathbf{R}}$ for MIMO systems, or of F, R, Q, $\hat{Q}$, and $\hat{R}$ for SISO systems, then, depending on which plots are used:

$$N_X = N_F \quad (\mathbf{F} = \mathbf{T} + \mathbf{I} = (\mathbf{Q} + \mathbf{H}^{-1})\mathbf{H}) \tag{14.17b}$$

$$N_X = N_Q - N_R \quad N_X = \hat{N}_R - \hat{N}_Q \quad (\hat{\mathbf{R}} = \hat{\mathbf{Q}} + \mathbf{H}) \tag{14.17c}$$

First the application to SISO systems will be discussed, both to classify the Nyquist criterion of Chapter 7 in this framework and to introduce the use of plots of inverse transfer functions.

14.4 INVERSE NYQUIST PLOTS FOR SISO SYSTEMS

The origin encirclements of $F = T + 1$ equal those of T about -1. Hence, as in Section 7.3:

> **Theorem 14.4.1.**   The system is stable if and only if the Nyquist diagram of T encircles point -1 p_0 times counterclockwise.

Alternatively, since $T = QH$, if $H = h =$ constant:

> **Theorem 14.4.2.**   The system is stable if and only if the Nyquist diagram of Q encircles point $-1/h$ p_0 times counterclockwise.

The *Nyquist diagram* is the plot as s travels once clockwise around D. For open-loop stable systems ($p_0 = 0$), only the polar plot, with s along the imaginary axis from $\omega \to 0^+$ to $\omega \to +\infty$, is needed and, as in Section 7.5:

Theorem 14.4.3. An open-loop stable system is stable if and only if point -1 ($-1/h$) lies to the left of the polar plot of $T(Q)$.

Chapter 7 provides numerous examples.

Inverse plots are frequently used in the design of H for SISO systems, because it appears in $\hat{R} = \hat{Q} + H$ as a simple term. Consider first the construction of a plot of

$$\hat{Q} = \frac{a_n s^n + \cdots + a_1 s + a_0}{b_m s^m + \cdots + b_1 s + b_0} \tag{14.18}$$

$$\hat{Q}(\omega = 0^+) = \frac{a_0}{b_0} \qquad \hat{Q}(\omega \to +\infty) = \frac{a_n}{b_m}(j\omega)^{n-m} \tag{14.19}$$

Usually, $n > m$ and $\hat{Q}$ is on the real axis for $\omega = 0^+$, and it tends to infinity for $\omega \to +\infty$. Since each factor $j\omega$ contributes $+90°$, the plot of $\hat{Q}$ tends to infinity along the positive imaginary axis for $n - m = 1$, the negative real axis for $n - m = 2$, and the negative imaginary axis for $n - m = 3$. The behavior between 0^+ and $+\infty$ can be calculated, or sketched by use of the vector diagrams in Fig. 7.6. The solid curves in Fig. 14.3 show typical *inverse polar plots*.

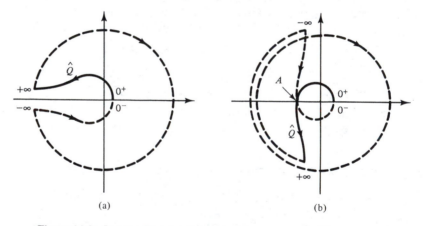

(a)   (b)

Figure 14.3 Inverse Nyquist diagrams: (a) $n - m = 2$; (b) $n - m = 3$.

To extend these into *inverse Nyquist diagrams,* the plots for the frequency range 0^- to $-\infty$ are the mirror images of the inverse polar plots. On the Nyquist contour D, the vector s rotates π radians clockwise at large radius R between $+\infty$ and $-\infty$ on the imaginary axis. Since $R \to \infty$, $\hat{Q} \to (a_n/b_m)s^{n-m}$, so vector $\hat{Q}$ rotates $(n - m)\pi$ radians clockwise at large radius for this rotation of s. This means 2π radians rotation in Fig. 14.3(a), and 3π in Fig. 14.3(b).

Because $\hat{R} = \hat{Q} + h$, the origin encirclements $\hat{N}_R$ in (14.17c) equal those of $\hat{Q}$ about $-h$. Hence the Theorem 14.3.1 when expressed in terms of inverse transfer functions becomes:

Theorem 14.4.4.   The system is stable if and only if the difference of the clockwise encirclements of $\hat{Q}$ about $-h$ and the origin equals $-p_0$.

For open-loop stable systems ($p_0 = 0$), $-h$ and the origin must be encircled an equal number of times, so they must lie on the same side of the inverse polar plot. If Fig. 14.3 represents open-loop stable systems, (a) is stable for all gains h, and (b) only if $-h$ lies to the right of point A.

Note that, appearances to the contrary, the origin encirclements for one complete circuit are zero in Fig. 14.3(b) as well as in (a).

14.5 INVERSE AND DIRECT NYQUIST ARRAY TECHNIQUES

The difficulty in applying the Stability Theorem 14.3.1 to multivariable systems is that determination of N_X requires finding the origin encirclements of plots of the determinants of matrices as defined in (14.17). Rosenbrock's Nyquist array techniques overcome this problem if the matrices are sufficiently close to diagonal, in a well-defined sense.

If in Fig. 14.1, $\mathbf{Q} = \mathbf{GK}$ is diagonal, the system, with $\mathbf{H} = \text{diag}\{h_i\}$, will consist of a number of independent SISO feedback control loops, each of which can be designed independently by classical techniques. Also, from (14.4) and (14.8), $\mathbf{R}$ would be diagonal and the system noninteracting or decoupled. However, aside from conditions on its existence, the decoupling controller $\mathbf{K}$ required to achieve this tends to be of the same order of complexity as the plant $\mathbf{G}$ itself.

The important Nyquist array (NA) design techniques originated by Rosenbrock are based on the principle of *diagonal dominance*. This represents a condition under which interaction has been reduced sufficiently that the overall control can still be obtained by independent SISO design of diagonal loops. Moreover, as discussed later, often the $\mathbf{K}$ needed to achieve this dominance can consist of constant-gain elements.

Definition.   An $m \times m$ matrix $\mathbf{Z}(s) = \{z_{ij}(s)\}$ is diagonally dominant on the Nyquist contour D if for all s on D and for all i either

$$|z_{ii}(s)| > d_{ir}(s) = \sum_{j=1, \neq i}^{m} |z_{ij}(s)| \quad \text{(row dominance)} \quad (14.20a)$$

or

$$|z_{ii}(s)| > d_{ic}(s) = \sum_{j=1, \neq i}^{m} |z_{ji}(s)| \quad \text{(column dominance)} \quad (14.20b)$$

Thus the magnitude of the diagonal element z_{ii} is larger than the sum d_{ir} or d_{ic} of the magnitudes of the off-diagonal elements in the row or the column. A graphical interpretation of this condition is based on the use of the *Gershgorin bands,* which are constructed as follows:

1. Construct Nyquist plots of the diagonal elements $z_{ii}(s)$, as illustrated in Fig. 14.4 for one element.
2. For each, draw circles of radii d_{ir} or d_{ic} calculated from (14.20), with their centers on the plot of z_{ii} at the corresponding frequency. The bands swept out by these circles, illustrated in Fig. 14.4 for z_{ii}, are the Gershgorin bands.

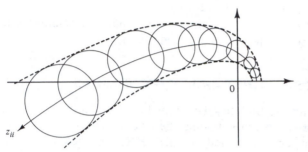

Figure 14.4   Gershgorin band.

Evidently, if for all i these bands exclude the origin, then (14.20) is satisfied and **Z** is diagonally dominant. It is noted that interactive computer graphics are used for computation and display of such plots.

*****

The exploitation of this concept depends on the following theorem, given here without proof.

Theorem 14.5.1.   For diagonally dominant matrices **Z**, the origin encirclements N_Z of det **Z** as s travels once clockwise around Nyquist contour D equal the sum of the encirclements N_{zi} of the diagonal elements z_{ii}:

$$N_Z = \sum_{i=1}^{m} N_{zi} \tag{14.21}$$

This is evident for diagonal matrices, for which the determinant equals the product of the diagonal elements.

Rosenbrock's *inverse Nyquist array (INA) technique* is used extensively for design, and is based on inverse Nyquist plots. If $\hat{\mathbf{Q}}$ and $\hat{\mathbf{R}} = \hat{\mathbf{Q}} + \mathbf{H}$ are diagonally dominant, the condition $N_X = \hat{N}_R - \hat{N}_Q = -p_0$ in (14.17c) of the basic stability theorem becomes $\Sigma_i (\hat{N}_{ri} - \hat{N}_{qi}) = -p_0$, where $\hat{N}_{ri}$ and $\hat{N}_{qi}$ are the clockwise origin encirclements of the elements $\hat{r}_{ii}$ and $\hat{q}_{ii}$ of $\hat{\mathbf{R}}$ and $\hat{\mathbf{Q}}$. Since $\hat{r}_{ii} = \hat{q}_{ii} + h_i$, $\hat{N}_{ri}$ equals the encirclements of $\hat{q}_{ii}$ about $-h_i$, so that the stability condition can be written: Σ_i (clockwise encirclements of $\hat{q}_{ii}$ about $-h_i$ minus those around the origin) $= -p_0$.

Figure 14.5 shows $\hat{q}_{11}$ and $\hat{q}_{22}$ with their Gershgorin bands as given by Rosenbrock for an example. $\hat{\mathbf{Q}}$ is apparently dominant. Since $\hat{r}_{ij} = \hat{q}_{ij}$, $i \neq j$,

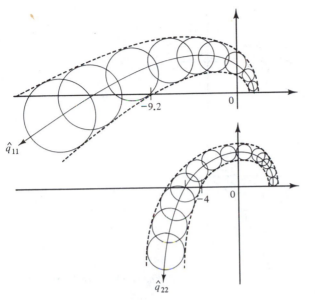

Figure 14.5 INA example.

the Gershgorin circle radii of $\hat{Q}$ and $\hat{R}$ are the same. For $\hat{R}$ to be dominant, the bands centered on $\hat{r}_{ii} = \hat{q}_{ii} + h_i$ must exclude the origin, so those centered on $\hat{q}_{ii}$ must exclude $-h_i$. Hence follows the

> **INA Stability Theorem 14.5.2.**   If the Gershgorin bands centered on the $\hat{q}_{ii}$ exclude the origin and the points $-h_i$ (i.e., if $\hat{Q}$ and $\hat{R}$ are dominant), the system is stable if and only if
>
> $$\sum_i \text{(clockwise encirclements of } \hat{q}_{ii} \text{ about } -h_i \text{ minus}$$
> $$\text{those around the origin)} = -p_0 \qquad (14.22)$$

The example in Fig. 14.5 is given to be open-loop stable, so $p_0 = 0$ and stability is guaranteed if the bands encircle the origin and $-h_i$ an equal number of times. This will be the case if the origin and $-h_i$ lie on the same side of the bands, so for all gain combinations $0 \leq h_1 < 9.2$, $0 \leq h_2 < 4.0$, ignoring negative gains. Dominance of $\hat{Q}$ and of $\hat{R}$ are sufficient conditions, so it may not be concluded that the system will be unstable for values of $-h_i$ inside the bands.

The *direct Nyquist array (DNA) technique* uses the condition $N_X = N_F = -p_0$ in (14.17b) of the basic stability theorem, with $\mathbf{F} = (\mathbf{Q} + \mathbf{H}^{-1})\mathbf{H}$. Since $\det \mathbf{F} = \det (\mathbf{Q} + \mathbf{H}^{-1}) \det \mathbf{H}$, for constant $\mathbf{H}$ the encirclements N_F of $\det \mathbf{F}$ equal those of $\det (\mathbf{Q} + \mathbf{H}^{-1})$. The condition for the DNA technique is that $(\mathbf{Q} + \mathbf{H}^{-1})$ is diagonally dominant. The encirclements of $(q_{ii} + h_i^{-1})$ about the origin equal those of q_{ii} about $-h_i^{-1}$, and the stability condition then becomes: Σ_i (clockwise encirclements of q_{ii} about $-h_i^{-1}$) $= -p_0$.

Figure 14.6 shows q_{11} and q_{22} with their Gershgorin bands for a two-by-

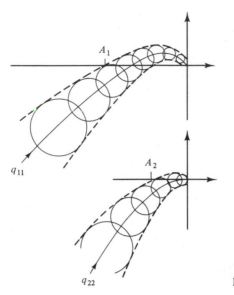

Figure 14.6 DNA example.

two system. Since $\mathbf{H}^{-1}$ is diagonal, the Gershgorin circle radii of $(\mathbf{Q} + \mathbf{H}^{-1})$
are the same as those of $\mathbf{Q}$. For dominance, the bands based on $(q_{ii} + h_i^{-1})$
should exclude the origin, so those based on q_{ii} should exclude $-h_i^{-1}$. Assuming
open-loop stability ($p_0 = 0$), this ensures stability if $-1/h_1$ and $-1/h_2$ lie to the
left of A_1 and A_2, respectively. The sufficient condition that $(\mathbf{Q} + \mathbf{H}^{-1})$ be
dominant is then satisfied, and stability, in direct extension of Theorem 14.4.3
for SISO systems, is guaranteed because these points lie to the left of the polar
plots.

> **DNA Stability Theorem 14.5.3.**   If the Gershgorin bands centered on the
> q_{ii} exclude the points $-h_i^{-1}$, then the system is stable if and only if
>
> $$\sum_i \text{(clockwise encirclements of } q_{ii} \text{ about } -h_i^{-1}) = -p_0 \qquad (14.23)$$

The gains h_i in the individual feedback loops for INA and DNA can be
"moved to" the corresponding forward paths, ahead of $\mathbf{K}$, for implementation.
This does not affect stability because it does not change the loop-gain matrix

$$\mathbf{T} = \mathbf{GKH} = \mathbf{QH}.$$

In fact, it is useful to reformulate the DNA approach in terms of $\mathbf{T}$ to see more
clearly the links with classical SISO design based on the loop gain function.
$\mathbf{F} = \mathbf{T} + \mathbf{I}$ is dominant if the Gershgorin bands centered on $f_{ii} = t_{ii} + 1$ exclude
the origin, so if those centered on t_{ii} exclude -1. Since $f_{ij} = t_{ij}$, $i \neq j$, the
circle radii of $\mathbf{F}$ and $\mathbf{T}$ are the same. Hence, equivalent to the DNA theorem:

> If the bands centered on the t_{ii} exclude -1, the system is stable if and only if
>
> $$\sum_i \text{(clockwise encirclements of } t_{ii} \text{ about } -1) = -p_0 \qquad (14.24)$$

In the open-loop stable case, this is equivalent to the -1 point lying to the left of each polar plot t_{ii}, a direct extension of the condition for SISO systems.

K should really be considered as a product $\mathbf{K}_a\mathbf{K}_d$, where $\mathbf{K}_a$ is designed to produce dominance as discussed later, and $\mathbf{K}_d$ is a diagonal matrix representing the controllers designed independently by classical SISO techniques for each of the diagonal loops after dominance has been established.

It may be that a system can be made closed-loop dominant but not open-loop dominant. For $\mathbf{H} = \mathbf{I}$, the closed-loop transfer function matrix $\mathbf{R} = (\mathbf{I} + \mathbf{Q})^{-1}\mathbf{Q}$. For high loop gains, when the elements of $\mathbf{Q}$ are large, this approaches the ideal diagonal matrix $\mathbf{Q}^{-1}\mathbf{Q} = \mathbf{I}$, regardless of the form of $\mathbf{Q}$. In the inverse domain the condition (14.17c) can then be written as $\Sigma_i \hat{N}_{ri} - \hat{N}_Q = -p_0$.

Design of **K** to achieve dominance will be discussed after a note on the characteristic locus method.

14.6 A NOTE ON THE CHARACTERISTIC LOCUS (CL) METHOD

Characteristic loci are the basis of the CL method and of an intensive effort originated by MacFarlane and coworkers to generalize concepts such as the Nyquist criterion and root loci.

Condition (14.17b) gives the stability condition $N_F = -p_0$, where N_F is the number of clockwise origin encirclements of det **F**. To find a novel expression for N_F, it is recalled that the eigenvalues $\lambda_1, \ldots, \lambda_n$, assumed distinct, of a constant matrix **A** are the roots of $|\mathbf{A} - \lambda_i\mathbf{I}| = 0$. A transformation **T** of which the columns are the eigenvectors of **A** transforms **A** to the diagonal form $\mathbf{\Lambda} = \text{diag} \{\lambda_i\} = \mathbf{T}^{-1}\mathbf{A}\mathbf{T}$. Hence $|\mathbf{T}^{-1}|\,|\mathbf{A}|\,|\mathbf{T}| = |\mathbf{A}| = |\mathbf{\Lambda}| = \lambda_1\lambda_2\cdots\lambda_n \equiv \Pi_{i=1}^n \lambda_i$, or

$$\det \mathbf{A} = \prod_{i=1}^{n} \lambda_i$$

Now, for any value s_1 of s, $\mathbf{F}(s_1)$ is also a constant matrix, be it one with generally complex elements. Its eigenvalues $f_i(j\omega_a)$, $i = 1, \ldots, m$, at a selected frequency ω_a can be found using a standard algorithm after first calculating $\mathbf{F}(j\omega_a)$. These eigenvalues change with s_1, and the *characteristic loci* $f_i(s)$, $i = 1, \ldots, m$, are their plots as s travels once clockwise around the Nyquist contour D. Analogous to the relation for det **A**:

$$\det \mathbf{F}(s) = \prod_{i=1}^{m} f_i(s) \tag{14.25}$$

Therefore, the condition $N_F = -p_0$ yields

Theorem 14.6.1.   The system is stable if and only if $\Sigma_{i=1}^m N_{fi} = -p_0$, where $N_{fi} =$ number of clockwise origin encirclements of $f_i(s)$.

By the eigenvalue shift theorem of matrix theory, because $\mathbf{F} = \mathbf{T} + \mathbf{I}$, the characteristic loci $f_i(s)$ and $t_i(s)$ of $\mathbf{F}$ and $\mathbf{T}$ are related by

$$f_i(s) = t_i(s) + 1 \qquad (14.26)$$

From this and Theorem 14.6.1 follows

> **Theorem 14.6.2. (Generalized Nyquist Criterion).** The system is stable if and only if $\sum_{i=1}^{m} N_{ti} = -p_0$, where N_{ti} = number of clockwise encirclements of $t_i(s)$ about -1.

Note that this generalization is in terms of the characteristic loci of the loop gain matrix. CL design is concerned with modifying these loci in a desired direction, and uses also the eigenvectors of $\mathbf{T}(s)$.

14.7 DESIGN TO ACHIEVE DIAGONAL DOMINANCE

Numerous techniques exist for design of $\mathbf{K}$ such that $\mathbf{GK}$ will be dominant. Often, as will be assumed here, only constant gains are required. In addition to a technique using elementary matrix operations indicated below, methods are available to minimize a measure of the ratios of d_{ir} or d_{ic} to z_{ii} in (14.20). These can be numerical hillclimbing procedures or pseudodiagonalization schemes which require the solution of eigenvalue/eigenvector problems.

These techniques generally allow for a *postcompensator* $\mathbf{L}$, indicated in Fig. 14.7, in addition to the *precompensator* $\mathbf{K}$, to facilitate achieving dominance of $\mathbf{Q} = \mathbf{LGK}$, or of $\hat{\mathbf{Q}} = \hat{\mathbf{K}}\hat{\mathbf{G}}\hat{\mathbf{L}}$. Considering 2 by 2 systems, let

$$\mathbf{Q} = \begin{bmatrix} l_{11} & l_{12} \\ l_{21} & l_{22} \end{bmatrix} \begin{bmatrix} g_{11} & g_{12} \\ g_{21} & g_{22} \end{bmatrix} \begin{bmatrix} k_{11} & k_{12} \\ k_{21} & k_{22} \end{bmatrix} \qquad (14.27)$$

If $\mathbf{L} = \mathbf{I}$, only the jth column of $\mathbf{K}$ appears in the jth column of $\mathbf{Q}$. Thus design is simplified by considering column dominance since this permits the columns of $\mathbf{K}$ to be found independently. For $\hat{\mathbf{Q}} = \hat{\mathbf{K}}\hat{\mathbf{G}}$, row-dominance design makes the rows of $\hat{\mathbf{K}}$ independent. $\mathbf{K}$ performs column operations on $\mathbf{LG}$ (i.e., the columns of $\mathbf{Q}$ are linear combinations of those of $\mathbf{LG}$). $\mathbf{L}$ performs row operations on $\mathbf{GK}$. $\mathbf{K}$ is a *permutation matrix* if $k_{11} = k_{22} = 0$ and $k_{12} = k_{21} = 1$. This interchanges the columns of $\mathbf{LG}$, an *elementary column operation*. A permutation matrix $\mathbf{L}$ interchanges the rows of $\mathbf{GK}$.

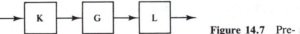

Figure 14.7 Pre- and postcompensation.

For a perspective on this, it is important to observe that $\mathbf{G}$ is just a matrix of transfer functions relating input and output variables numbered in specified orders and expressed in given physical units. One is evidently free to change these orders or to use other units. If input 1 affects output 2 most strongly, and input 2 similarly affects output 1, dominance may result by using a permutation matrix $\mathbf{K}$ to interchange the columns of $\mathbf{G}$. This renumbers the inputs to achieve the desired pairing of inputs and outputs. A permutation $\mathbf{L}$, interchanging rows and renumbering the outputs, has the same effect and gives the same column dominance ratios.

A diagonal matrix $\mathbf{K}$ multiplies the first column of $\mathbf{G}$ by k_{11} and the second by k_{22}. This is equivalent to expressing the inputs in different units, and does not change the column dominance ratios. A diagonal $\mathbf{L}$ rescales the outputs and can be very useful. If, say, the elements in the second row of $\mathbf{G}$ are much smaller than those in the first row, then $\mathbf{K}$, which only combines columns of $\mathbf{G}$, is not well suited to achieve dominance of the second column. But the first column is likely to be highly dominant, and $\mathbf{L}$ permits this to be shared with the second column. If $\mathbf{K} = \mathbf{I}$ and $l_{11} = 1$, the dominance ratios are $l_{22}g_{21}/g_{11}$ in column one and $g_{12}/(l_{22}g_{22})$ in column two. Increasing $l_{22} > 1$ improves the second column at the expense of the first.

It should be noted that $\mathbf{L}$ is in the plant output. A permutation or diagonal $\mathbf{L}$ can be implemented here, but a general form of $\mathbf{L}$ cannot, and must therefore be moved around the loop after design to be combined with $\mathbf{K}$. This does not affect stability, but does change interactions.

14.8 VECTOR DIAGRAM TECHNIQUE FOR DESIGN OF K

As mentioned, several routines are available for design of $\mathbf{K}$ to improve column dominance in $\mathbf{Q} = \mathbf{GK}$. These include interactive routines with computer graphics used to produce Nyquist arrays, that is, Nyquist plots of all elements of $\mathbf{Q}$, or $\hat{\mathbf{Q}}$.

In this section a technique for 2 by 2 and 3 by 3 systems is discussed which, while it has been implemented as an interactive routine with graphics, is much less dependent on such aids and has been applied to numerous examples without them. For this reason it may be more appropriate in the context of this introductory chapter.

The system configuration considered in this section and the next is the very common one with unity feedback in Fig. 14.8a. For 2 by 2 systems, let

$$\mathbf{G} = \begin{bmatrix} \dfrac{n_{11}}{d_{11}} & \dfrac{n_{12}}{d_{12}} \\[2mm] \dfrac{n_{21}}{d_{21}} & \dfrac{n_{22}}{d_{22}} \end{bmatrix} \qquad \mathbf{K} = \begin{bmatrix} k_{11} & k_{12} \\ k_1 & k_2 \end{bmatrix} \qquad \mathbf{Q} = \mathbf{GK}$$

$$\mathbf{K}_d = \begin{bmatrix} k_{d1}(s) & 0 \\ 0 & k_{d2}(s) \end{bmatrix} \qquad \begin{aligned} D_i &= k_{di}q_{ii} \\ k_{11}, k_{12} &= 1 \text{ or } 0 \end{aligned} \tag{14.28}$$

Here $n_{ij}(s)$ and $d_{ij}(s)$ are polynomials. In $\mathbf{K}$, if k_{11} and/or k_{12} is not chosen to be zero, it can be taken to equal 1 without loss of generality since only the ratios in the columns of $\mathbf{K}$ affect column dominance. $\mathbf{K}_d$ is a diagonal matrix of controller transfer functions $k_{di}(s)$ identified in the explicit form of Fig. 14.8(a) shown in

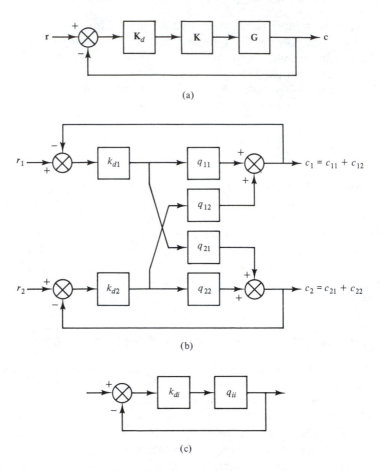

(a)

(b)

(c)

Figure 14.8 System configuration.

Fig. 14.8(b). After $\mathbf{K}$ has been found, these are designed independently, on the basis of the loop gain function D_i of the ith diagonal loop shown in Fig. 14.8(c) as discussed in the next section. If a polynomial matrix $\hat{\mathbf{Q}} = \{\hat{q}_{ij}\}$ is defined by

$$q_{ij} = \frac{\hat{q}_{ij}}{d_{i1}d_{i2}} \qquad \hat{q}_{ij} = k_{1j}n_{i1}d_{i2} + k_{j}n_{i2}d_{i1} \qquad (14.29)$$

then D_i and the column dominance ratio r_{cj} in column j of $\mathbf{Q}$ can be written

$$D_i = k_{di}q_{ii} = \frac{k_{di}\hat{q}_{ii}}{d_{i1}d_{i2}}$$

$$r_{cj} = \frac{q_{ij}}{q_{jj}} = \frac{\hat{q}_{ij}d_{j1}d_{j2}}{\hat{q}_{jj}d_{i1}d_{i2}} \qquad (14.30)$$

Column j of $\mathbf{K}$ should be designed to minimize r_{cj}, subject to the constraint that, if possible, $\hat{q}_{jj}$ is Hurwitz (i.e., has only left-half s-plane roots). Otherwise, D_j

would be a nonminimum-phase transfer function and, as discussed in Section 8.7, performance might be seriously constrained.

The technique is based on the use of Bode plots and a simple type of vector diagram and will be discussed by means of an example.

Example 14.8.1   Design of K

$$n_{11} = 1, \quad d_{11} = s + 1 \qquad\qquad n_{12} = 5, d_{12} = s + 5 \tag{14.31}$$
$$n_{21} = 0.1, d_{21} = s + 0.5 \qquad n_{22} = 4, d_{22} = s + 2$$

$$q_{1j} = \frac{\hat{q}_{1j}}{(s + 1)(s + 5)} \qquad \hat{q}_{1j} = k_{1j}(s + 5) + 5k_j(s + 1)$$
$$\tag{14.32}$$
$$q_{2j} = \frac{\hat{q}_{2j}}{(s + 0.5)(s + 2)} \qquad \hat{q}_{2j} = 0.1k_{1j}(s + 2) + 4k_j(s + 0.5)$$

At very low ($s \to 0$) and very high ($s \to \infty$) frequencies:

$$q_{1jl} = k_{1j} + k_j \qquad\qquad sq_{1jh} = k_{1j} + 5k_j$$
$$\tag{14.33}$$
$$q_{2jl} = 0.2k_{1j} + 2k_j \qquad sq_{2jh} = 0.1k_{1j} + 4k_j$$

Here subscript *l* corresponds to $s \to 0$, and *h* to $s \to \infty$. Figure 14.9(a) shows each constructed as a sum of vectors, for $k_{1j} = 1$ and an arbitrary but simple choice of k_j, here $k_j = -1$. Vector q_{1jl} is the sum of $+1$ (to the right) and -1 (to the left, and offset for clarity). For $k_{1j} = 0$ the first vectors are zero, and for values of k_j other than -1 the corresponding vectors are scaled proportionally, including a change of direction for $k_j > 0$.

The choice of k_j and k_{1j} (zero or 1) is made from Fig. 14.9(a) to minimize r_{cj} at both low and high frequencies. To minimize r_{c1}, inspection suggests the choices $k_{11} = 1$ and k_1 near zero, because this makes q_{21} small relative to q_{11} at both ends of the frequency range. This order-of-magnitude choice is refined by vector-length measurements, which can be done by counting plot divisions, and simple ratio calculations, yielding $k_1 = -0.05$. For r_{c2}, k_{12} and k_2 must be chosen to minimize q_{12} relative to q_{22}. Here the low-frequency diagram suggests $k_{12} = 1$, $k_2 = -1$, and that for high frequencies suggests $k_{12} = 1$, $k_2 = -0.2$. However, this second choice would give bad results at low frequencies. Refinement of the first choice does not improve the ratio, and this set is selected even though r_{c2h} will slightly exceed 1. The elements of **K** and the corresponding column dominance ratios at low and high frequencies which result from this provisional selection are now as follows:

$$k_{11} = 1, k_1 = -0.05 \qquad\qquad k_{12} = 1, k_2 = -1.0 \tag{14.34}$$
$$r_{c1l} = 0.105, r_{c1h} = -0.133 \qquad r_{c2l} = 0, r_{c2h} = 1.026$$

Substituting into (14.32), the polynomials $\hat{q}_{ij}$ and their roots (in parentheses) are found to be

$$\hat{q}_{11} = 0.75s + 4.75 \quad (-6.333) \qquad \hat{q}_{21} = -0.1s + 0.1 \quad (+1.0)$$
$$\tag{14.35}$$
$$\hat{q}_{22} = -3.9s - 1.8 \quad (-0.462) \qquad \hat{q}_{12} = -4.0s \quad (0.0)$$

As desired, $\hat{q}_{11}$ and $\hat{q}_{22}$ are Hurwitz. If they had not been, the pole–zero patterns of $\hat{q}_{ij}$ shown in Fig. 14.9(b) are useful. From (14.32), the root of $\hat{q}_{1j}$ is -1 for

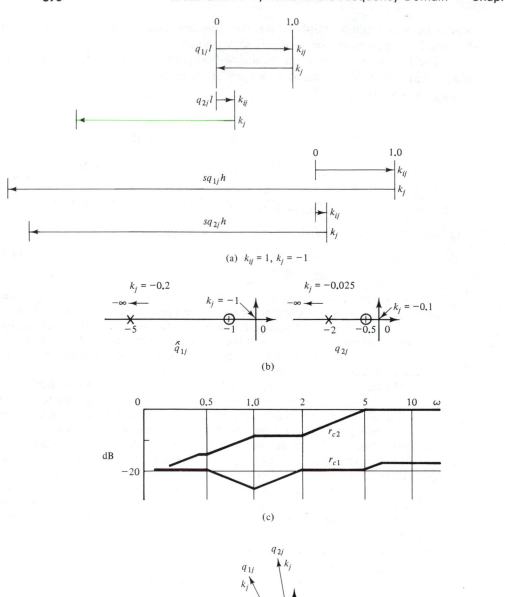

(a) $k_{ij} = 1$, $k_j = -1$

(b)

(c)

(d) $\omega = 10$, $k_{ij} = 1$, $k_j = -1$

Figure 14.9 Design of K for Example 14.8.1.

$k_{1j} = 0$, and for $k_{1j} = 1$, $\hat{q}_{1j}$ can be written

$$\hat{q}_{1j} = (s + 5)\left[1 + \frac{5k_j(s + 1)}{(s + 5)}\right] \tag{14.36}$$

From the root locus technique, the form in brackets means that the roots of $\hat{q}_{1j}$ lie along loci for varying root locus gain $5k_j$ which could be constructed from an open loop pole–zero pattern with a pole at -5 and a zero at -1. Figure 14.9(b) shows these patterns for $\hat{q}_{2j}$ as well as $\hat{q}_{1j}$. Superimposing roots of $\hat{q}_{jj}$ calculated numerically for contemplated values of k_{1j} and k_j onto these patterns will, without constructing the complete loci, suggest directions of change of k_j to achieve a Hurwitz result. It is noted that loci for negative gains were discussed in Sections 6.2 and 6.7.

Bode plots of r_{cj} are constructed to examine dominance over the frequency range, and determine whether the design of K based on low and high frequencies is satisfactory. Using (14.32) and (14.35) yields

$$r_{c1} = \frac{(-0.1s + 0.1)Z}{0.75s + 4.75} \qquad r_{c2} = \frac{-4.0s}{(-3.9s - 1.8)Z} \tag{14.37}$$

$$Z \equiv \frac{(s + 1)(s + 5)}{(s + 0.5)(s + 2)}$$

With the roots in (14.35), these give the Bode plots in Fig. 14.9(c). The low-frequency asymptote 0.105 of r_{c1} is also given in (14.34), and that for r_{c2} is obtained from (14.37) for $s \rightarrow 0$:

$$r_{c2l} = \frac{-4.0s}{-1.8 \times 5} = 0.444s$$

Column one is highly dominant, so that dominance sharing via diagonal L, implemented by multiplying the rows of G by the corresponding diagonal elements of L, could be used to improve r_{c2}.

This design was based on low and high frequencies alone. If such Bode plots of r_{c1} or r_{c2} show considerable peaking over a range of frequencies, vector diagrams for one or more frequencies in this range can be used to consider improvements. Figure 14.9(d) shows these diagrams for $\omega = 10$, calculated by substituting $s = 10j$ into (14.32):

$$q_{1j} = \frac{k_{1j}(5 + 10j) + 5k_j(1 + 10j)}{(1 + 10j)(5 + 10j)}$$

$$= (0.01 - 0.1j)k_{1j} + (0.2 - 0.4j)k_j \tag{14.38}$$

$$q_{2j} = \frac{0.1k_{1j}(2 + 10j) + 4k_j(0.5 + 10j)}{(2 + 10j)(0.5 + 10j)}$$

$$= (0.0005 - 0.01j)k_{1j} + (0.077 - 0.385j)k_j$$

Inspection again shows the difficulty of making q_{12}/q_{22} small except by dominance sharing.

One additional consideration in the design of K, which links the choice of its columns, will be discussed in Section 14.9.

The technique has been applied to 2 by 2 and 3 by 3 systems using only a hand-held calculator with program modules for multiplying polynomials and for finding their roots. For routine use it has also been implemented for interactive computer design with graphics.

14.9 CLASSICAL DESIGN OF 2 BY 2 SYSTEMS

The design of $\mathbf{K}_d$ in Fig. 14.8 is considered next. As in the case of $\mathbf{K}$, the technique benefits from computer implementation but does not depend on it. The purpose is the design of at least 2 by 2 systems using only simple concepts and familiar tools, as an extension of SISO frequency-domain design.

The closed-loop transfer function matrix $\mathbf{R}$ is, from (14.4) and (14.8) with $\mathbf{H} = \mathbf{I}$,

$$\mathbf{R} = (\mathbf{I} + \mathbf{QK}_d)^{-1}\mathbf{QK}_d \tag{14.39}$$

From (14.29), $\hat{\mathbf{Q}}$ can be expressed as

$$\hat{\mathbf{Q}} = \mathbf{EQ} \qquad \mathbf{E} = \text{diag}\{d_{11}d_{12}, d_{21}d_{22}\} \tag{14.40}$$

Hence $\mathbf{R} = (\mathbf{I} + \mathbf{QK}_d)^{-1}\mathbf{E}^{-1}\mathbf{EQK}_d = [\mathbf{E}(\mathbf{I} + \mathbf{QK}_d)]^{-1}\mathbf{EQK}_d$, or

$$\mathbf{R} = (\mathbf{E} + \hat{\mathbf{Q}}\mathbf{K}_d)^{-1}\hat{\mathbf{Q}}\mathbf{K}_d \tag{14.41}$$

$$\hat{\mathbf{Q}}\mathbf{K}_d = \begin{bmatrix} k_{d1}\hat{q}_{11} & k_{d2}\hat{q}_{12} \\ k_{d1}\hat{q}_{21} & k_{d2}\hat{q}_{22} \end{bmatrix}$$

$$(\mathbf{E} + \hat{\mathbf{Q}}\mathbf{K}_d)^{-1} = \frac{1}{\hat{\Delta}}\begin{bmatrix} d_{21}d_{22} + k_{d2}\hat{q}_{22} & -k_{d2}\hat{q}_{12} \\ -k_{d1}\hat{q}_{21} & d_{11}d_{12} + k_{d1}\hat{q}_{11} \end{bmatrix}$$

where $\hat{\Delta} = \det(\mathbf{E} + \hat{\mathbf{Q}}\mathbf{K}_d)$. Also,

$$\hat{q}_{11}\hat{q}_{22} - \hat{q}_{12}\hat{q}_{21} = |\hat{\mathbf{Q}}| = |\mathbf{EQ}| = |\mathbf{EGK}| = |\mathbf{EG}|\,|\mathbf{K}|$$
$$= (k_{11}k_2 - k_{12}k_1)\delta \tag{14.42a}$$
$$\delta = |\mathbf{EG}| = n_{11}n_{22}d_{12}d_{21} - n_{12}\,n_{21}\,d_{11}\,d_{22} \tag{14.42b}$$

where δ is called the plant zeros polynomial, and its roots the *plant zeros*.

Straightforward matrix multiplication now yields

$$\mathbf{R} = \frac{1}{\Delta}\begin{bmatrix} k_{d1}N_{11} & k_{d2}N_{12} \\ k_{d1}N_{21} & k_{d2}N_{22} \end{bmatrix} \tag{14.43}$$

$$\Delta = d_{i1}d_{i2}B_i + k_{di}N_{ii} \qquad i = 1 \text{ or } 2$$

$$N_{ji} = \hat{q}_{ji}d_{i1}d_{i2} \qquad B_i = d_{j1}d_{j2} + k_{dj}\hat{q}_{jj} \qquad i \neq j \tag{14.44a}$$

$$N_{ii} = \hat{q}_{ii}d_{j1}d_{j2} + k_{dj}(k_{11}k_2 - k_{12}k_1)\,\delta \qquad i \neq j \tag{14.44b}$$

In Fig. 14.8(a) and (b), the input–output relations are

$$\mathbf{c} = \mathbf{Rr} \qquad \mathbf{r} = \begin{bmatrix} r_1 \\ r_2 \end{bmatrix} \qquad \mathbf{c} = \begin{bmatrix} c_1 \\ c_2 \end{bmatrix} = \begin{bmatrix} c_{11} + c_{12} \\ c_{21} + c_{22} \end{bmatrix} \tag{14.45}$$

c_{ij} = component of output c_i due to input r_j

With equations (14.43) and (14.44), this gives

$$\frac{c_{ii}}{r_i} = \frac{k_{di}N_{ii}}{d_{i1}d_{i2}B_i + k_{di}N_{ii}} \qquad \frac{c_{ji}}{c_{ii}} = \frac{N_{ji}}{N_{ii}} \tag{14.46}$$

The *diagonal response* c_{ii}/r_i is recognized as the closed-loop transfer function of the equivalent loop for input i (EL_i) shown in Fig. 14.10. The *interaction*

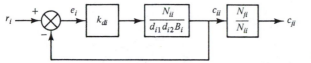

Figure 14.10 Equivalent system.

ratio c_{ji}/c_{ii}, which gives the ratio of off-diagonal to diagonal responses for input r_i, is also represented.

$$*****$$

The loop gain functions D_i of the diagonal loop in Fig. 14.8(c), as given by (14.30), and A_i in Fig. 14.10 are

$$D_i = \frac{k_{di}\hat{q}_{ii}}{d_{i1}d_{i2}} \qquad A_i = \frac{k_{di}N_{ii}}{d_{i1}d_{i2}B_i} \tag{14.47}$$

It is noted that the closed-loop poles in Fig. 14.10 are the same for $i = 1$ and $i = 2$, and are the roots of Δ in (14.43).

The controllers k_{di} can be designed sequentially or independently on the basis of the diagonal loops D_i. In the independent design, account is taken of the performance, stability, and interaction of the actual system, and these properties are verified for the actual loops A_i. In sequential design, k_{dj} is designed first, as above, on the basis of D_j. This determines N_{ii} and B_i and allows k_{di} to be designed directly on the basis of A_i in Fig. 14.10 for stability, diagonal response performance, and interaction. An important design parameter is defined first.

Definition. The *interaction factor* r of $\mathbf{Q}$ is the ratio of its determinant to the product of its diagonal elements.

This is a measure of the effective degree of diagonality, and is unity for diagonal matrices. Using (14.42a) gives

$$r = \frac{|\mathbf{Q}|}{q_{11}q_{22}} = \frac{q_{11}q_{22} - q_{12}q_{21}}{q_{11}q_{22}}$$

$$= 1 - r_{c1}r_{c2} = \frac{|\hat{\mathbf{Q}}|}{\hat{q}_{11}\hat{q}_{22}} \tag{14.48a}$$

$$= \frac{(k_{11}k_2 - k_{12}k_1)\delta}{\hat{q}_{11}\hat{q}_{22}} \tag{14.48b}$$

Equations (14.44) now readily yield

$$B_i = d_{j1}d_{j2}(1 + D_j) \qquad N_{ii} = \hat{q}_{ii}d_{j1}d_{j2}(1 + D_j r) \tag{14.49}$$

and from (14.47) and (14.46), using (14.44a) and (14.30),

$$A_i = \frac{D_i(1 + D_j r)}{1 + D_j}$$

$$\frac{c_{ji}}{c_{ii}} = \frac{r_{ci}}{1 + D_j r} \qquad i \neq j \qquad (14.50)$$

If

$$D_j, D_j r \gg 1: \quad A_i \approx D_i r \qquad \frac{c_{ji}}{c_{ii}} \approx \frac{r_{ci}}{D_j r} \qquad (14.51a)$$

$$D_j, D_j r \ll 1: \quad A_i \approx D_i \qquad \frac{c_{ji}}{c_{ii}} \approx r_{ci} \qquad (14.51b)$$

Usually, (14.51a) and (14.51b) apply at frequencies well inside and well outside the bandwidth, respectively.

*****

$A_i \approx D_i$ if r is near 1. Desired loop gains A_i can then be realized by independent design of diagonal loops D_i. Also, from (14.51a), to avoid a loss of gain in A_i over the bandwidth range, r should not be far below unity in this range. Hence:

> If possible, **K** should be designed such that $r = 1 - r_{c1} r_{c2}$ is near 1 and not far below unity magnitude, that is, such that at least one of r_{c1} and r_{c2} is small if the other is not. (14.52)

If, say, k_1 favors r_{c1} at low frequencies, k_2 should favor high frequencies. This links design of the two columns of **K**, but does not prevent one of r_{c1} and r_{c2} from exceeding 1, provided that, to limit interactions, they do not become large.

These considerations must be incorporated in the design of **K** in the preceding section, and a Bode plot of r, constructed after design of **K** from the last of (14.48), is available during the design of $\mathbf{K}_d$.

*****

D_j is designed to be stable, so that B_i in (14.49) will be Hurwitz, and A_i in (14.47) or (14.50) open-loop stable if the plant is open-loop stable. This promotes the simplicity of k_{di}. But (14.50) and (14.51) show that in addition to the D_j loop a loop with loop gain function $D_j r$ should be considered:

1. Its closed-loop poles are zeros of $(1 + D_j r)$, so of A_i, and should if possible be stable to avoid the constraints of nonminimum-phase systems.
2. By (14.51a), its accuracy should be satisfactory.
3. Within its bandwidth, interaction ratio c_{ji}/c_{ii} for input r_i is reduced by increasing D_j, $j \neq i$. (14.53)

If r is near 1, then from (14.49), $N_{ii} \approx \hat{q}_{ii} B_i$, so condition 1 of (14.53) is probably satisfied if $\hat{q}_{ii}$ is Hurwitz. The inherent performance limitations of nonminimum-phase plants, when δ of (14.42b) has right-half-plane roots, show in that the conditions that r_{ci} be minimized and $\hat{q}_{ii}$ be Hurwitz turn out to oppose each other. The Bode plot of $D_j r$, the sum of those of D_j and of r, is usually not plotted, but the features (14.53) are estimated visually.

*****

The choice of bandwidth of D_j involves a number of considerations. It must realize a desired bandwidth for A_j, which according to (14.51) may fall between those of D_j and $D_j r$. Also, if c_{ji}/c_{ii} must be suppressed over a frequency range where r_{ci} is not small, D_j, $j \neq i$, over this range must be adequately large. A higher crossover frequency of D_j may be needed to achieve this. Control of interactions has a strong effect on the choice. Figure 14.11 illustrates a system where r_{c1} and r_{c2} both approach 1 inside the desired bandwidths. From (14.50) and (14.51b), the interaction ratios c_{ji}/c_{ii} will approach the same values as the r_{ci}, at frequencies of the order of magnitude of the crossover frequencies of $D_j r$, $j \neq i$. The diagram is quite useful to predict interactions. Output c_{12} for input r_2 will be small because c_{12}/c_{22} is small at and below the crossover frequency of $D_2 r$ and only approaches 1 at frequencies where the diagonal response c_{22} has become small. But c_{21} will be large, since c_{21}/c_{11} is near 1 over a wide range of frequencies below the crossover frequency of $D_1 r$. Increasing the crossover frequency of D_2 or reducing that of D_1 will reduce c_{21}, at the expense of c_{12}. This would not be the case if c_{12}/c_{22} were small at high frequencies, so if condition (14.52) were satisfied.

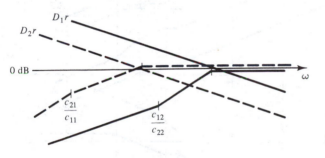

Figure 14.11   Interaction prediction.

   This discussion suggests a sequential approach in which D_j is designed first and the resulting plot of c_{ji}/c_{ii} is used in the choice of crossover frequency for D_i. This approach is also the basis of design for stability. If D_j is known, N_{ii} and B_i in (14.49) can be found, so the plant $N_{ii}/(B_i d_{i1} d_{i2})$ in Fig. 14.10 is available. Then k_{di} can be designed directly for this system, and hence guarantee stability. It is usually more convenient, however, to design k_{di} first, in the manner of k_{dj}, on the basis of D_i, and modify it, if necessary, on the basis of Fig. 14.10. The techniques will be discussed by application to Example 14.8.1.

14.10   DESIGN OF K_d FOR EXAMPLE 14.8.1

Table 14.10.1 summarizes data from Section 14.8 and shows other results, concerning the interaction factor r, which normally are part of the design of **K**. The

TABLE 14.10.1 DESIGN OF **K**

$d_{11} = s + 1$, $d_{12} = s + 5$, $d_{21} = s + 0.5$, $d_{22} = s + 2$
$\hat{q}_{11} = 0.75s + 4.75$ (-6.333); $\hat{q}_{21} = -0.1s + 0.1$ $(+1.0)$
$\hat{q}_{22} = -3.9s - 1.8$ (-0.462); $\hat{q}_{12} = -4.0s$ (0.0)
$k_{11}k_2 - k_{12}k_1 = -0.95$; $\delta = 3.5s^2 + 20.5s + 9$ $(-0.478, -5.379)$
$N_{21} = -0.1(s - 1)(s + 1)(s + 5)$; $N_{12} = -4.0s(s + 0.5)(s + 2)$

plant zeros polynomial δ, obtained from (14.42b), is used to express $r = (k_{11}k_2 - k_{12}k_1)\delta/(\hat{q}_{11}\hat{q}_{22})$. The Bode plot of r in Fig. 14.12 is close to 1 at all frequencies. Since, therefore, **K** appears satisfactory, N_{12} and N_{21} as calculated from (14.44a) are also shown in the table.

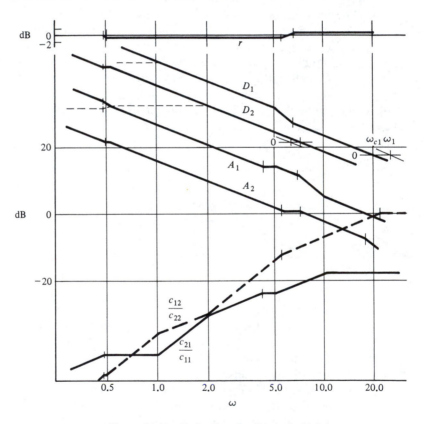

Figure 14.12 Bode plots for Example 14.8.1.

The controllers k_{di} are taken to be of the form

$$k_{di} = K_i\left(1 + \frac{a_i}{s}\right)G_{ci} \qquad G_{ci} = 1 \quad \text{or} \quad \frac{1 + s/z_i}{1 + s/p_i} \qquad (14.54)$$

G_{ci} is included if needed to improve stability. The PI control forces each output to become equal to the corresponding input under static conditions, and therefore also ensures static decoupling. From (14.30), then,

$$D_i = \frac{K_i(s + a_i)\hat{q}_{ii}G_{ci}}{sd_{i1}d_{i2}} \tag{14.55}$$

Bode plots of D_1 and D_2 for $k_{di} = K_i$ are shown dashed in Fig. 14.12, with the 0 dB axis levels as yet undecided. They suggest

$$a_1 = 1 \qquad a_2 = 2$$

The solid lines then show D_1 and D_2 for $G_{ci} = 1$.

From inspection of Fig. 14.9(c), D_1 is designed first, for a crossover frequency $\omega_{c1} = 20$. This should suppress c_{12}/c_{22} adequately at frequencies of the order of 5 rad/sec. If D_i of (14.55) must have a crossover frequency ω_{ci}, the 0 dB axis must be where D_i passes ω_{ci}. To find K_i, the frequency ω_i at which the extension of the low-frequency asymptote crosses 0 dB is found by construction. This low-frequency asymptote ($s \to 0$) is $K_i a_i \hat{q}_{iil}/(sd_{i1l}d_{i2l})$, where the subscript l identifies the constant terms. Since its magnitude must equal 1 for $s = j\omega_i$,

$$K_i = \frac{\omega_i d_{i1l} d_{i2l}}{a_i \hat{q}_{iil}} \tag{14.56}$$

For D_1: $\omega_{c1} = 20$, $\omega_1 = 25$, $K_1 = 26.316$.

Inspection of Fig. 14.12 suggests a high degree of stability for the D_1 and D_1r loops. The roots of B_2 and N_{22} in Table 14.10.2, obtained from (14.44), confirm this. The interaction ratio $c_{12}/c_{22} = N_{12}/N_{22}$ can now also be expressed, and is given in terms of the low-frequency asymptote, in brackets, and the poles (p) and zeros (z) in order of increasing frequency. These data yield the Bode plot of c_{12}/c_{22} in Fig. 14.12.

The plot of c_{12}/c_{22} is used to choose for D_2 a crossover frequency $\omega_{c2} = 7$, where c_{12}/c_{22} is -10 dB. Since r_{c1} is small, it does not impede this choice. Following the procedure used for D_1 gives

$$\omega_{c2} = 7 \qquad \omega_2 = 6.2 \qquad K_2 = -1.7222$$

TABLE 14.10.2 DESIGN OF $\mathbf{K}_d$

$sB_2 = (s + 1)(s^2 + 24.737s + 125.0)$; roots: $-1, -7.079, -17.658$
$sN_{22} = (s + 1)[s(s + 5)(-3.9s - 1.8) - 26.316 \times 0.95(3.5s^2 + 20.5s + 9)]$; roots: -1, $-0.478, -5.503, -21.916$
c_{12}/c_{22}: $[0.01778s^2]$; $-0.478(p), -0.5(z), -1(p), -2(z), -5.503(p), -21.916(p)$
$sB_1 = (s + 2)(s^2 + 7.2166s + 3.10)$; roots: $-0.459, -2, -6.758$
$sN_{11} = (s + 2)[s(s + 0.5)(0.75s + 4.75) + 1.7222 \times 0.95(3.5s^2 + 20.5s + 9)]$; roots: -2, $-0.476, -4.217, -9.775$
c_{21}/c_{11}: $[0.01698s]$; $-0.476(p), -1(z), +1(z), -2(p), -4.217(p), -5(z), -9.775(p)$
A_1: $[25/s]$; $-0.459(p); -0.476(z); -4.217(z); -5(p); -6.758(p); -9.775(z)$
A_2: $[6.2/s]$; $-0.478(z); -0.5(p); -5.503(z); -7.079(p); -17.658(p); -21.916(z)$
Δ roots: $-1, -2, -0.477, -4.132; -13.672 \pm 3.133j$, $(\omega_n, \zeta = 14.027, 0.97)$

The roots of B_1 and N_{11} in Table 14.10.2 confirm the high degree of stability suggested by the plots of D_2 and $D_2 r$.

To determine whether, as discussed in Section 14.9, redesign of k_{d2} is necessary, the data in the table are used to express A_2 from $A_i = k_{di} N_{ii}/(B_i d_{i1} d_{i2})$ in terms of its low-frequency asymptote and poles and zeros. These results are given in Table 14.10.2, and the Bode plot of A_2 in Fig. 14.12. This plot suggests a high degree of system stability. The closed-loop eigenvalues, given in Table 14.10.2 as the roots of Δ of (14.43), confirm this. Thus redesign of k_{d2} is not necessary.

However, to verify the design, A_1 is also expressed in Table 14.10.2 and plotted in Fig. 14.12. While the closed-loop poles for this loop are the same as for A_2, the zeros are not, and in general the loop gain and phase margin must be verified to be satisfactory, as they are in this case.

These plots of A_1 and A_2 are also needed to judge the interactions by comparison with those of c_{21}/c_{11} and c_{12}/c_{22} as discussed in Section 14.9. Interaction c_{21} is expected to be small, and c_{12} for input r_2 moderate. Step responses to unit step inputs r_1 and r_2 are shown in Fig. 14.13 and show the expected behavior. The faster response of c_{11} originates in the larger bandwidth of D_1. Such response curves, used to verify the design, are usually computed digitally, but can also be calculated from (14.43) to (14.45) by the usual partial fraction expansion techniques, using the roots in Table 14.10.2.

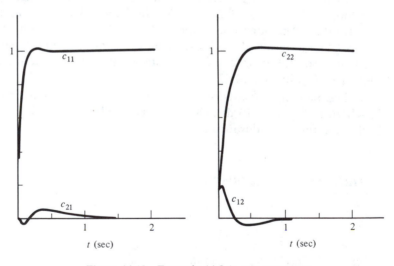

Figure 14.13   Example 14.8.1: step responses.

14.11 CONCLUSION

Frequency-domain design techniques have been introduced, with considerable attention to the stability theorems associated with the well-known Nyquist array and characteristic locus methods. The application of these techniques requires the use of computer-aided graphics.

As an alternative, thus far limited to 2 by 2 and 3 by 3 systems, a classical technique has been discussed. The technique uses only Bode plots, pole–zero patterns, and a simple type of vector diagram. Interactive computer programs with graphics have been developed for routine work, but certainly for 2 by 2 systems application does not depend on them. The technique uses only familiar plots and exploits all classical insights. The application to a simple 2 by 2 example illustrates its use in design for selected bandwidths and controlled interactions, and hence shows how it serves to improve understanding of multivariable system behavior generally. This provides a useful basis for further study of this area of control engineering.

PROBLEMS

14.1. In the configuration of Fig. 14.1, let

$$\mathbf{H} = \mathbf{I} \quad \mathbf{K} = \begin{bmatrix} 1.0 & 1.0 \\ -0.05 & -1.0 \end{bmatrix} \quad \mathbf{G} = \begin{bmatrix} \dfrac{1}{s+1} & \dfrac{5}{s+5} \\ \dfrac{0.1}{s+0.5} & \dfrac{4}{s+2} \end{bmatrix}$$

(a) Formulate **Q**, **T**, and **F**.
(b) Find the characteristic polynomial and determine system stability. (Remember that a necessary condition is that all coefficients have the same sign.)

14.2. In the configuration of Fig. 14.1, let

$$\mathbf{H} = \mathbf{I} \quad \mathbf{K} = \begin{bmatrix} 5 & 0 \\ 0 & 4 \end{bmatrix} \quad \mathbf{G} = \begin{bmatrix} \dfrac{1}{s+2} & \dfrac{0.25}{s+2} \\ \dfrac{0.2}{s+1} & \dfrac{1}{s+1} \end{bmatrix}$$

(a) Find the matrices **Q**, **T**, and **F**.
(b) Determine the open-loop and closed-loop characteristic polynomials and the closed-loop eigenvalues.
(c) Find the closed-loop transfer function matrix.

14.3. Use the results in Problem 14.2 to calculate the responses of outputs c_1 and c_2 to a unit step input of r_1.

14.4. In Problem 14.2:
(a) Find the inverse matrices $\hat{\mathbf{Q}}$ and $\hat{\mathbf{R}}$.
(b) Determine the open-loop and closed-loop characteristic polynomials from part (a).
(c) Verify the closed-loop matrix found in Problem 14.2(c).

14.5. Sketch both direct and inverse Nyquist diagrams for the loop gain function $T(s)$ = $2/(s + 1)$.

14.6. (a) Repeat Problem 14.5 for $T(s) = 2/[s(s + 1)]$.

    (b) Detail the inverse plot near the origin 0 to determine the position of 0 relative to the plot. (Note that a small semicircular indentation on the Nyquist contour is used to avoid a pole at the origin.)

14.7. (a) Use stability theorems associated with each of the plots in Problems 14.5 and 14.6 to determine the stability of the closed-loop systems.

    (b) If it is appropriate to do so, also examine the stability in terms of the polar plots (for $\omega = 0$ to $\omega = +\infty$).

14.8. Repeat Problems 14.5 and 14.7 for $T(s) = 2/(s - 1)$.

14.9. Repeat Problem 14.8 for $T(s) = 2/[s(s - 1)]$.

14.10. Investigate the stability of a system with loop gain function $T(s) = 4/[s(s + 1)(s + 2)]$ using:

    (a) Direct polar plots.

    (b) Inverse polar plots.

14.11. Repeat Problem 14.10 for $T(s) = 1/[s^2(s + 1)]$.

14.12. Determine the combinations of feedback gains h_1 and h_2 for which the two systems given in Fig. P14.12 in terms of inverse and direct polar plots of the diagonal elements with their Gershgorin bands will be stable. The systems are known to be open-loop stable.

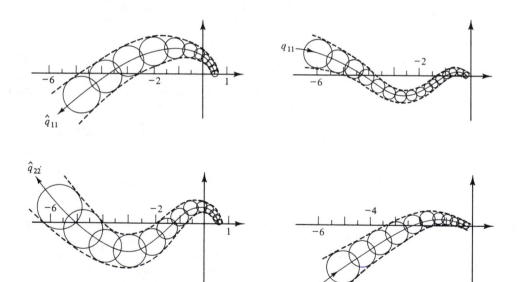

(a)   (b)

Figure P14.12

14.13. In Fig. P14.13, obtain the polar plot of the diagonal element g_{11} of **G** together with its Gershgorin band in terms of the column dominance ratio. (It is sufficient to consider the frequencies $\omega = 4$, 6, and 8.)

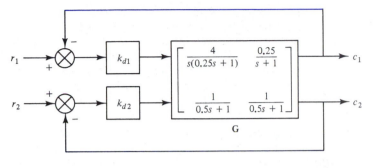

Figure P14.13

14.14. In Problem 14.13:
   (a) Also plot g_{22} with its Gershgorin band.
   (b) What type of matrix **K** is represented by the constant gain elements k_{d1} and k_{d2}? Are the column dominance ratios of $\mathbf{Q} = \mathbf{GK}$ different from those of **G**?
   (c) Is **G**, or **Q**, column diagonally dominant, and is this needed for application of an appropriate stability theorem?
   (d) Determine the ranges of values of k_{d1} and k_{d2} for which this theorem guarantees stability.

14.15. In Fig. 14.1, let

$$\mathbf{G} = \begin{bmatrix} 1 & 6 \\ 5 & 2 \end{bmatrix}$$

with **K** to be designed for dominance.
   (a) Show that for dominance of $\mathbf{Q} = \mathbf{GK}$ the use of column diagonal dominance simplifies design of **K**, by setting up the pertinent ratios for both row and column dominance in terms of the elements of **K**.
   (b) Similarly, show that row dominance is simpler for design of $\hat{\mathbf{K}}$ to achieve dominance of $\hat{\mathbf{Q}} = \hat{\mathbf{K}}\hat{\mathbf{G}}$.

14.16. In Problem 14.15:
   (a) Design simple matrices **K** in 15(a) and $\hat{\mathbf{K}}$ in 15(b) of which only two elements are nonzero to achieve the desired types of dominance.
   (b) What is the type of matrix **K** for Problem 14.15(a) designed above, and what is its physical significance?

14.17. In Fig. P14.17, mass flow rate and temperature of a flow are controlled by valve openings in "hot" and "cold" supply lines.
   (a) What are the dominance ratios in the columns of **G**?
   (b) Give the forms of **G** and the corresponding column dominance ratios for each of the following:

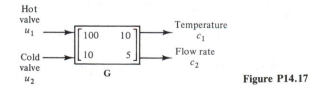

Figure P14.17

(i) If the order of the inputs is reversed.

(ii) If the order of the outputs is reversed.

(iii) If u_2 is measured in units 0.1th of those in Fig. P14.17.

(iv) If c_2 is measured in units one-fifth of those in Fig. P14.17.

(c) Give the pre- and/or postcompensators **K** and **L** which represent the actions of part (b) with the form of **G** as given in Fig. P14.17.

(d) Choose **K** and **L** for column diagonal dominance of **Q** = **LGK** and note conditions under which this is difficult to achieve using **K** alone.

14.18. (a) Noting that in general the outputs of **G** cannot be manipulated, how is a permutation matrix **L** implemented?

(b) To implement a diagonal **L**, is it in effect sufficient to multiply the rows of **G** by the diagonal elements of **L**?

(c) What are the units of the dominance ratio in, say, column one of **G** in Fig. P14.17? In view of its dependence on the choice of units and of order of inputs and outputs, does the ratio have a physical significance or is it strictly a numerical value useful for a criterion concerning closed-loop stability?

14.19. For the system of Fig. P14.13:

(a) Make asymptotic Bode plots of the column dominance ratios $r_{cj} = g_{ij}/g_{jj}$ to examine dominance of **G** over the entire frequency range.

(b) Also make Bode plots of all elements g_{ij} and determine how the r_{cj} may be found from these plots.

14.20. (a) Verify that Fig. P14.20 could be a refinement of Fig. P14.17, with allowance for the fact that temperature generally responds more slowly than flow rate, and with c_1 and c_2 expressed in units, respectively, 10 and 5 times those in Fig. P14.17.

(b) Plot the low- and high-frequency vector diagrams for design of **K**.

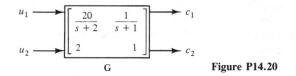

$$u_1 \longrightarrow \begin{bmatrix} \dfrac{20}{s+2} & \dfrac{1}{s+1} \\ 2 & 1 \end{bmatrix} \begin{matrix} \longrightarrow c_1 \\ \\ \longrightarrow c_2 \end{matrix}$$

G   **Figure P14.20**

14.21. In Problem 14.20:

(a) Use the vector diagrams to choose **K** so that **Q** = **GK** will be diagonal at low and high frequencies, with $\hat{q}_{11}$ Hurwitz.

(b) Make a Bode plot of r_{c2} to determine whether column two is dominant at all frequencies.

14.22. In Problem 14.21:

(a) To lower the peak values in the Bode plot of r_{c2}, draw the vector diagram for frequency $\omega = 1.5$ and use it to choose k_2 and k_{12} to achieve an approximate minimum of r_{c2} at $\omega = 1.5$.

(b) Draw the Bode plot of r_{c2} for this choice and compare it with that in Problem 14.21(b).

14.23. In Fig. 14.7, use elementary column and/or row operations to obtain a **K** and **L** such that **Q** = **LGK** will be column diagonally dominant. The numerator and denominator polynomials n_{ij} and d_{ij} of the elements g_{ij} of **G** are as follows: $n_{11} =$

2, $d_{11} = s + 2$; $n_{12} = 1$, $d_{12} = s + 1$; $n_{21} = 0.4$, $d_{21} = s + 5$; $n_{22} = 0.01$, $d_{22} = s + 0.5$.

14.24. In Fig. 14.8, let **K** have been designed for a given plant **G** such that the numerator and denominator polynomials n_{ij} and d_{ij} of the elements q_{ij} of **Q** = **GK** are as follows: $n_{11} = 1$, $d_{11} = s + 1$; $n_{12} = 3$, $d_{12} = s + 2$; $n_{21} = 1.5$, $d_{21} = s + 1$; $n_{22} = 1.5$, $d_{22} = s + 2$.
  (a) Determine whether **Q** is column diagonally dominant.
  (b) Determine whether the condition for application of the DNA stability theorem is satisfied.

14.25. For a system as in Fig. 14.8, known to be open-loop stable, design of **K** and $\mathbf{K}_d$ resulted in the plot shown in Fig. P14.25, analogous to that in Fig. 14.11.
  (a) Predict the severity of the interactions expected for inputs r_1 and r_2.
  (b) Suggest how the interactions may be reduced while maintaining good relative stability and without loss of speed of response.

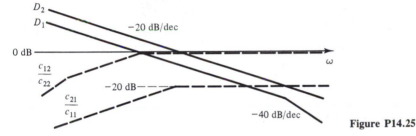

Figure P14.25

14.26. The polynomials n_{ij} and d_{ij} of equations (14.28) for the plant **G** in Fig. 14.8 are as follows: $n_{11} = 1$, $n_{12} = 5$, $n_{21} = 0.75$, $n_{22} = 4$; $d_{11} = s + 1$, $d_{12} = s + 5$, $d_{21} = s + 0.5$, $d_{22} = s + 2$. For this plant, the following matrix **K** has been designed:
$$k_{11} = k_{12} = 1 \qquad k_1 = -0.18 \qquad k_2 = -1$$
  (a) Find the plant zeros polynomial and the plant zeros, and determine whether the plant is minimum phase.
  (b) Express the elements $\hat{q}_{ij}$ of $\hat{\mathbf{Q}}$, find their roots and determine whether $\hat{q}_{11}$ and $\hat{q}_{22}$ are Hurwitz.
  (c) Express the interaction factor r.

14.27. For Problem 14.26:
  (a) Express the column dominance ratios in **Q**.
  (b) Plot these ratios and the interaction factor r.
  (c) Is **Q** diagonally dominant?

14.28. In Problems 14.26:
  (a) Express the diagonal loop gain functions D_i, and plot them if for both the crossover frequency of the asymptotic plot is chosen as 5 rad/sec, and k_{di} is constant.
  (b) Modify these plots if PI controllers $k_{d1} = K_1(1 + 1/s)$, $k_{d2} = K_2(1 + 2/s)$ are included to achieve zero steady-state errors and static decoupling and to raise dynamic accuracy at lower frequencies.

14.29. For the system with PI control considered in Problems 14.26 to 14.28, and using only the plots already obtained:

(a) Estimate whether the diagonal loops are stable.

(b) Estimate whether the single-loop equivalent systems in Fig. 14.10 are open-loop stable.

(c) Estimate whether these systems are minimum phase.

(d) Reason why the interactions for system input r_2 are likely to be appreciable, and considerably larger than those for r_1.

(e) Suggest how the interactions for r_2 may be reduced without those for r_1 being likely to change appreciably.

(f) Note the frequency ranges that are likely to dominate in the interactions, and hence estimate which interactions are likely to decay most rapidly following step inputs.

APPENDIX A

VECTORS, MATRICES, AND DETERMINANTS

A.1 VECTORS AND MATRICES

The $m \times n$ matrix A in (A.1) is a rectangular array of mn elements, where a_{ij} is the element in row i and column j. The notation $\{a_{ij}\}$ is often used to identify the matrix. If $m = n$, the matrix is square, and n dimensional. For $n = 1$ A is a column vector, such as $\mathbf{x}$ in (A.1), and for $m = 1$ it is a row vector, such as $\mathbf{y}$ in (A.1).

$$\mathbf{A} = \{a_{ij}\} = \begin{bmatrix} a_{11} & a_{12} & \cdots & a_{1n} \\ a_{21} & a_{22} & & a_{2n} \\ \vdots & & & \vdots \\ a_{m1} & a_{m2} & \cdots & a_{mn} \end{bmatrix} \qquad \mathbf{x} = \begin{bmatrix} x_1 \\ x_2 \\ \vdots \\ x_n \end{bmatrix}$$

$$\mathbf{y} = [y_1 \quad y_2 \quad \cdots \quad y_n] \tag{A.1}$$

Some important definitions:

1. *Diagonal matrix:* A square matrix of which all elements not on the main diagonal are zero: $a_{ij} = 0$, $i \neq j$.
2. *Unit matrix or identity matrix I:* A diagonal matrix of which all elements on the main diagonal are equal to 1.
3. *Zero matrix or null matrix 0:* A matrix of which all elements are zero.
4. *Symmetrical matrix:* $a_{ij} = a_{ji}$ for all $i \neq j$.
5. *Transpose matrix A' of A:* Obtained by interchanging the rows and columns of **A**. The transpose of the column vector **x** in (A.1) is the row vector

$\mathbf{x}' = [x_1 \;\; x_2 \;\; \cdots \;\; x_n]$. $(\mathbf{A}')' = \mathbf{A}$, and a matrix is symmetrical if $\mathbf{A}' = \mathbf{A}$.

6. *Trace of a square matrix* A: Is the sum of the elements on the main diagonal.

$$\text{tr } \mathbf{A} = a_{11} + a_{22} + \cdots + a_{nn} \tag{A.2}$$

7. *Partitioned matrices:* A matrix can be partitioned into submatrices or vectors. Equations (A.3) show the $m \times n$ matrices $\mathbf{A} = \{a_{ij}\}$ and $\mathbf{B} = \{b_{ij}\}$ partitioned into, respectively, column vectors and row vectors.

$$\mathbf{A} = [\mathbf{a}_1 \;\; \cdots \;\; \mathbf{a}_n]$$

$$\mathbf{b}'_i = [b_{i1} \;\; \cdots \;\; b_{in}] \qquad \mathbf{B} = \begin{bmatrix} \mathbf{b}'_1 \\ \vdots \\ \mathbf{b}'_m \end{bmatrix} \qquad \mathbf{a}_i = \begin{bmatrix} a_{1i} \\ \vdots \\ a_{mi} \end{bmatrix} \tag{A.3}$$

A.2 VECTOR AND MATRIX OPERATIONS

For matrices $\mathbf{A} = \{a_{ij}\}$, $\mathbf{B} = \{b_{ij}\}$, or vectors $\mathbf{x} = \{x_i\}$, $\mathbf{y} = \{y_i\}$, some important properties and operations are the following.

1. $\mathbf{A} = \mathbf{B}$ ($\mathbf{x} = \mathbf{y}$) if and only if $a_{ij} = b_{ij}$ ($x_i = y_i$) for all i, j. The matrices (vectors) are then equal.

2. Multiplication by a scalar h, $h\mathbf{A} = \{ha_{ij}\}$, $h\mathbf{x} = \{hx_i\}$. Each element is multiplied by the scalar.

3. Addition and subtraction, $\mathbf{C} = \mathbf{A} \pm \mathbf{B}$, $\mathbf{z} = \mathbf{x} \pm \mathbf{y}$. Corresponding elements are added or subtracted: $c_{ij} = a_{ij} \pm b_{ij}$, $z_i = x_i \pm y_i$. Some properties:

   $\mathbf{A} + \mathbf{B} = \mathbf{B} + \mathbf{A}$ $(\mathbf{A} + \mathbf{B}) + \mathbf{C} = \mathbf{A} + (\mathbf{B} + \mathbf{C})$ $(\mathbf{A} + \mathbf{B})'$

   $\qquad = \mathbf{A}' + \mathbf{B}'$ $(h_1 + h_2)\mathbf{x} = h_1\mathbf{x} + h_2\mathbf{x}$ $\mathbf{A} + \mathbf{0} = \mathbf{A}$ $\mathbf{A} - \mathbf{A} = \mathbf{0}$

4. Multiplication, $\mathbf{C} = \mathbf{AB}$. The number of columns of $\mathbf{A}$ must equal the number of rows of $\mathbf{B}$. If $\mathbf{A}$ is $m \times n$ and $\mathbf{B}$ is $n \times p$, then element c_{ij} of the $m \times p$ matrix $\mathbf{C} = \{c_{ij}\}$ is

$$c_{ij} = \sum_{k=1}^{n} a_{ik} b_{kj} \tag{A.4}$$

That is, element k in row i of $\mathbf{A}$ is multiplied by element k in column j of $\mathbf{B}$, and these products are added for all k in this row i and column j to obtain c_{ij}. Multiplication using partitioned matrices gives the same result:

$$\mathbf{C} = \mathbf{AB} = \begin{bmatrix} \mathbf{a}'_1 \\ \vdots \\ \mathbf{a}'_m \end{bmatrix} [\mathbf{b}_1 \;\; \cdots \;\; \mathbf{b}_p]$$

$$c_{ij} = \mathbf{a}_i' \mathbf{b}_j = [a_{i1} \quad \cdots \quad a_{in}] \begin{bmatrix} b_{1j} \\ \vdots \\ b_{nj} \end{bmatrix} = \sum_{k=1}^{n} a_{ik} b_{kj}$$

Some properties: In general, $\mathbf{AB} \neq \mathbf{BA}$ (i.e., matrix multiplication is not commutative). If $\mathbf{A}$ and $\mathbf{B}$ commute, then $\mathbf{AB} = \mathbf{BA}$.

$$\mathbf{AI} = \mathbf{IA} = \mathbf{A} \qquad \mathbf{III} = \mathbf{I}^3 = \mathbf{I} \qquad (\mathbf{AB})\mathbf{C} = \mathbf{A}(\mathbf{BC}) = \mathbf{ABC}$$

$$\mathbf{A}(\mathbf{B} + \mathbf{C}) = \mathbf{AB} + \mathbf{AC} \qquad \mathbf{A0} = \mathbf{0A} = \mathbf{0}$$

Note that in the matrix case $\mathbf{AB} = \mathbf{0}$ need not imply that $\mathbf{A}$ or $\mathbf{B}$ is a null matrix. For example:

$$\begin{bmatrix} a & b \\ 0 & 0 \end{bmatrix} \begin{bmatrix} b & 0 \\ -a & 0 \end{bmatrix} = \begin{bmatrix} 0 & 0 \\ 0 & 0 \end{bmatrix}$$

5. *$(AB)' = B'A'$:* The transpose of a product is the product of the transposes in reverse order.

6. *Linear transformations and algebraic equations:* Alternative forms of the same set of simultaneous linear algebraic equations are as follows:

$$a_{11}x_1 + \cdots + a_{1n}x_n = y_1$$
$$\vdots \qquad\qquad\qquad \vdots \qquad\qquad y_i = \sum_{j=1}^{n} a_{ij}x_j \qquad i = 1, \ldots, m \tag{A.5}$$
$$a_{m1}x_1 + \cdots + a_{mn}x_n = y_m \qquad \mathbf{y} = \mathbf{Ax} \qquad \mathbf{A} = \{a_{ij}\}$$

The matrix representation may be verified by applying the general multiplication rule to the matrix-vector product, which results in a vector. Equations (A.5) may represent a transformation of variables, used extensively to simplify problem formulations. It can also be a set of equations to be solved for $\mathbf{x}$. If $\mathbf{y} = \mathbf{0}$, the set is homogeneous, and a necessary condition for a nontrivial solution, that is, a solution $\mathbf{x}$ of which not all elements are zero, is known to be that the determinant of the matrix $\mathbf{A}$ of the coefficients be zero.

7. *Quadratic forms $x'Ax$:*

$$\mathbf{x}'\mathbf{Ax} = [x_1 \quad \cdots \quad x_n] \begin{bmatrix} a_{11} & \cdots & a_{1n} \\ \vdots & & \vdots \\ a_{n1} & \cdots & a_{nn} \end{bmatrix} \begin{bmatrix} x_1 \\ \vdots \\ x_n \end{bmatrix} = \sum_{i=1}^{n} x_i y_i \tag{A.6}$$

where y_i is as defined in (A.5). Note that the product is a scalar number.

8. *Vector products and orthogonality:* The scalar product, or dot product, or inner product, of two vectors $\mathbf{x}$ and $\mathbf{y}$ is a scalar number.

$$\mathbf{x'y} = [x_1 \cdots x_n] \begin{bmatrix} y_1 \\ \vdots \\ y_n \end{bmatrix} = \sum_{i=1}^{n} x_i y_i = \mathbf{y'x}$$

(A.7)

$$\mathbf{x'x} = \sum_{i=1}^{n} x_i^2 \qquad |\mathbf{x}| = \sqrt{x_1^2 + \cdots + x_n^2} = \sqrt{\mathbf{x'x}}$$

As an extension of three-dimensional concepts, $|\mathbf{x}|$ identifies the length of vector $\mathbf{x}$. A vector is said to be normalized if $|\mathbf{x}| = 1$, and is then called a unit vector. Two vectors are orthogonal if their inner product is zero: $\mathbf{x'y} = \mathbf{y'x} = 0$. This is a generalization of the mutually perpendicular coordinate axes for rectangular coordinate systems in three-dimensional space. The axes are given by the unit vectors $\mathbf{e}_1' = [1 \quad 0 \quad 0]$, $\mathbf{e}_2' = [0 \quad 1 \quad 0]$, and $\mathbf{e}_3' = [0 \quad 0 \quad 1]$, with zero inner products.

A.3 DETERMINANTS, THE INVERSE, AND THE RANK OF A MATRIX

The determinant $|\mathbf{A}|$ or det $\mathbf{A}$ of an $n \times n$ matrix $\mathbf{A}$ is a scalar number or function. It is found via the use of minors and cofactors. The minor m_{ij} of element a_{ij} is the determinant of a matrix of order $n - 1$ obtained from $\mathbf{A}$ by removing the row and column containing a_{ij}. The cofactor c_{ij} of a_{ij} is $c_{ij} = (-1)^{i+j} m_{ij}$. The determinant of $\mathbf{A}$ is then

$$|\mathbf{A}| = \sum_{j=1}^{n} a_{ij} c_{ij} = \sum_{i=1}^{n} a_{ij} c_{ij}$$

(A.8)

As this implies, $|\mathbf{A}|$ may be found by expansion in this manner along any row or column of $\mathbf{A}$. For example, if $\mathbf{A}$ is 3 by 3, expansion along the first row gives

$$|\mathbf{A}| = a_{11}(a_{22}a_{33} - a_{23}a_{32}) - a_{12}(a_{21}a_{33} - a_{23}a_{31}) + a_{13}(a_{21}a_{32} - a_{22}a_{31}) \quad \text{(A.9)}$$

Some properties are that $|\mathbf{A'}| = |\mathbf{A}|$, $|h\mathbf{A}| = h^n|\mathbf{A}|$ for scalars h, $|\mathbf{AB}| = |\mathbf{A}| \, |\mathbf{B}|$. Interchanging two columns or two rows of $\mathbf{A}$ changes the sign of $|\mathbf{A}|$, and $|\mathbf{A}| = 0$ if $\mathbf{A}$ has two equal rows or columns.

The inverse matrix $\mathbf{A}^{-1}$ of $\mathbf{A}$ is now found by first defining the adjoint matrix $adj(\mathbf{A})$ of $\mathbf{A}$. It is the transpose of the matrix obtained by replacing each element of $\mathbf{A}$ by its cofactor. From (A.8), it is readily verified that $\mathbf{A} \, adj(\mathbf{A}) = adj(\mathbf{A})\mathbf{A} = |\mathbf{A}|\mathbf{I}$. The inverse $\mathbf{A}^{-1}$ of $\mathbf{A}$ is defined by the relations $\mathbf{A}^{-1}\mathbf{A} = \mathbf{A}\mathbf{A}^{-1} = \mathbf{I}$, and is therefore equal to

$$\mathbf{A}^{-1} = \frac{adj(\mathbf{A})}{|\mathbf{A}|}$$

(A.10)

If $|\mathbf{A}| = 0$, the matrix $\mathbf{A}$ is said to be singular, and the inverse does not exist. Some properties are that $(\mathbf{A}^{-1})^{-1} = \mathbf{A}$, $(\mathbf{A}^{-1})' = (\mathbf{A}')^{-1}$, and $(\mathbf{AB})^{-1} = \mathbf{B}^{-1}\mathbf{A}^{-1}$.

The last equation says that the inverse of a product is equal to the product of the inverses in reverse order.

The rank $R(\mathbf{A})$ of $\mathbf{A}$ is defined to be the dimension of the largest nonsingular matrix, that is, square and with a nonzero determinant, contained in $\mathbf{A}$. Equivalently, it is the maximum number of linearly independent rows or columns of $\mathbf{A}$. A set of vectors $\mathbf{x}_1, \ldots, \mathbf{x}_n$ is linearly dependent if there are constants $c_1, \ldots, c_n$ which are not all zero such that $c_1\mathbf{x}_1 + c_2\mathbf{x}_2 + \cdots + c_n\mathbf{x}_n = \mathbf{0}$. This is equivalent to $\mathbf{Ac} = \mathbf{0}$, where $\mathbf{A} = [\mathbf{x}_1 \cdots \mathbf{x}_n]$ and $\mathbf{c}' = [c_1 \ c_2 \cdots c_n]$. $|\mathbf{A}| = 0$ is the necessary condition for a nontrivial solution, so $\mathbf{A}$ must be nonsingular for linear independence. Some rank properties are that $R(\mathbf{A}') = R(\mathbf{A})$ and $R(\mathbf{A}) = R(\mathbf{A}'\mathbf{A}) = R(\mathbf{A}\mathbf{A}')$.

A.4 MATRIX CALCULUS

The derivative or integral of a vector or a matrix is the vector or matrix consisting of the derivatives or integrals of the elements. For example, if $\mathbf{A}(t) = \{a_{ij}(t)\}$, then

$$\frac{d}{dt}(\mathbf{A}(t)) = \left\{\frac{d}{dt}(a_{ij}(t))\right\}$$

Some properties are the following:

$$\frac{d}{dt}(\mathbf{A}\,\mathbf{B}) = \mathbf{A}\frac{d}{dt}(\mathbf{B}) + \frac{d}{dt}(\mathbf{A})\mathbf{B} \tag{A.11}$$

$$\frac{d}{dt}(h\mathbf{A}) = h\frac{d}{dt}(\mathbf{A}) + \frac{dh}{dt}\mathbf{A} \qquad h \text{ scalar}$$

APPENDIX B

COMPUTER AIDS FOR ANALYSIS AND DESIGN

B.1 INTRODUCTION

Programs for computer-assisted analysis and design, with examples of their use, are given for the following:

1. Finding the roots of a polynomial
2. Plotting the transient response for unit step inputs
3. Plotting root loci
4. Constructing polar plots
5. Constructing Bode plots

The programs are written in FORTRAN, and for an interactive mode of operation with computer graphics. The graphics package used is called PLTLIB, developed in the Department of Mechanical Engineering of the University of Toronto. For ease of use, the calling commands have been made largely the same as those of the CALCOMP basic software package. The programs of PLTLIB used in the computer programs are

> PLOTST, PLOT, SYMBOL, RECT, LABEL, LINE, NUMBER, SCALE, AXIS, LGAXS, LLGRID, LGLIN, DLGLIN, and PLOTND.

B.2 ROOTS OF A POLYNOMIAL: ROOT

The ROOT program determines the roots of a polynomial with real coefficients. It is based on the Lin–Bairstow method [1], in which an iterative routine is used to find and divide out successive quadratic factors of the polynomial. Their roots, plus that of any remaining first-order factor, constitute the solution.

The program listing is shown in Table B.1 and demonstrates the interactive mode of use. Following the instructions which appear on the screen, the coefficients are typed in, separated by spaces or carriage returns. The calculated roots appear on the selected output device as two columns showing the real and imaginary parts, respectively.

TABLE B.1 PROGRAM FOR THE ROOTS OF A POLYNOMIAL

```
C*********************************************************************
C*                                                                  *
C*    THIS PROGRAM READS THE COEFFICIENTS OF A POLYNOMIAL AND        *
C*    CALLS THE SUBROUTINE PROOT TO FIND THE ROOTS OF THE            *
C*    POLYNOMIAL                                                     *
C*                                                                  *
C*********************************************************************
C*
      REAL A(30)
      COMPLEX ROOT(29)
10    WRITE (5,*) 'Enter the order+1 of the polynomial'
      READ (5,*) NA
      WRITE (5,*) 'Enter the coeff. of the polynomial'
      WRITE (5,*) 'in the order of ascending power'
      READ (5,*) (A(I),I=1,NA)
      WRITE (5,*) 'Enter choice of output: screen=5, printer=3'
      READ (5,*) NW
      IF (NW .EQ. 3) OPEN (UNIT=3,STATUS='NEW',FILE='ROOT.DAT')
      CALL PROOT (A,NA,ROOT,NR)
      CALL WRITEM (A,NA,ROOT,NR,NW)
      IF (NW .EQ. 3) CLOSE (UNIT=3)
      WRITE (5,*) 'Do you want to run again? Y=1'
      READ (5,*) I
      IF (I .EQ. 1) GO TO 10
      IF (NW .EQ. 3) WRITE (5,*) 'Please print ROOT.DAT for output'
      STOP
      END
```

The subprograms used are listed in Section B.7:

(1) PROOT   (2) QUAD   (3) WRITEM

Example B.2.1

The roots of the polynomial $10X^9 + 9X^8 + 8X^7 + 7X^6 + 6X^5 + 5X^4 + 4X^3 + 3X^2 + 2X + 1$ are:

Real	Imaginary
0.1892543	0.7576993
0.1892543	−0.7576993
0.6212025	0.5367069
0.6212025	−0.5367069
−0.6175181	0.4261032
−0.6175181	−0.4261032
−0.2694115	0.7112944
−0.2694115	−0.7112944
−0.7470546	0.0

B.3 STEP RESPONSE OF A TRANSFER FUNCTION MODEL: TRAN1

TRAN1 is based on the partial fraction expansion technique [2]. The transform $C(s)$ of the output is a ratio of polynomials $P(s)$ and $Q(s)$. It is assumed that $Q(s)$ is of higher degree than $P(s)$ and that the roots R_n of $Q(s)$ are all distinct. The residues K_n in the partial fraction expansion

$$C(s) = \frac{P(s)}{Q(s)} = \sum_{n=1}^{m} \frac{K_n}{(s - R_n)}$$

are then known to be given by

$$K_n = \lim_{s \to R_n} (s - R_n) \frac{P(s)}{Q(s)}$$

Since $Q(s) = (s - R_1) \cdots (s - R_n) \cdots (s - R_m)$, it may be seen that (L'Hopital's rule)

$$\lim_{s \to R_n} \frac{s - R_n}{Q(s)} = \lim_{s \to R_n} \frac{1}{Q'(s)} = \frac{1}{Q'(R_n)}$$

where $Q'(s)$ is the derivative of $Q(s)$. Hence

$$K_n = \frac{P(R_n)}{Q'(R_n)}$$

and inverse transformation gives the response

$$c(t) = \sum_{n=1}^{m} \frac{P(R_n)}{Q'(R_n)} e^{R_n t}$$

This is valid for both real and complex distinct roots.

The interactive program listing is shown in Table B.2. A subroutine ING, common to this and the following programs, is used to enter the coefficients of $P(s)$ and $Q(s)$. $P(s)$, in particular, is frequently available in the form of a product of several polynomials. Therefore, ING asks first how many polynomial factors are contained in each of $P(s)$ and $Q(s)$. For each in turn it will then repeat as often as necessary the prompts to enter the order plus one of the polynomial factor, and its coefficients in order of ascending power, separated by spaces or carriage returns as in the ROOT program. Note that only the transfer function data are entered. The pole at the origin due to the step input is added by the program, which also checks to ensure that all poles are distinct. The screen prompts the user to choose the numerical values to be shown on the screen or printed on paper. The program allows for manual scaling or automatic scaling of the plot.

The subprograms used are listed in Section B.7:

(1) ING (2) RESPON (3) EQROOT (4) MUTPOL (5) DIFF
(6) PLTLIB

TABLE B.2 PROGRAM FOR TRANSIENT RESPONSE

```fortran
C***************************************************************
C*                                                             *
C*   THIS PROGRAM CALCULATES THE TRANSIENT RESPONSE OF A       *
C*   TRANSFER FUNCTION FOR UNIT STEP INPUT USING HEAVISIDE'S   *
C*   EXPANSION FORMULA                                         *
C*                                                             *
C***************************************************************
C*
      REAL ZERO(20),POLE(20),Y(202),X(202),TEMP1(20),TEMP2(20)
      COMPLEX ROOT(20)
C
C  --> TO READ THE COEFFICIENTS OF THE NUMERATOR AND DENOMINATOR
C      OF THE TRANSFER FUNCTION BY CALLING SUBROUTINE ING
C
      CALL ING (ZERO,IZ,POLE,IP,TEMP1,TEMP2)
C
C  --> TO ADD THE POLE AT 0 TO THE DENOMINATOR, FOR THE STEP INPUT
C
      IP = IP + 1
      DO 90 I = 2,IP
      POLE(IP-I+2) = POLE(IP-I+1)
90    CONTINUE
      POLE(1) = 0.
C
C  --> TO ENTER THE PLOTTING PARAMETERS
C
100   WRITE (5,*) 'Enter no. of points (max. 200)'
      READ (5,*) NPOINT
      WRITE (5,*) 'Enter time interval'
      READ (5,*) TINT
C
      WRITE (5,*) 'Enter choice of output of numerical values!'
      WRITE (5,*) 'screen=5, printer=3'
      READ (5,*) NW
      IF (NW .EQ. 3) OPEN(UNIT=3,STATUS='NEW',FILE='TRAN1.DAT')
C
      CALL PROOT (POLE,IP,ROOT,IR)
      WRITE (NW,*) 'Roots of the denominator including 1/S are:'
      WRITE (NW,*) (ROOT(I),I=1,IR)
C
C  --> TO CHECK FOR EQUAL ROOTS IN THE DENOMINATOR.
C      IF THERE ARE EQUAL ROOTS, PROGRAM STOPS
C
      CALL EQROOT (ROOT,IR,IFLAG)
      IF (IFLAG .NE. 1) GO TO 110
      WRITE (5,*) 'Denominator has equal roots, program stops'
      STOP
C
110   CALL RESPON (POLE,IP,ZERO,IZ,ROOT,IR,NPOINT,TINT,Y)
C
      WRITE (NW,120)
120   FORMAT (/' ',10X,'TIME',15X,'RESPONSE'/
     +        ' ',7X,10('-'),11X,10('-')/)
C
      DO 130 I = 1, NPOINT
      X(I) = TINT * FLOAT(I-1)
      WRITE (NW,140) X(I),Y(I)
130   CONTINUE
140   FORMAT (' ',5X,F10.5,5X,'Response =',F10.5)
      IF (NW .EQ. 3) CLOSE (UNIT=3)
C
C  --> TO PLOT THE RESULT
C
      WRITE (5,*) 'Do you want to plot the response? Y=1'
      READ (5,*) I
      IF (I .NE. 1) GO TO 150
      CALL PLOT1 (X,Y,NPOINT)
C
150   WRITE (5,*) 'Do you want to run again with different no.'
      WRITE (5,*) 'of points and time interval? Y=1'
      READ (5,*) I
      IF (I .EQ. 1) GO TO 100
      IF (NW .EQ. 3) WRITE (5,*) 'Please print TRAN1.DAT for output'
      STOP
      END
C
C
      SUBROUTINE  PLOT1  (X,Y,NPOINT)
C
C***************************************************************
C*                                                             *
C*   THIS SUBROUTINE PLOTS THE RESULT                          *
C*   X(I) -- AN ARRAY OF TIME                                  *
C*   Y(I) -- AN ARRAY OF TIME RESPONSE                         *
C*   NPOINT -- NO. OF POINTS                                   *
C*                                                             *
C***************************************************************
C
      DIMENSION     X(I),Y(I)
      INTEGER       IPBUFF (256)
      M = NPOINT + 1
      N = NPOINT + 2
C
C  --> TO CHOOSE AUTOMATIC SCALING OR MANUAL SCALING
C
100   WRITE (5,*) 'Do you want manual scaling? Y=1'
      READ (5,*) I
      IF (I .EQ. 1) GO TO 110
C
C  --> AUTOMATIC SCALING
C
      CALL SCALE (X,6,,NPOINT,1)
      CALL SCALE (Y,4,,NPOINT,1)
      GO TO 120
C
```

TABLE B.2 (cont.)

```
C
C   --> MANUAL SCALING
C
110    WRITE (5,*) 'Enter starting value & change in value'
       WRITE (5,*) 'per inch of x-axis'
       READ (5,*) X(M),X(N)
       WRITE (5,*) 'Enter starting value & change in value'
       WRITE (5,*) 'per inch of y-axis'
       READ (5,*) Y(M),Y(N)
C
120    CONTINUE
       WRITE (5,*) 'Do you want a reference line? Y=1'
       READ (5,*) IREF
       IF (IREF .NE. 1) GO TO 130
C
C   --> TO CALCULATE THE POSITION OF THE REFERENCE LINE
C
       WRITE (5,*) 'Enter value of reference line'
       READ (5,*) REFL
       ZEROL = (REFL-Y(M)) / Y(N)
       WRITE (5,*) 'What do you want ? dash line=1, solid line=2'
       READ (5,*) DSL
       IF (ZEROL .LE. Y(M)) WRITE (5,*)
     +            'Reference line below axis, won t plot'
C
C   --> TO PLOT THE RESULTS
C
130    CALL PLOTST (256,IPBUFF)
       CALL PLOT   (-.5,0.25,-3)
       CALL RECT   (0.0,0.0,6.5,9.0,0.0,3)
       CALL LABEL  (0.75,0.7,8.25,0.7,'TRANSIENT RESPONSE',18,
     +              0.16,1.0.0)
       CALL PLOT   (1.5,1.5,-3)
       CALL AXIS   (0.0,0.0,'TIME (SEC)',-10,6.,0.,X(M),X(N))
       CALL AXIS   (0.0,0.0,'RESPONSE',8,4.,90.0,Y(M),Y(N))
       IF (ZEROL .LE. Y(M) .OR. IREF .NE. 1) GO TO 140
       CALL PLOT   (0.0,ZEROL,3)
       IF (DSL .EQ. 1.) CALL DASHP  (6.0,ZEROL,.1)
       IF (DSL .EQ. 2.) CALL PLOT   (6.0,ZEROL,2)
       CALL PLOT   (0.0,0.0,3)
140    CALL LINE   (X,Y,NPOINT,1,0,1)
       CALL PLOTND
C
       WRITE (5,*) 'Do you want to plot again with different'
       WRITE (5,*) 'plotting parameters? Y=1'
       READ (5,*) I
       IF (I .EQ. 1) GO TO 100
       RETURN
       END
```

Example B.3.1

Figure B.1 shows the step response of

$$G(s) = \frac{1}{s^2 + s + 1}$$

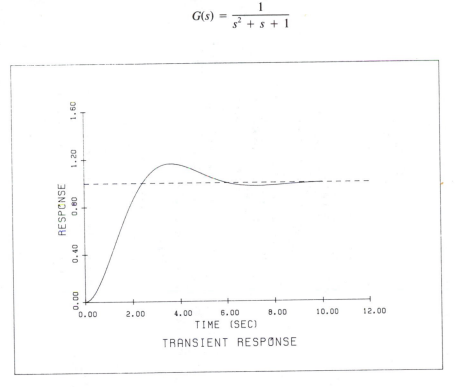

Figure B.1   Example of transient response plot.

B.4 *ROOT LOCUS PLOTS: RLOCI*

RLOCI plots the loci of the closed-loop poles of a system with loop gain function $G(s) = KN(s)/D(s)$ for varying K. $N(s)$ and $D(s)$ are polynomials and the poles are the roots of the polynomial

$$D(s) + KN(s)$$

which are calculated for specified values of K.

The interactive program listing is shown in Table B.3. The polynomials $N(s)$ and $D(s)$ can again be entered in terms of their factors as described for the TRAN1 program. The closed-loop poles are calculated a specified number of times with a specified positive or negative increment of gain K. The screen prompts the user to choose the numerical values to be shown on the screen or printed on paper. Manual scaling is allowed for to permit a part of the plot to be enlarged. The vertical axis can be moved as desired for a more efficient plot.

TABLE B.3 PROGRAM FOR ROOT LOCI

```fortran
C************************************************
C*                                             *
C*  THIS PROGRAM PLOTS THE ROOT LOCUS FOR A LOOP GAIN FUNCTION  *
C*                                             *
C************************************************
C*
      REAL ZERO(20),POLE(20),X1(20),X2(20),
     +     Y1(20),Y2(20),POLY(40),TEMP(4),
     +     TEMP1(20),TEMP2(20)
      COMPLEX ROOT1(20),ROOT2(20),ROOT3(40)
      INTEGER IPBUFF(256)
C
C --> TO READ THE COEFFICIENTS OF THE NUMERATOR AND DENOMINATOR
C     OF THE TRANSFER FUNCTION BY CALLING SUBROUTINE ING
C
      CALL ING (ZERO,IZ,POLE,IP,TEMP1,TEMP2)
C
C --> TO ENTER THE PLOTTING PARAMETERS
C
100   WRITE (5,*) 'Enter no. of times the closed loop poles'
      WRITE (5,*) 'must be calculated'
      READ (5,*) N1
      WRITE (5,*) 'Enter the starting gain'
      READ (5,*) RK
      WRITE (5,*) 'Enter the increment'
      READ (5,*) RK1
C
      WRITE (5,*) 'Enter choice of output of numerical values :'
      WRITE (5,*) 'screen=5, printer=3'
      READ (5,*) NW
      IF (NW .EQ. 3) OPEN (UNIT=3,STATUS='NEW',FILE='RLOCI.DAT')
C
      WRITE (5,*) 'Do you want to see all the closed loop poles'
      WRITE (5,*) 'at each gain? Y=1'
      READ (5,*) NPR
C
      WRITE (5,*) 'Do you want to move the vertical axis'
      WRITE (5,*) 'from the centre of the plot? Y=1'
      READ (5,*) IM
      IF (IM .NE. 1) GO TO 110
      WRITE (5,*) 'Which side do you want to move? right=1 left=2'
      READ (5,*) IMD
      WRITE (5,*) 'How many inches?'
      READ (5,*) SINCH
      IF (IMD .EQ. 2) SINCH = -SINCH
C
110   WRITE (5,*) 'Do you want to put a star for every nth gain'
      WRITE (5,*) 'on the root locus? Y=1'
      READ (5,*) IDASH
      IF (IDASH .NE. 1) GO TO 120
      WRITE (5,*) 'Enter the value n of the nth gain'
      READ (5,*) NINC
C
C --> TO CALCULATE OPEN LOOP POLES AND ZEROS
C
120   CALL PROOT (ZERO,IZ,ROOT1,NR1)

      WRITE (NW,130) 'ROOTS, OF THE NUMERATOR'
130   FORMAT (//A30)
      WRITE (NW,*) (ROOT1(I),I=1,NR1)
      CALL PROOT (POLE,IP,ROOT2,NR2)
      WRITE (NW,130) 'ROOTS OF THE DENOMINATOR'
      WRITE (NW,*) (ROOT2(I),I=1,NR2)
C
C --> TO FIND THE MAXIMUM VALUE FOR SCALING
C
      BIG = 0.
C
      DO 140 I = 1, NR1
      X1(I) = REAL(ROOT1(I))
      IF (ABS(X1(I)) .GT. BIG)  BIG = ABS(X1(I))
      Y1(I) = AIMAG(ROOT1(I))
      IF (ABS(Y1(I)) .GT. BIG)  BIG = ABS(Y1(I))
140   CONTINUE
C
      DO 150 I = 1, NR2
      X2(I) = REAL(ROOT2(I))
      IF (ABS(X2(I)) .GT. BIG)  BIG = ABS(X2(I))
      Y2(I) = AIMAG(ROOT2(I))
      IF (ABS(Y2(I)) .GT. BIG)  BIG = ABS(Y2(I))
150   CONTINUE
C
      WRITE (5,*) 'Do you want manual scaling? Y=1'
      READ (5,*) I
      IF (I .EQ. 1) GO TO 160
C
C --> AUTOMATIC SCALING
C
      TEMP(1) = 0.
      TEMP(2) = BIG
      CALL SCALE (TEMP,2.,2.,1)
      RMAX = TEMP(4) * 2.
      GO TO 170
C
C --> ENTER MANUAL SCALING VALUE
C
160   WRITE (5,*) 'Enter maximum value for scaling'
      READ (5,*) RMAX
C
170   WRITE (5,*) 'Enter no. of decimal places in the axis no.'
      READ (5,*) IDIG
      VMAX = -RMAX
C
      CALL PLOTST (256,IPBUFF)
      CALL PLOT   (-0.5,0.25,-3)
      CALL RECT   (0.0,0.0,6.5,9.,0.,3)
      CALL LABEL  (0.75,0.4,8.25,0.4,'ROOT LOCUS PLOT',15,0.16,1,0.0)
      CALL SYMBOL (7.75,1.,0.14,61,0.,-1)
      CALL SYMBOL (999.,1.02,0.07,' O.L. POLES',0.,11)
      CALL SYMBOL (7.75,0.75,0.14,36,0.,-1)
      CALL SYMBOL (999.,0.77,0.07,' O.L. ZEROS',0.,11)
      CALL SYMBOL (7.8,0.5,0.02,11,0.,-1)
      CALL SYMBOL (999.,0.47,0.07,' C.L. POLES',0.,12)
```

TABLE B.3 (cont.)

```fortran
        CALL PLOT     (4.5,-5.,-3)
        IF (IM .EQ. 1) CALL PLOT (SINCH,0.,-3)
        CALL PLOT     (0.,-2.,3)
        CALL PLOT     (0.,2.,2)
        CALL SYMBOL   (0.,-2.,0.15,13,90.,-1)
        CALL SYMBOL   (0.,-1.,0.15,13,90.,-1)
        CALL SYMBOL   (0.,1.,0.15,13,90.,-1)
        CALL SYMBOL   (0.,2.,0.15,13,90.,-1)
        CALL NUMBER   (-0.08,2.2,0.1,RMAX,0.,IDIG)
        CALL NUMBER   (-0.13,-2.28,0.1,VMAX,0.,IDIG)
        IF (IM .EQ. 1) GO TO 180
        CALL PLOT     (-2.,0.,3)
        CALL PLOT     (2.,0.,2)
        CALL SYMBOL   (-2.,0.,0.15,13,0.,-1)
        CALL SYMBOL   (-1.,0.,0.15,13,0.,-1)
        CALL SYMBOL   (1.,0.,0.15,13,0.,-1)
        CALL SYMBOL   (2.,0.,0.15,13,0.,-1)
        CALL NUMBER   (-2.2,-0.2,0.1,VMAX,0.,IDIG)
        CALL NUMBER   (1.9,-0.2,0.1,RMAX,0.,IDIG)
        GO TO 200
180     IF (IMD .EQ. 2) GO TO 190
        CALL PLOT     (-2.,0.,3)
        CALL PLOT     (0.,0.,2)
        CALL SYMBOL   (-2.,0.,0.15,13,0.,-1)
        CALL SYMBOL   (-1.,0.,0.15,13,0.,-1)
        CALL NUMBER   (-2.2,-0.2,0.1,VMAX,0.,IDIG)
        GO TO 200
190     CALL PLOT     (0.,0.,3)
        CALL PLOT     (2.,0.,2)
        CALL SYMBOL   (1.,0.,0.15,13,0.,-1)
        CALL SYMBOL   (2.,0.,0.15,13,0.,-1)
        CALL NUMBER   (1.9,-0.2,0.1,RMAX,0.,IDIG)
200     CALL SYMBOL   (0.0,0.0,0.07,25,0.,-1)
C
C ---> TO PLOT THE OPEN LOOP ZEROS
C
        DO 210 I = 1, NR1
        X = X1(I) * 2. / RMAX
        Y = Y1(I) * 2. / RMAX
        CALL SYMBOL   (X,Y,0.21,36,0.,-1)
210     CONTINUE
C
C ---> TO PLOT THE OPEN LOOP POLES
C
        DO 220 I = 1, NR2
        X = X2(I) * 2.0 / RMAX
        Y = Y2(I) * 2.0 / RMAX
        CALL SYMBOL   (X,Y,0.21,61,0.,-1)
220     CONTINUE
C
C ---> TO PLOT THE CLOSED LOOP POLES
C
        DO 270 I = 1, N1
        CALL SPCADD (IP,POLE,IZ,ZERO,POLY,1.,RK,NP)
        CALL REDUCE (POLY,NP)
        CALL PROOT (POLY,NP,ROOT3,NR3)
        IF (NPR .NE. 1) GO TO 240
C
        WRITE (NW,130) 'CLOSED-LOOP POLES FOR :'
        WRITE (NW,230) RK
230     FORMAT (' GAIN = ',F14.7)
        WRITE (NW,*) (ROOT3(MM),MM=1,NR3)
C
240     DO 260 J=1,NR3
        X = REAL(ROOT3(J)) * 2. / RMAX
        Y = AIMAG(ROOT3(J)) * 2. / RMAX
        IF (IDASH .EQ. 1) GO TO 250
        CALL SYMBOL (X,Y,0.02,11,0.,-1)
        GO TO 260
250     IF (MOD(I,NINC) .NE. 0) CALL SYMBOL (X,Y,0.02,11,0.,-1)
        IF (MOD(I,NINC) .EQ. 0) CALL SYMBOL (X,Y,0.05,11,0.,-1)
260     CONTINUE
        RK = RK + RK1
270     CONTINUE
        CALL PLOTND
C
        IF (NW .EQ. 3) CLOSE (UNIT=3)
C
        WRITE (5,*) 'Do you want to try other gains ? Y=1'
        READ (5,*) I
        IF (I .EQ. 1) GO TO 100
        IF (NW .EQ. 3) WRITE (5,*) 'Please print RLOCI.DAT for output'
        STOP
        END
```

The subprograms used are listed in Section B.7:

(1) ING 2) PROOT (3) REDUCE (4) SPCADD (5) PLTLIB

Example B.4.1

Figure B.2 shows the root locus plot with K increasing at increments 0.1 from 0 to 40, for the loop gain function

$$G(s) = \frac{K(s + 3)}{(s + 1)(s + 2)}$$

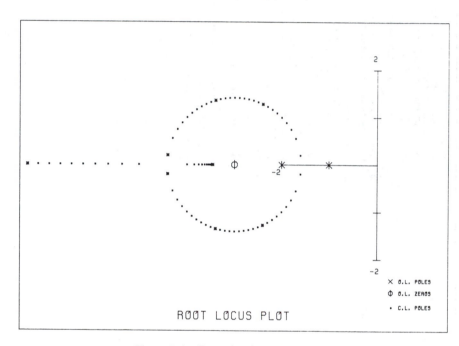

Figure B.2　Example of root locus plot.

B.5 POLAR PLOTS: POLAR

POLAR constructs a polar plot of the transfer function

$$G(s) = \frac{KN(s)}{D(s)}$$

where K is a constant and $N(s)$ and $D(s)$ are polynomials. The program is design oriented in that it allows for separate entry of a dynamic compensator and provides a polar plot of the product. The plot is obtained by calculating magnitude and phase angle of the transfer function for $s = j\omega$ a specified number of times with a specified increment of ω.

The interactive program listing is shown in Table B.4. $N(s)$ and $D(s)$ can

TABLE B.4 PROGRAM FOR POLAR PLOTS

```fortran
C*********************************************************************
C*                                                                  *
C*  THIS PROGRAM PLOTS THE POLAR PLOT OF A TRANSFER FUNCTION         *
C*                                                                  *
C*********************************************************************
C*
      REAL ZERO(20),POLE(20),TZERO(20),TPOLE(20),X(202),Y(202),
     +     COMN(10),COMD(10),TEMP1(20),TEMP2(20)
      COMPLEX VZERO,VPOLE,PMAG
C
C --> TO READ THE COEFFICIENTS OF THE NUMERATOR AND DENOMINATOR
C     OF THE TRANSFER FUNCTION BY CALLING SUBROUTINE ING
C
      CALL ING (ZERO,IZ,POLE,IP,TEMP1,TEMP2)
C
100   WRITE (5,*)  'Enter gain'
      READ (5,*)  GAIN
      WRITE (5,*)  'Do you want dynamic compensators? Y=1'
      READ (5,*)  I
      IF (I .EQ. 1) GO TO 130
C
      DO 110 I = 1, IP
      TPOLE(I) = POLE(I)
110   CONTINUE
      DO 120 I = 1, IZ
      TZERO(I) = ZERO(I) * GAIN
120   CONTINUE
      IP2 = IP
      IZ2 = IZ
      GO TO 150
C
C --> TO ENTER THE DYNAMIC COMPENSATOR
C
130   CONTINUE
      WRITE (5,*)  'Enter the order+1 of the compensator NUMERATOR'
      READ (5,*)  ICOMN
      WRITE (5,*)  'Enter the coeff. of the compensator NUMERATOR'
      WRITE (5,*)  'in the order of ascending power'
      READ (5,*)  (COMN(I),I=1,ICOMN)
      WRITE (5,*)  'Enter the order+1 of the compensator DENOMINATOR'
      READ (5,*)  ICOMD
      WRITE (5,*)  'Enter the coeff. of the compensator DENOMINATOR'
      WRITE (5,*)  'in the order of ascending power'
      READ (5,*)  (COMD(I),I=1,ICOMD)
C
      DO 140 I = 1, ICOMN
      COMN(I) = COMN(I) * GAIN
140   CONTINUE
      CALL MUTPOL (ZERO,COMN,IZ,ICOMN,TZERO,IZ2)
      CALL MUTPOL (POLE,COMD,IP,ICOMD,TPOLE,IP2)
C
C --> TO ENTER THE PLOTTING DATA
C
150   WRITE (5,*)  'Enter no. of times the magnitudes'
      WRITE (5,*)  'must be calculated (maximum 200)'
      READ (5,*)  NPOINT
      WRITE (5,*)  'Enter the starting frequency'
      READ (5,*)  WS
      WRITE (5,*)  'Enter the frequency increment'
      READ (5,*)  WINC
      WRITE (5,*)  'Do you want to see the magnitude at each'
      WRITE (5,*)  'frequency? Y=1'
      READ (5,*)  NPR
      IF (NPR .NE. 1) GO TO 170
C
      WRITE (5,*)  'Enter choice of output of numerical values:'
      WRITE (5,*)  'screen=5, printer=3'
      READ (5,*)  NW
      IF (NW .EQ. 3) OPEN (UNIT=3,STATUS='NEW',FILE='POLAR.DAT')
      WRITE (NW,160)
160   FORMAT (' ',10X,'FREQUENCY',14X,'REAL',14X,'IMAGINARY'/
     +        ' ',9X,11('-'),10X,10('-'),10X,11('-')/)
C
170   DO 180 I = 1, NPOINT
      CALL CALCPX (TZERO,IZ2,WS,VZERO)
      CALL CALCPX (TPOLE,IP2,WS,VPOLE)
      IF (VPOLE .EQ. (0.,0.)) GO TO 200
      PMAG = VZERO / VPOLE
      X(I) = REAL (PMAG)
      Y(I) = AIMAG (PMAG)
      IF (NPR .EQ. 1) WRITE (NW,190) WS,X(I),Y(I)
      WS = WS + WINC
180   CONTINUE
190   FORMAT (' ',3X,'Freq=',F10.5,5X,'Real=',F10.5,6X,'Imag=',F10.5)
      GO TO 220
C
C --> TO INPUT THE STARTING FREQUENCY AGAIN IF THE MAGNITUDE
C     AT THE CHOSEN STARTING FREQUENCY IS INFINITY
C
200   WRITE (5,210)
210   FORMAT (' ','Starting magnitude is infinity,'/' ',
     +        'Please enter another starting frequency')
      READ (5,*)  WS
      GO TO 170
C
220   WRITE (5,*)  'Do you want to plot the result? Y=1'
      READ (5,*)  I
      IF (I .NE. 1) GO TO 240
C
C --> TO PLOT THE RESULT
C
      WRITE (5,*)  'Do you want to add pt. (0.,0.) as the last pt.? Y=1'
      READ (5,*)  I
      IF (I .NE. 1) GO TO 230
      NPOINT = NPOINT + 1
      X(NPOINT) = 0.
      Y(NPOINT) = 0.
230   CALL PLOT2 (X,Y,NPOINT)
C
240   IF (NW .EQ. 3) CLOSE (UNIT=3)
      WRITE (5,*)  'Do you want to try other compensators? Y=1'
      READ (5,*)  I
      IF (I .EQ. 1) GO TO 100
      IF (NW .EQ. 3) WRITE (5,*)  'Please print POLAR.DAT for output'
      STOP
      END
```

405

```
      SUBROUTINE   PLOT2 (X,Y,NPOINT)
      DIMENSION    X(1),Y(1),TEMP(4)
      INTEGER      IPBUFF(256)
C
C***********************************************************
C*                                                         *
C*       THIS SUBROUTINE PLOTS THE POLAR PLOT.             *
C*                                                         *
C*       X(1)  -- THE REAL PART OF THE MAGNITUDES STORED   *
C*                IN AN ARRAY                              *
C*       Y(1)  -- THE IMAGINARY PART OF THE MAGNITUDES STORED *
C*                IN AN ARRAY                              *
C*       NPOINT -- NO. OF POINTS                           *
C***********************************************************
C
C
C --> TO CHOOSE AUTOMATIC SCALING OR MANUAL SCALING
C
100   WRITE (5,*) 'Do you want manual scaling? Y=1'
      READ (5,*) I
      IF (I .EQ. 1) GO TO 300
C
C --> AUTOMATIC SCALING
C
      BIG = 0.
      DO 200 I=1,NPOINT
      IF (ABS(X(I)) .GT. BIG) BIG = ABS(X(I))
      IF (ABS(Y(I)) .GT. BIG) BIG = ABS(Y(I))
200   CONTINUE
      TEMP(1) = BIG
      TEMP(2) = 0.
      CALL SCALE (TEMP,2.,2,1)
      GO TO 400
C
C --> MANUAL SCALING
C
300   WRITE (5,*) 'Enter change in value per inch of axes'
      READ (5,*) TEMP(4)
C
400   CONTINUE
      WRITE (5,*) 'Enter no. of decimal places in the axis no.'
      READ (5,*) IDIG
      RMAX = ABS(TEMP(4)*2.)
      VMAX = -RMAX
      X(NPOINT+1) = 0.
      X(NPOINT+2) = TEMP(4)
      Y(NPOINT+1) = 0.
      Y(NPOINT+2) = TEMP(4)
C
      CALL PLOTST (256,IPBUFF)
      CALL PLOT   (-0.5,0.25,-3)
      CALL RECT   (0.0,0.0,6.5,9.,0.,3)
      CALL LABEL  (0.75,0.65,8.25,0.65,'POLAR PLOT',10,0.16,1,0.0)
      CALL PLOT   (2.5,3.5,-3)
      CALL PLOT   (4.,0.,2)
      CALL PLOT   (2.,-2.,3)


      CALL PLOT    (2.,2.,2)
      CALL SYMBOL  (0.,0.,0.14,13,0.,-1)
      CALL SYMBOL  (1.,0.,0.14,13,0.,-1)
      CALL SYMBOL  (3.,0.,0.14,13,0.,-1)
      CALL SYMBOL  (4.,0.,0.14,13,0.,-1)
      CALL SYMBOL  (2.,-2.,0.14,13,90.,-1)
      CALL SYMBOL  (2.,-1.,0.14,13,90.,-1)
      CALL SYMBOL  (2.,1.,0.14,13,90.,-1)
      CALL SYMBOL  (2.,2.,0.14,13,90.,-1)
      CALL NUMBER  (-.5,-0.2,0.1,VMAX,0.,IDIG)
      CALL NUMBER  (4.1,-0.2,0.1,RMAX,0.,IDIG)
      CALL NUMBER  (1.95,2.2,0.1,RMAX,0.,IDIG)
      CALL NUMBER  (1.9,-2.28,0.1,VMAX,0.,IDIG)
      CALL PLOT    (2.,0.,-3)
C
C --> TO CALCULATE AND TO PLOT THE POSITION OF THE -1 POINT
C
      IF (X(NPOINT+2)*-2. .GE. -1.) GO TO 500
      RM1 = -1. / X(NPOINT+2)
      CALL SYMBOL  (RM1,0.,0.07,13,0.,-1)
      POS = RM1 - 0.1
      CALL SYMBOL  (POS,0.15,0.1,'-1',0.,2)
C
500   CALL LINE    (X,Y,NPOINT,1,0,1)
      CALL PLOTND
      WRITE (5,*)  'Do you want to plot again with different'
      WRITE (5,*)  'plotting parameters? Y=1'
      READ (5,*) I
      IF (I .EQ. 1) GO TO 100
      RETURN
      END
```

again be entered in terms of their polynomial factors. A warning is given if at the starting frequency the magnitude is infinite, with a request to enter another starting frequency. For a strictly proper transfer function, the magnitude approaches the origin when the frequency tends to infinity. A prompt is provided to allow the origin to be made the last point of the plot. For purposes of design, the screen prompts the user whether he or she wants to enter a different dynamic compensator. The user can choose the numerical values to be shown on the screen or printed on paper.

The subprograms used are listed in Section B.7:

(1) ING   (2) CALCPX   (3) MUTPOL   (4) PLTLIB

Example B.5.1

Figure B.3 shows the polar plot for frequencies from 0 to 100 with increment 0.1 of the transfer function

$$G(s) = \frac{s + 3}{(s + 1)(s + 2)}$$

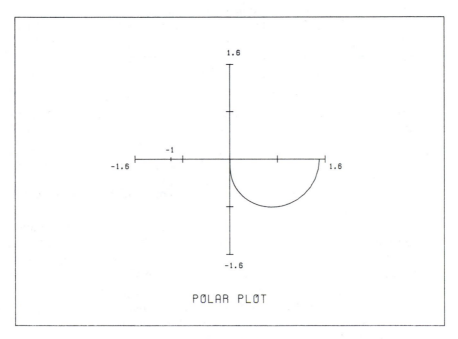

Figure B.3   Example of polar plot.

B.6 *BODE PLOTS: DEBODE*

DEBODE plots the asymptotic Bode magnitude plot and the phase angle curve for a transfer function

$$G(s) = \frac{A(s)}{s^n B(s)}$$

where $A(s)$ and $B(s)$ are polynomials with nonzero constant terms and n is a, possibly zero, integer constant. $A(s)$ and $B(s)$ may again be entered in terms of their polynomial factors, and the program is oriented to design in that a dynamic compensator can be entered separately. The screen prompts the user to choose either a PI-controller of the form PI = GAIN$(1 + RI/s)$ or a lead or lag compensator of the form $(As + B)/(Cs + D)$, or both, if dynamic compensation is desired by the user.

To plot $G(s)$, the roots of $A(s)$ and $B(s)$ are found, and these roots are stored in a vector X in the order of increasing frequency. Here the frequency corresponding to a complex root $c + jd$ is calculated as $\sqrt{c^2 + d^2}$. If there are N such break frequencies, the vector X is of dimension $N + 2$, with $X(1)$ and $X(N + 2)$ identifying the lowest and highest frequencies for which the plot is to be made.

The asymptotic magnitude plot is constructed by flagging the roots of $A(s)$ to cause a $+20$ dB/dec change of slope and those of $B(s)$ for a -20 dB/dec change of slope. These flags are stored in a vector Flag of dimension $N + 1$, where Flag(1) gives the slope of the low-frequency asymptote, Flag(1) $= -20n$. The asymptotic magnitudes for the frequencies stored in X can now be calculated and are stored in a vector Y of dimension $N + 2$. The starting magnitude is

$$Y(1) = 20 \log K + \text{Flag}(1) \log X(1)$$

where K is the ratio of the constant terms of $A(s)$ and $B(s)$. The other elements of Y are, with $i = 2, \ldots, N + 2$,

$$Y(i) = Y(i - 1) + \left[\sum_{j=1}^{i-1} \text{Flag}(j) \right] [\log X(i) - \log X(i - 1)]$$

The X and Y vectors form a set of points which are connected by lines to generate the plot.

For nonminimum-phase transfer functions an option is included to allow the phase margin to be still visualized from the magnitude plot by changing the slope at nonminimum-phase zeros by -20 instead of $+20$ dB/dec. It is safer, however, to use the phase angle curve for this purpose.

The phase angles are calculated at Np frequencies, stored in a vector $X1$, where $X1(1)$ is the lowest frequency for the plot and $X1(Np)$ the highest. The phase angles are stored in a vector $Y1$ of dimension Np, of which the ith element is found from

$$Y1(i) = \sum_{j=1}^{Na} \arg [X1(i) - RN(j)] - 180SK$$

$$- \sum_{j=1}^{Nb} \arg [X1(i) - RD(j)] - 90n$$

where Na, Nb = order of polynomial $A(s)$, $B(s)$

RN, RD = complex vector consisting of the roots of $A(s)$, $B(s)$

SK = 0, 1 if $(-1)^{NPR}$ times the ratio of the coefficients of the highest powers of $A(s)$ and $B(s)$ is > 0, < 0

NPR = total number of right-half plane roots of RN and RD

The $X1$ and $Y1$ vectors form a set of points which are connected by lines to generate the phase-angle curve. If its smoothness is inadequate, Np should be increased.

The interactive program listing is shown in Table B.5. Screen prompts allow the user to choose printing of numerical values. The subprograms used are listed in Section B.7:

(1) ING (2) REDUCE (3) PROOT (4) WRITEM (5) CHECKN
(6) RCHECK (7) MUTPOL (8) BODPLT (9) BODE (10) SORT
(11) FMAX (12) PLTLIB

Example B.6.1

Figures B.4 and B.5 show Bode magnitude and phase angle curves for the transfer function

$$G(s) = \frac{s + 3}{(s + 1)(s + 2)}$$

Figure B.4   Example of asymptotic Bode magnitude plot.

TABLE B.5 PROGRAM FOR BODE PLOTS

410

```fortran
C*****************************************************
C*
C*   THIS PROGRAM USES BODE PLOT TO DESIGN A SINGLE INPUT     *
C*   AND SINGLE OUTPUT CONTROL SYSTEM                         *
C*                                                            *
C*****************************************************
C*
      REAL    FD(2),S(2),PIN(2),PID(2),
     +        DIN(20),D1D(20),ZERO(20),POLE(20),
     +        TEMP1(20),TEMP2(20)
      COMPLEX ROOT1(19),ROOT2(19)
C
C --> TO READ THE COEFFICIENTS OF THE NUMERATOR AND DENOMINATOR
C     OF THE TRANSFER FUNCTION BY CALLING SUBROUTINE ING
C
      CALL ING (ZERO,IZ,POLE,IP,TEMP1,TEMP2)
C
C --> TO CHOOSE VALUES PRINTED ON SCREEN OR PAPER
C
100   WRITE (5,*) 'Enter choice of output of numerical values :'
      WRITE (5,*) 'screen=5, printer=3'
      READ (5,*) NW
      IF (NW .EQ. 3) OPEN (UNIT=3,STATUS='NEW',FILE='DEBODE.DAT')
C
C --> TO GENERATE THE COMPENSATOR
C
      WRITE (5,*) 'Do you want PI compensator? Y=1'
      READ (5,*) I
      IF (I .NE. 1) GO TO 110
      WRITE (5,*) 'PI = GAIN * (1 + RI/S)'
      WRITE (5,*) 'Enter value of GAIN & RI'
      READ (5,*) GAIN, RI
      FD(1) = GAIN * RI
      FD(2) = GAIN
      S(1) = 0.
      S(2) = 1.
      GO TO 120
C
110   WRITE (5,*) 'Enter gain'
      READ (5,*) GAIN
      FD(1) = GAIN
      FD(2) = 0.
      S(1) = 1.
      S(2) = 0.
      RI = 0.
C
120   CONTINUE
C
C --> TO ENTER LEAD/LAG COMPENSATOR
C
      WRITE (5,*) 'Do you want lead/lag compensator? Y=1'
      READ (5,*) ICP
      IF (ICP .EQ. 1) GO TO 130
      PIN(1) = 1.
      PIN(2) = 0.
      PID(1) = 1.
      PID(2) = 0.
      GO TO 140

130   WRITE (5,*) '(A*S+B)/(C*S+D)'
      WRITE (5,*) 'Enter A,B,C & D'
      READ (5,*) PIN(2), PIN(1), PID(2), PID(1)
C
140   CONTINUE
C
C --> TO WRITE THE COMPENSATOR VALUES
C
      WRITE (NW,150) GAIN, RI
150   FORMAT (' GAIN =',F10.5,6X,' RI =',F10.5/)
      IF (ICP .EQ. 1) WRITE (NW,160) PIN(2),PIN(1),PID(2),PID(1)
160   FORMAT (' (A*S+B)/(C*S+D)'/
     +        '    A =',F10.5,5X,' B =',F10.5/
     +        '    C =',F10.5,5X,' D =',F10.5//)
C
C --> TO CALCULATE THE OVERALL TRANSFER FUNCTION AND ITS ROOTS
C
      CALL MUTPOL (FD,ZERO,2,IZ,TEMP1,NT1)
      CALL MUTPOL (TEMP1,PIN,NT1,2,DIN,NDIN)
      WRITE (NW,170) ' NUMERATOR'
170   FORMAT (//A20)
      CALL REDUCE (DIN,NDIN)
      CALL PROOT (DIN,NDIN,ROOT1,NR1)
      CALL WRITEM (DIN,NDIN,ROOT1,NR1,NW)
      CALL CHECKN (ROOT1,NR1,NO1)
      CALL MUTPOL (S,PID,2,2,TEMP2,NT2)
      CALL MUTPOL (POLE,TEMP2,IP,NT2,D1D,ND1D)
      WRITE (NW,170) ' DENOMINATOR'
      CALL REDUCE (D1D,ND1D)
      CALL PROOT (D1D,ND1D,ROOT2,NR2)
      CALL WRITEM (D1D,ND1D,ROOT2,NR2,NW)
      CALL CHECKN (ROOT2,NR2,NO2)
C
C --> TO DETERMINE THE GAIN SIGN
C
      D1K = 1.
      DO 190 I = 1, NR2
      IF (I .GT. NR1) GO TO 180
      IF (REAL(ROOT1(I)) .GT. 0.) D1K = -D1K
180   IF (REAL(ROOT2(I)) .GT. 0.) D1K = -D1K
190   CONTINUE
      IF ((DIN(NDIN)/D1D(ND1D)) .LT. O.) D1K = -D1K
C
C --> TO PLOT THE BODE PLOT
C
      WRITE (5,*) 'Do you want to see the Bode plot? Y=1'
      READ (5,*) I
      IF (I .NE. 1) GO TO 200
      CALL RCHECK (NR1,ROOT1,NR2,ROOT2,NM1,NW)
      CALL DLPGN (ROOT1,NR1,NO1,DIN(NO1+1),ROOT2,NR2,NO2,D1D(NO2+1),
     +            0,NM1,O,D1K)
200   WRITE (5,*) 'DO YOU WANT TO REDESIGN THE LOOP? Y=1'
      READ (5,*) I
      IF (NW .EQ. 3) CLOSE (UNIT=3)
      IF (I .EQ. 1) GO TO 100
      IF (NW .EQ. 3) WRITE (5,*) 'Please print DEBODE.DAT for output'
      STOP
      END
```

TABLE B.5 (cont.)

```fortran
      SUBROUTINE DLPGN (RQA,NA,NORDA,COEFA,RD,ND,NORDD,COEFD,
     +NS,NM,NDL,SIGN)
      DIMENSION  FREQ(120),X(120),Y(120),FLAG20(120),RQA(1),RD(1)
      COMPLEX    RQA,RD,ZZZ(120)
      INTEGER    IPBUFF(256)
C
C**********************************************************
C*                                                        *
C*    THIS SUBROUTINE PLOTS THE BODE MAGNITUDE AND PHASE   *
C*    PLOTS.                                               *
C*    RQA    --   ROOTS OF NUMERATOR                       *
C*    RD     --   ROOTS OF DENOMINATOR                     *
C*    NA     --   NO. OF ROOTS OF NUMERATOR                *
C*    ND     --   NO. OF ROOTS OF DENOMINATOR              *
C*    NORDA  --   LOWEST ORDER IN NUMERATOR                *
C*    NORDD  --   LOWEST ORDER IN DENOMINATOR              *
C*    COEFA  --   COEFF. OF LOWEST POWER IN NUMERATOR      *
C*    COEFD  --   COEFF. OF LOWEST POWER IN DENOMINATOR    *
C*    NDL    --   0  SOLID LINES FOR BODE PLOT             *
C*                1  DASH  LINES FOR BODE PLOT             *
C*    NM     --   CORRECTION FACTOR FOR +VE ZEROS          *
C*    SIGN   --   GAIN SIGN FOR PHASE ANGLE PLOT           *
C*                                                         *
C**********************************************************
C
C  --> ENTER PLOTTING PARAMETERS FOR MAGNITUDE PLOT (INTERACTIVELY)
C
      WRITE (5,*) 'Input the following data for magnitude plot'
      WRITE (5,*) 'Starting frequency:'
      READ  (5,*) XFIRS1
      WRITE (5,*) 'Ending frequency:'
      READ  (5,*) XLAST1
      WRITE (5,*) 'Starting magnitude (db):'
      READ  (5,*) YFIRS1
      WRITE (5,*) 'Change in magnitude (db) per inch of axis:'
      READ  (5,*) YD1
      XL1 = 8.
      YL1 = 5.
      WRITE (5,*) 'Number of cycles on frequency axis:'
      READ  (5,*) NXDIV1
      WRITE (5,*) 'Number of divisions on magnitude axis:'
      READ  (5,*) NYDIV1
      XD1 = (NXDIV1 * 1.) / XL1
C
C  --> INITIALIZE PLOT BUFFER, LABEL PLOTS AND DRAW AXES
C
      CALL PLOTST (256,IPBUFF)
      CALL RECT   (-0.75,0.25,7.0,9.5,0.0,3)
      CALL PLOT   (-0.75,0.25,-3)
      CALL LABEL  (1.0,-0.5,8.25,0.5,'BODE PLOT',9,0.14,1.0,0.0)
      CALL PLOT   (0.75,1.25,-3)
      CALL RECT   (0.0,0.0,YL1,XL1,0.0,3)
      CALL LGAXS  (0.0,0.0,'FREQUENCY (RAD/SEC)',-19,XL1,0.0,XFIRS1,XD1)
      CALL AXIS   (0.0,0.0,'ABS VALUE IN DB',15,YL1,90.,YFIRS1,YD1)
      CALL LLGRID (0.0,0.0,XL1,YL1,'LOG ','LIN ',NXDIV1,NYDIV1)
C
C  --> GENERATE BODE MAGNITUDE PLOT
C
      N = NA + ND
      NT = N + 4
      NORD = NORDD - NORDA
      COEF = COEFA / COEFD
      CALL BODPLT (YFIRS1,YD1,XFIRS1,XD1,XLAST1,NS,NT,N,NA,ND,NORD,RQA,
     +             RD,COEF,X,Y,FREQ,FLAG20,ZZZ,NM,NDL)
C
      CALL PLOTND
C
C  --> TO INDICATE IF PHASE PLOT REQUIRED
C
      WRITE (5,*) 'Do you want to see the phase plot? Y=1'
      READ  (5,*) J
      IF (J .NE. 1) GO TO 999
C
C  --> ENTER PLOTTING PARAMETERS FOR PHASE PLOT (INTERACTIVELY)
C
      WRITE (5,*) 'Input the following data for the phase plot'
      WRITE (5,*) 'Starting phase angle:'
      READ  (5,*) YFIRS2
      WRITE (5,*) 'Change in phase angle per inch of axis:'
      READ  (5,*) YD2
      WRITE (5,*) 'Input no. of points per cycle in calculating angles:'
      WRITE (5,*) '(multiple of 9)'
      READ  (5,*) NDIV
C
C  --> INITIALIZE PLOT BUFFER, LABEL PLOTS AND DRAW AXES
C
      CALL PLOTST (256,IPBUFF)
      CALL RECT   (-0.75,0.25,7.0,9.5,0.0,3)
      CALL PLOT   (-0.75,0.25,-3)
      CALL LABEL  (1.0,-0.5,8.25,0.5,'BODE PLOT',9,0.14,1.0,0.0)
      CALL PLOT   (0.75,1.25,-3)
      CALL RECT   (0.0,0.0,YL1,XL1,0.0,3)
      CALL LGAXS  (0.0,0.0,'FREQUENCY (RAD/SEC)',-19,XL1,0.0,XFIRS1,XD1)
      CALL AXIS   (0.0,0.0,'PHASE ANGLE IN DEGREES',22,YL1,90.,YFIRS2,YD2)
      CALL LLGRID (0.0,0.0,XL1,YL1,'LOG ','LIN ',NXDIV1,NYDIV1)
C
C  --> TO GENERATE BODE PHASE PLOT
C
      CALL APHPLT (YFIRS2,YD2,XFIRS1,XD1,XLAST1,1,NA,ND,NORD,RQA,
     +             RD,NDIV,SIGN)
C
      CALL PLOTND
C
999   RETURN
      END
```

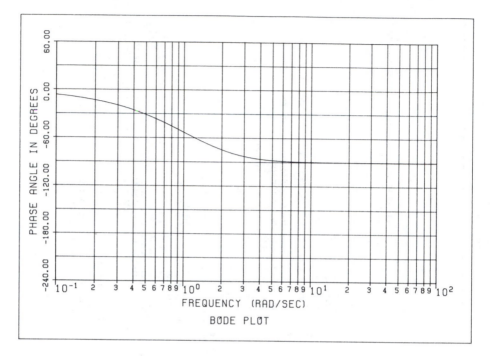

Figure B.5 Example of phase angle curve.

B.7 LISTINGS OF SUBPROGRAMS

Listings of the subroutines used in the programs of the preceding sections are given in Table B.6.

B.8 REFERENCES

1. CURTIS, F. G., *Applied Numerical Analysis,* 2nd ed. Reading, Mass.: Addison-Wesley Publishing Company, Inc., 1980.
2. CHURCHILL, R. V., *Modern Operational Mathematics in Engineering.* New York: McGraw-Hill Book Company, 1944, pp. 44–45.

TABLE B.6 PROGRAMS OF SUBROUTINES

```fortran
C**************************************************************
C*                                                           *
C*    THIS SUBROUTINE PLOTS A SINGLE BODE GRAPH ON EACH CALL: *
C*                                                           *
C*    NN    - ROOTSN(NN); THE NUMBER OF ROOTS IN THE NUMERATOR   *
C*    ND    - ROOTSD(ND); THE NUMBER OF ROOTS IN THE DENOMINATOR *
C*    NDIV  - NO. OF POINTS PER CYCLE IN CALCULATING ANGLES      *
C*            (MULTIPLE OF 9)                                    *
C*    NORDER - THE ORDER OF THE S MULTIPLYING THE DENOMINATOR    *
C*    SIGN   - GAIN SIGN FOR PHASE ANGLE PLOT                    *
C*                                                           *
C**************************************************************
C*
      SUBROUTINE APHPLT(YFIRST,YD,XFIRST,XD,XLAST,ISHAPE,NN,ND,
     +              NORDER,ROOTSN,ROOTSD,NDIV,SIGN)
      DIMENSION X(100),Y(100)
      COMPLEX ROOTSN(1),ROOTSD(1),WJ,TEMP
C
      NCYCLE = IFIX(ALOG10(XLAST)-ALOG10(XFIRST))
      NT = NCYCLE * NDIV + 1
      IF (NT .LT. 100) GO TO 10
C
      WRITE (5,*) 'no. of points exceeds array size,'
      WRITE (5,*) 'no phase angle plot'
      RETURN
C
10    SGAIN = 0.
      IF (SIGN .LT. 0.) SGAIN = 1.
      X(1) = XFIRST
      K = 1
      PI = 3.141593
C
C --> TO CALCULATE X VECTOR
C
      DO 20 I = 1, NCYCLE
      DO 20 J = 1, NDIV
      K = K + 1
      X(K) = XFIRST * 10**(I-1) * (1.+9.*FLOAT(J)/FLOAT(NDIV))
20    CONTINUE
C
C --> TO CALCULATE Y VECTOR
C
      AYN = 0.
      DO 30 J = 1, NN
      TEMP = WJ - ROOTSN(J)
      IF (ABS(REAL(TEMP)) .LE. 1.E-5) GO TO 25
      ANGN = ATAN(AIMAG(TEMP)/REAL(TEMP)) * 180./PI
      GO TO 28
25    ANGN = 90. * 180. / PI
      IF (AIMAG(TEMP) .LT. 0.) ANGN = -ANGN
28    AYN = AYN + ANGN
30    CONTINUE
C
      AYD = 0.
      DO 40 J = 1, ND
      TEMP = WJ - ROOTSD(J)
      IF (ABS(REAL(TEMP)) .LE. 1.E-5) GO TO 45
      ANGD = ATAN(AIMAG(TEMP)/REAL(TEMP)) *180./PI
      GO TO 48
45    ANGD = 90. * 180. / PI
      IF (AIMAG(TEMP) .LT. 0.) ANGD = - ANGD
48    AYD = AYD + ANGD
40    CONTINUE
      Y(I) = AYN - AYD - 90.*FLOAT(NORDER) - SGAIN * 180.
50    CONTINUE
C
      Y(NT+1) = YFIRST
      Y(NT+2) = YD
      X(NT+1) = XFIRST
      X(NT+2) = XD
C
C --> TO PLOT THE RESULT
C
      CALL LGLIN (X,Y,NT,1,0,ISHAPE,-1)
      RETURN
      END


C**************************************************************
C*                                                           *
C*    THIS SUBROUTINE SETS UP THE X AND Y VECTORS FOR PLOTTING WITHIN  *
C*    THE ALLOWABLE FREQUENCY RANGE                                    *
C*                                                           *
C**************************************************************
C*
      SUBROUTINE BODE(XFIRST,XLAST,NT,N,NORDER,COEFF,X,Y,FREQ,FLAG2O,Z,
     +           NEWNT)
      DIMENSION X(1),Y(1),FREQ(1),FLAG2O(1)
      COMPLEX Z
      DIMENSION Z(1)
C
C --> COMPUTE THE FREQUENCY OF EACH ROOT:
C
      DO 10 I = 1, N
      FREQ(I) =  CABS (Z(I))
10    CONTINUE
C
C --> SORT THE ROOTS INTO ASCENDING ORDER; AND SORT FLAG2O
C     IN THE SAME ORDER SIMULTANEOUSLY:
C
      CALL SORT (N,FREQ,FLAG2O)
C
C --> SET X(1) TO THE LOWER OF:
C
      X(1) = FREQ(1) / 10.0
      IF (XFIRST .LT. FREQ(1)) X(1) = XFIRST / 10.0
```

TABLE B.6 (cont.)

```fortran
C*************************************************************************
C*                                                                     *
C*        THIS SUBROUTINE PLOTS A SINGLE BODE GRAPH ON EACH CALL        *
C*                                                                     *
C*        NN      - ROOTSN (NN);  THE NUMBER OF ROOTS IN THE NUMERATOR  *
C*        ND      - ROOTSD (ND);  THE NUMBER OF ROOTS IN THE DENOMINATOR*
C*        N       - NN+ND; THE TOTAL NUMBER OF ROOTS IN THE NUMERATOR AND*
C*                  DENOMINATOR                                          *
C*        NT      - N+4; THE FOLLOWING ARE PLOTTED: N ROOTS, 1 STARTING  *
C*                  POINT, 1 ENDING POINT; AND THE STARTING FREQUENCY AND *
C*                  DELTA FREQUENCY WHICH MAKE UP THE LAST TWO ENTRIES   *
C*                  X (NT), Y (NT); THESE ARE THE FREQUENCY AND MAGNITUDE*
C*                  VECTORS                                              *
C*        NORDER  - THE ORDER OF THE S MULTIPLYING THE DENOMINATOR       *
C*        COEFF   - THE GAIN COMPUTED FROM ZERO ORDER TERMS IN THE NUMERATOR*
C*                  AND THE DENOMINATOR POLYNOMIALS, AND ANY GAIN        *
C*                  MULTIPLYING THE OVERALL TRANSFER FUNCTION            *
C*        NM      - = 1 FOR NON-MINIMUM PHASE CORRECTION FACTOR          *
C*                  = 0 FOR MINIMUM PHASE                                *
C*        NL      - = 1 FOR DASH LINE                                    *
C*                  = 0 FOR SOLID LINE                                   *
C*                                                                     *
C*************************************************************************
C*
      SUBROUTINE BODPLT(YFIRST,YD,XFIRST,XD,XLAST,ISHAPE,NT,N,NN,ND,
     +              NORDER,ROOTSN,ROOTSD,COEFF,X,Y,FREQ,FLAG20,
     +              Z,NM,NL)
C
      DIMENSION X(1),Y(1),FREQ(1),FLAG20(1)
      COMPLEX Z,ROOTSN,ROOTSD
      DIMENSION Z(1),ROOTSN(1),ROOTSD(1)
C
C -->  SET UP Z VECTOR AND FLAG20 VECTOR:
C
      IF (NN .EQ. 0) GO TO 20
      DO 10 J = 1, NN
      Z(J) = ROOTSN(J)
      FLAG20(J) = 20.0
      IF (NM.EQ.1 .AND. (REAL(ROOTSN(J)).GT.0.0) FLAG20(J) = -20.0
10    CONTINUE
20    IF (ND .EQ. 0) GO TO 40
      DO 30 J = 1, ND
      K = NN + J
      Z(K) = ROOTSD(J)
      FLAG20(K) = -20.0
30    CONTINUE
40    CONTINUE
C
C -->  PREPARE X AND Y VECTORS FOR PLOTTING:
C
      CALL BODE(XFIRST,XLAST,NT,N,NORDER,COEFF,X,Y,FREQ,FLAG20,Z,NEWNT)
C
      Y(NEWNT - 1) = YFIRST
      Y(NEWNT) = YD
      X(NEWNT - 1) = XFIRST
      X(NEWNT) = XD
      NOW = NEWNT - 2

C
C -->  SET UP Y(1) ACCORDING TO GAIN AND S**NORDER:
C
      Y(1) = 20.0 * ALOG10(ABS(COEFF)) - NORDER * 20.0 * ALOG10(X(1))
C
C -->  SET X(N+2) TO THE LARGER OF THE FOLLOWING:
C
      LAST = N + 2
      X(LAST) = FREQ(N) * 10.0
      IF (XLAST .GT. FREQ(N)) X(LAST) = XLAST * 10.0
C
C -->  SET UP REMAINING ROOTS FOR PLOTTING:
C
      ORDER = -NORDER * 20.0
      DO 30 I = 2, LAST
      J = I - 1
      IF (J .GT. N) GO TO 20
      X(I) = FREQ(J)
20    Y(I) = Y(J) + (ALOG10(X(I)) - ALOG10(X(J))) * ORDER
      IF (J .GT. N) GO TO 30
      ORDER = ORDER + FLAG20(J)
30    CONTINUE
C
C -->  NOW REDUCE THE X VECTOR TO FIT WITHIN THE CONFINES OF THE PLOT:
C
      DO 40 J = 1, LAST
      JFIRST = J
      IF (X(J) .GT. XFIRST) GO TO 50
40    CONTINUE
C
50    DO 60 J = 1, LAST
      JLAST = J
      IF (X(J) .GT. XLAST) GO TO 70
60    CONTINUE
C
70    CONTINUE
      YFF = Y(JFIRST-1)+((Y(JFIRST)-Y(JFIRST-1))/(ALOG10(X(JFIRST)
     +     -ALOG10(X(JFIRST-1))))*(ALOG10(XFIRST)-ALOG10(X(JFIRST-
     +     1)))
C
      YLL = Y(JLAST-1)+((Y(JLAST)-Y(JLAST-1))/(ALOG10(X(JLAST))
     +     -ALOG10(X(JLAST-1))))*(ALOG10(XLAST)-ALOG10(X(JLAST-1)))
      KLAST = JLAST - JFIRST + 1
C
      DO 80 J = 2, KLAST
      K = JFIRST + J - 2
      XX = X(K)
      YY = Y(K)
      X(J) = XX
      Y(J) = YY
80    CONTINUE
C
      X(1) = XFIRST
      Y(1) = YFF
      X(KLAST+1) = XLAST
      Y(KLAST+1) = YLL
      NEWNT = KLAST + 3
      RETURN
      END
```

TABLE B.6 (cont.)

```
C
C --> PLOT THE RESULTING GRAPH:
C
      IF (NL .EQ. 0) CALL LGLIN (X,Y,NOW,1,0,ISHAPE,-1)
      IF (NL .EQ. 1) CALL DLGLIN (X,Y,NOW,1,0,ISHAPE,-1)
C
      RETURN
      END
C*******************************************************
C*                                                     *
C*    THIS SUBROUTINE CALCULATES THE VALUE OF A        *
C*    POLYNOMIAL AT A GIVEN COMPLEX VALUE.             *
C*    POLY(1) -- GIVEN POLYNOMIAL WITH COEFFICIENTS    *
C*               STORED IN THE ORDER OF ASCENDING POWER*
C*    IP     -- ORDER + 1 OF GIVEN POLYNOMIAL          *
C*    FREQ   -- GIVEN COMPLEX VALUE                    *
C*    FRVECT -- RESULT IN COMPLEX                      *
C*                                                     *
C*******************************************************
C
      SUBROUTINE CALCPX (POLY,IP,FREQ,FRVECT)
      COMPLEX FRVECT
      REAL    POLY(1)
C
      X = POLY(1)
      Y = POLY(2) * FREQ
C
      J = 1
      DO 10 I = 3, IP, 2
      X = X + FREQ**(I-1) * (-1)**J * POLY(I)
      IF (I .EQ. IP) GO TO 20
      Y = Y + FREQ**(I) * (-1)**J * POLY(I+1)
      J = J + 1
10    CONTINUE
C
20    FRVECT = CMPLX(X,Y)
      RETURN
      END
C*******************************************************
C*                                                     *
C*    THIS SUBROUTINE CHECKS THE ROOTS WHICH ARE EQUAL *
C*    TO ZERO. THE ZERO ROOTS WILL BE TAKEN OUT        *
C*    FROM THE ARRAY.                                  *
C*    NORDER = NO. OF ZERO ROOTS                       *
C*                                                     *
C*******************************************************
C*
      SUBROUTINE CHECKN (ROOT,NR,NORDER)
      COMPLEX ROOT(1)
      NORDER = 0
      J = 0
C
      DO 10 I = 1, NR
      A = REAL(ROOT(I))
      B = AIMAG(ROOT(I))
      IF (ABS(A).GE.1.E-6 .OR. ABS(B).GE.1.E-6) GO TO 20
      NORDER = NORDER + 1
      GO TO 10
C
20    CONTINUE
      J = J + 1
      ROOT(J) = ROOT(I)
10    CONTINUE
C
      NR = NR - NORDER
      RETURN
      END
C*******************************************************
C*                                                     *
C*    THIS SUBROUTINE DIFFERENTIATES A POLYNOMIAL ONCE *
C*                                                     *
C*******************************************************
C*
      SUBROUTINE DIFF (N,A,M,B)
      DIMENSION A(1),B(1)
      M = N - 1
      DO 10 I = 1, M
      B(I) = A(I+1) * FLOAT(I)
10    RETURN
      END
C*******************************************************
C*                                                     *
C*    THIS SUBROUTINE CHECKS THE EXISTENCE OF EQUAL ROOTS*
C*    IN AN ARRAY.  IN RETURN :                        *
C*    IFLAG = 0  --- NO EQUAL ROOT                     *
C*    IFLAG = 1  --- EQUAL ROOTS EXIST                 *
C*                                                     *
C*******************************************************
C*
      SUBROUTINE EQROOT (ROOT,IR,IFLAG)
      COMPLEX ROOT(1)
      IFLAG = 0
      IR1 = IR - 1
      DO 10 I = 1, IR1
      K = I + 1
      DO 10 J = K, IR
      DIFFI = ABS((AIMAG(ROOT(I))) - (AIMAG(ROOT(J))))
```

```fortran
      IF (DIFFI .GT. 1.E-6) GO TO 10
      DIFFR = REAL(ROOT(I)) - REAL(ROOT(J))
      IF (ABS(DIFFR) .GT. 1.E-6) GO TO 10
      IFLAG = 1
      GO TO 99
10    CONTINUE
99    RETURN
      END

C************************************************************
C*                                                         *
C*  THIS SUBROUTINE FINDS THE LARGEST VALUE OF A VECTOR X AND THE  *
C*  CORRESPONDING VALUE OF VECTOR FLAG2O AND STORES THEM IN THE   *
C*  VARIABLES ARGE AND GUT                                 *
C*                                                         *
C************************************************************

      SUBROUTINE FMAX (N,X,J,ARGE,FLAG2O,GUT)
      DIMENSION X(1),FLAG2O(1)
      ARGE = X(1)
      GUT = FLAG2O(1)
      J = 1
      IF (N .EQ. 1) GO TO 20
      DO 10 I = 2, N
      IF (X(I) .LE. ARGE) GO TO 10
      ARGE = X(I)
      GUT = FLAG2O(I)
      J = I
10    CONTINUE
20    RETURN
      END

C************************************************************
C*                                                         *
C*  THIS SUBROUTINE READS THE VALUES OF A TRANSFER FUNCTION *
C*                                                         *
C************************************************************

      SUBROUTINE ING (ZERO,IZ,POLE,IP,TEMP1,TEMP2)
      REAL ZERO(1),POLE(1),TEMP1(1),TEMP2(1)
C
C --> TO ENTER THE TRANSFER FUNCTION INTERACTIVELY
C
      WRITE (5,*) 'Please enter the transfer function :'
      WRITE (5,*) 'How many polynomials in the NUMERATOR'
      READ (5,*) NPN
      WRITE (5,*) 'How many polynomials in the DENOMINATOR'
      READ (5,*) NPD
C
C --> TO ENTER THE NUMERATOR
C
      I2 = 1
      TEMP2(1) = 1.
      DO 40 I = 1, NPN
      WRITE (5,10) I
10    FORMAT (' ','Enter the order+1 of the no. ',I2,
     +         ' NUMERATOR polynomial')
      READ (5,*) I1
      WRITE (5,20)
20    FORMAT (' ','Enter the coeff. of the no. ',I2,
     +         ' NUMERATOR polynomial'/
     +         ' ','in the order of ascending power')
      READ (5,*) (TEMP1(J), J = 1, I1)
      CALL MUTPOL (TEMP1,TEMP2,I1,I2,ZERO,IZ)
C
      IF (I .EQ. NPN) GO TO 50
      I2 = IZ
      DO 30 K = 1, IZ
      TEMP2(K) = ZERO(K)
30    CONTINUE
40    CONTINUE
C
C --> TO ENTER THE DENOMINATOR
C
50    I2 = 1
      TEMP2(1) = 1.
      DO 90 I = 1, NPD
      WRITE (5,60) I
60    FORMAT (' ','Enter the order+1 of the no. ',I2,
     +         ' DENOMINATOR polynomial')
      READ (5,*) I1
      WRITE (5,70)
70    FORMAT (' ','Enter the coeff. of the no. ',I2,
     +         ' DENOMINATOR polynomial'/
     +         ' ','in the order of ascending power')
      READ (5,*) (TEMP1(J), J = 1, I1)
      CALL MUTPOL (TEMP1,TEMP2,I1,I2,POLE,IP)
C
      IF (I .EQ. NPD) GO TO 100
      I2 = IP
      DO 80 K = 1, IP
      TEMP2(K) = POLE(K)
80    CONTINUE
90    CONTINUE
C
100   RETURN
      END
```

TABLE B.6 (cont.)

```
C*************************************************
C*                                              *
C*      THIS SUBROUTINE MULTIPLIES POLYNOMIALS  *
C*      A & B WITH PRODUCT C                    *
C*                                              *
C*      A, B & C -- POLYNOMIALS WITH COEFFICIENTS*
C*                  STORED IN THE ORDER OF      *
C*                  ASCENDING POWER             *
C*                                              *
C*      M, N & K -- ORDER+1 OF A, B & C RESPECTIVELY *
C*                                              *
C*************************************************
C*
      SUBROUTINE MUTPOL (A,B,M,N,C,K)
      DIMENSION A(1),B(1),C(1)
      K = N + M - 1
C
      DO 10 I = 1, K
      C(I) = 0.
10    CONTINUE
C
      DO 20 I = 1, M
      DO 20 J = 1, N
      TEMP = A(I) * B(J)
      C(I+J-1) = C(I+J-1) + TEMP
20    CONTINUE
C
      RETURN
      END
```

```
C*************************************************
C*                                              *
C*      THIS SUBROUTINE USES LIN-BAIRSTOW METHOD TO FIND THE  *
C*      ROOTS OF A POLYNOMIAL.  THE PROGRAM IS WRITTEN TO AVOID *
C*      THE CHANCE OF OVERFLOW.  HOWEVER, IF OVERFLOW OCCURS,  *
C*      THE PROGRAM CANNOT SOLVE FOR THE ROOTS.                *
C*                                              *
C*************************************************
C*
      SUBROUTINE PROOT (A,NA,ROOT,NR)
      REAL A(1),B(30),C(30),D(30)
      COMPLEX ROOT(1),RQUAD(2)
      IF (NA .GT. 0) GO TO 10
      NR = 0
      WRITE (5,*) 'NO COEFF. GIVEN, NO ROOT'
      RETURN
C
10    IF (NA .GT. 1) GO TO 20
      NR = 0
      RETURN
C
20    B(1) = 0.
      B(2) = 0.
      C(1) = 0.
      C(2) = 0.
      NC = NA - 1
      NR = NC
      DO 30 I = 2, NA
      D(I+2) = A(NA-I+1) / A(NA)
      D(3) = 1.
      N3 = NC + 3
C
30
C
      J = 1
40    IF (NC-2) 50, 60, 70
50    ROOT(J) = CMPLX (-D(4), 0.)
      RETURN
60    CALL QUAD (1.,D(4),D(5),RQUAD)
      ROOT(J) = RQUAD(1)
      ROOT(J+1) = RQUAD(2)
      RETURN
C
70    R = 0.
      S = 0.
      TRY = 0.
      ITRY = 0
      TOL = 1.E-7
C
80    DO 110 IC = 1, 50
      DO 90 K = 3, N3
      B(K) = D(K) + R*B(K-1) + S*B(K-2)
90    C(K) = B(K) + R*C(K-1) + S*C(K-2)
C
      DENOM = C(NC+1)**2 - C(NC+2)*C(NC)
      IF (ABS(DENOM) .NE. 0.0) GO TO 100
      R = R + 1.
      S = S + 1.1
      GO TO 80
C
100   RNUM = -B(NC+2)*C(NC+1) + B(NC+3)*C(NC)
      SNUM = -C(NC+1)*B(NC+3) + C(NC+2)*B(NC+2)
      IF (ABS(DENOM).GT.1.E15 .OR. ABS(RNUM).GT.1.E15
     +  .OR. ABS(SNUM).GT.1.E15) GO TO 120
      DR = RNUM / DENOM
      DS = SNUM / DENOM
      R = R + DR
      S = S + DS
      IF (ABS(R).GT.1.E20 .OR. ABS(S).GT.1.E20) GO TO 120
      IF (ABS(DR).LT.TOL .AND. ABS(DS).LT.TOL) GO TO 140
110   CONTINUE
C
      TOL = TOL * 10.
120   IF (TOL .LE. 1.E-4) GO TO 80
      TRY = TRY + 0.5
      R = TRY
      S = -TRY
      ITRY = ITRY + 1
      TOL = 1.E-7
      IF (ITRY .LE. 40) GO TO 80
      WRITE (5,130)
130   FORMAT (' ','FAIL TO CONVERGE: ENTER TWO ESTIMATES TO CONTINUE,'/
     +  ' ','ENTER TWO 999 TO RETURN, ENTER TWO -999 TO STOP')
```

```
      READ (5,*) R,S
      IF (R.EQ.999 .OR. R.EQ.-999) NR=0
      IF (R .EQ. 999) RETURN
      IF (R .EQ. -999) STOP
      GO TO 80
140   CALL QUAD (1.,-R,-S,RQUAD)
      ROOT(J) = RQUAD(1)
      ROOT(J+1) = RQUAD(2)
      J = J + 2
      NC = NC - 2
      N3 = NC + 3
      D(I) = B(I)
150   DO 150 I = 3, N3
      GO TO 40
      END

C********************************************************
C*                                                    *
C*    THIS SUBROUTINE SOLVES THE ROOTS OF A QUADRATIC  *
C*    POLYNOMIAL IN THE FORM OF                        *
C*                                                    *
C*                 2                                   *
C*             XS  + YS + Z                            *
C*                                                    *
C********************************************************
C*
      SUBROUTINE QUAD (X,Y,Z,RQUAD)
      COMPLEX RQUAD(2)
      X2 = 2. * X
      DET = Y**2 - 4.*X*Z
      IF (DET) 10, 20, 30
10    RAD = SQRT (-DET)
      RQUAD(1) = CMPLX (-Y/X2,  RAD/X2)
      RQUAD(2) = CMPLX (-Y/X2, -RAD/X2)
      RETURN
20    RQUAD(1) = CMPLX (-Y/X2, 0.)
      RQUAD(2) = RQUAD(1)
      RETURN
30    RAD = SQRT (DET)
      RQUAD(1) = CMPLX ((-Y+RAD)/X2, 0.)
      RQUAD(2) = CMPLX ((-Y-RAD)/X2, 0.)
      RETURN
      END

C********************************************************
C*                                                    *
C*    THIS SUBROUTINE CHECKS THE EXISTENCE OF POSITIVE ROOTS *
C*    IN BOTH ZEROS AND POLES.  CORRECTION FACTOR FOR  *
C*    POSITIVE ZEROS CAN BE ASSIGNED IN OPTION.        *
C*                                                    *
C********************************************************
```

```
C*
      SUBROUTINE RCHECK (N1,R1,N2,R2,NM,NW)
      COMPLEX R1(1),R2(1)
      IZ = 0
      NM = 0
      DO 20 I = 1, N1
      TEMP = REAL (R1(I))
      IF (TEMP .LT. 0.0) GO TO 20
      WRITE (NW,10) TEMP
      IF (NW .NE. 5) WRITE (5,10) TEMP
10    FORMAT (' ONE ZERO IS +VE ROOT',F14.7)
      IZ = 1
20    CONTINUE
C
      DO 40 I = 1, N2
      TEMP = REAL (R2(I))
      IF (TEMP .LT. 0.0) GO TO 40
      WRITE (NW,30) TEMP
      IF (NW .NE. 5) WRITE (5,30) TEMP
30    FORMAT (' ONE POLE IS +VE ROOT',F14.7)
40    CONTINUE
C
      IF (IZ .NE. 1) RETURN
      WRITE (5,*) 'Do you want to use correction factor for +zero(s)?'
      WRITE (5,*) 'Y=1, N=0'
      READ (5,*) NM
      RETURN
      END

C********************************************************
C*                                                    *
C*    THIS SUBROUTINE REDUCES A POLYNOMIAL WHICH HAS   *
C*    ZERO COEFFICIENT IN THE HIGHEST POWER            *
C*                                                    *
C*    POL(1) -- POLYNOMIAL WITH COEFFICIENTS STORED    *
C*              IN THE ORDER OF ASCENDING POWER        *
C*    N -- ORDER+1 OF THE POLYNOMIAL                   *
C*                                                    *
C********************************************************
C*
      SUBROUTINE REDUCE (POL,N)
      DIMENSION POL(1)
      DO 10 I = 1, N
      IF (ABS (POL(N-I+1)) .GT. 1.E-6) GO TO 20
10    CONTINUE
      N = N - I + 1
20    RETURN
      END
```

TABLE B.6 (cont.)

```
C*********************************************************************
C*                                                                  *
C*    THIS SUBROUTINE CALCULATES THE INVERSE LAPLACE TRANSFORM      *
C*    OF A RATIO OF POLYNOMIALS WITH THE DENOMINATOR HAVING         *
C*    DISTINCT ROOTS                                                *
C*                                                                  *
C*    CIJN -- POLYNOMIAL OF THE NUMERATOR WITH THE                  *
C*            COEFFICIENTS STORED IN THE ORDER OF ASCENDING         *
C*            POWER                                                 *
C*                                                                  *
C*    CIJD -- POLYNOMIAL OF THE DENOMINATOR WITH THE                *
C*            COEFFICIENTS STORED IN THE ORDER OF ASCENDING         *
C*            POWER                                                 *
C*                                                                  *
C*    JJ, II -- ORDER+1 OF THE NUMERATOR AND THE DENOMINATOR        *
C*              RESPECTIVELY                                        *
C*                                                                  *
C*    ROOT -- ROOTS OF THE DENOMINATOR                             *
C*                                                                  *
C*    KK -- II - 1                                                  *
C*                                                                  *
C*********************************************************************
C*
      SUBROUTINE RESPON(CIJD,II,CIJN,JJ,ROOT,KK,NPOINT,TINT,CCIJ)
      DIMENSION CIJD(1),CIJN(1),DCIJD(30),CCIJ(1)
      COMPLEX ROOT(1),CIJ(202),DN,DD
C
C  --> TO DIFFERENTIAE THE DENOMINATOR
C
      CALL DIFF (II,CIJD,LL,DCIJD)
C
      IPOINT = NPOINT - 1
C
      DO 10 I = 1, IPOINT
10    CIJ(1) = CMPLX(0.,0.)
C
      DO 40 N = 1, IPOINT
      DO 40 I = 1, KK
      DN = CMPLX(CIJN(1),0.)
      DD = CMPLX(DCIJD(1),0.)
C
      DO 20 J = 2, JJ
20    DN = DN + CIJN(J) * ROOT(1) ** (J-1)
C
      DO 30 M = 2, LL
30    DD = DD + DCIJD(M) * ROOT(1) ** (M-1)
C
      CIJ(N) = (DN/DD) * CEXP(ROOT(1) * FLOAT(N) * TINT) + CIJ(N)
C
40    CONTINUE
C
      CCIJ(1) = 0.
      DO 50 I = 2, NPOINT
      CCIJ(1) = REAL (CIJ(I-1))
50    CONTINUE
C
      RETURN
      END
```

```
C*********************************************************************
C*                                                                  *
C*    THIS SUBROUTINE SORTS THE ENTRIES OF VECTOR A INTO ASCENDING ORDER  *
C*    AND SORTS THE ENTRIES OF THE VECTOR FLAG2O SIMULTANEOUSLY IN THE    *
C*    SAME WAY                                                      *
C*                                                                  *
C*********************************************************************
C*
      SUBROUTINE SORT(N,A,FLAG2O)
      DIMENSION A(1),FLAG2O(1)
      DO 10 I = 1, N
      M = N - I + 1
      CALL FMAX (M,A,J,ARGE,FLAG2O,GUT)
      A(J) = A(M)
      A(M) = ARGE
      FLAG2O(J) = FLAG2O(M)
      FLAG2O(M) = GUT
10    CONTINUE
      RETURN
      END
```

```
C*********************************************************************
C*                                                                  *
C*    THIS SUBROUTINE ADDS TWO POLYNOMIALS WITH                     *
C*    EACH POLYNOMIAL MULTIPLIES BY A CONSTANT.                     *
C*                                                                  *
C*********************************************************************
C*
      SUBROUTINE SPCADD (NNI,QI,NDI,DE,RPOL,FQI,FDE,MAX)
      DIMENSION QI(1),DE(1),RPOL(1)
      IX = NNI
      IY = NDI
      MAX = NNI
      IF (NNI .LT. NDI) MAX = NDI
      IF (NNI .GT. NNI) NNNI = NNI
      IF (NDI .GT. NNI) NNNI = NDI
C
      DO 40 K = 1, NNNI
      IF (IX .LE. 0) GO TO 20
      IF (IY .LE. 0) GO TO 10
      RPOL(K) = FQI * QI(K) + FDE * DE(K)
      GO TO 30
C
10    RPOL(K) = FQI * QI(K)
      GO TO 40
C
20    RPOL(K) = FDE * DE(K)
      GO TO 40
C
30    IX = NNI - K
      IY = NDI - K
40    CONTINUE
C
      RETURN
      END
```

TABLE B.6 (cont.)

```
C******************************************************
C*                                                   *
C*        THIS SUBROUTINE WRITES THE COEFFICIENTS    *
C*        AND THE ROOTS OF A POLYNOMIAL              *
C*                                                   *
C*        POLY --  POLYNOMIAL WITH THE COEFFICIENTS  *
C*                 STORED IN THE ORDER OF ASCENDING  *
C*                 POWER                             *
C*        NP   --  ORDER+1 OF THE POLYNOMIAL         *
C*        ROOT --  ROOTS OF THE POLYNOMIAL           *
C*        NR   --  NP - 1                            *
C*                                                   *
C******************************************************
C*
         SUBROUTINE WRITEM (POLY,NP,ROOT,NR,NW)
         DIMENSION POLY(1)
         COMPLEX ROOT(1)
         WRITE (NW,10)
10       FORMAT (' THE COEFFICIENTS ARE :'/
     +           ' X**N',30X,'COEFFICIENTS'/)
C
         DO 20 I = 1, NP
         J = I - 1
         WRITE (NW,30) J, POLY(I)
20       CONTINUE
30       FORMAT (' N=',I2,30X,E14.8)
         IF (NP .GT. 1) GO TO 50
C
         WRITE (NW,40)
40       FORMAT (' NO ROOT')
         RETURN
.
50       WRITE (NW,60)
60       FORMAT (' ','THE ROOTS ARE :'//
     +           ' ',12X,'REAL',14X,'IMAGINARY'/
     +           ' ',7X,14('-'),7X,14('-'))
C
         DO 80 I = 1, NR
         A = REAL (ROOT(I))
         B = AIMAG (ROOT(I))
         WRITE (NW,70) A, B
70       FORMAT (' ',7X,F14.7,7X,F14.7)
80       CONTINUE
C
90       WRITE (NW,90)
         FORMAT (//)
         RETURN
         END
```

REFERENCES

GENERAL

AUSLANDER, D. M., Y. TAKAHASHI, and M. J. RABINS, *Introducing Systems and Control.* New York: McGraw-Hill Book Company, 1974.

BREWER, J. W., *Control Systems—Analysis, Design and Simulation.* Englewood Cliffs, N.J.: Prentice-Hall, Inc., 1974.

CANFIELD, E. B., *Electromechanical Control Systems and Devices.* New York: John Wiley & Sons, Inc., 1965.

CANNON, R. H., JR., *Dynamics of Physical Systems.* New York: McGraw-Hill Book Company, 1967.

CHEN, C. F., and I. J. HAAS, *Elements of Control Systems Analysis.* Englewood Cliffs, N.J.: Prentice-Hall, Inc., 1968.

CHEN, C. T., *Analysis and Synthesis of Linear Control Systems.* New York: Holt, Rinehart and Winston, 1975.

D'AZZO, J. J., and C. H. HOUPIS, *Linear Control System Analysis and Design.* New York: McGraw-Hill Book Company, 1975.

DISTEFANO, J. J., III, A. R. STUBBERUD, and I. J. WILLIAMS, *Feedback and Control Systems* (Schaum's Outline Series). New York: Schaum Publishing Co., 1967.

DOEBELIN, E. O., *Dynamic Analysis and Feedback Control.* New York: McGraw-Hill Book Company, 1962.

———, *System Modeling and Response.* New York: John Wiley & Sons, Inc., 1980.

DORF, R. C., *Modern Control Systems,* 3rd ed. Reading, Mass.: Addison-Wesley Publishing Company, Inc., 1980.

DRANSFIELD, P., *Engineering Systems and Automatic Control.* Englewood Cliffs, N.J.: Prentice-Hall, Inc., 1968.

ELGERD, O. I., *Control Systems Theory*. New York: McGraw-Hill Book Company, 1967.

EVELEIGH, V. W., *Introduction to Control Systems Design*. New York: McGraw-Hill Book Company, 1972.

FORTMAN, T. E., and K. L. HITZ, *An Introduction to Linear Control Systems*. New York: Marcel Dekker, Inc., 1977.

GOLDBERG, J. H., *Automatic Controls: Principles of Systems Dynamics*. Boston: Allyn and Bacon, Inc., 1964.

GUPTA, S. C., and L. HASDORFF, *Fundamentals of Automatic Control*. New York: John Wiley & Sons, Inc., 1970.

HALE, F. J., *Introduction to Control System Analysis and Design*. Englewood Cliffs, N.J.: Prentice-Hall, Inc., 1973.

HARRISON, H. L., and J. G. BOLLINGER, *Introduction to Automatic Controls,* 2nd ed. Scranton, Pa.: International Textbook Company, 1969.

KELLER, R. E., *Statics and Dynamics of Components and Systems*. New York: John Wiley & Sons, Inc., 1971.

KUO, B. C., *Automatic Control Systems,* 4th ed. Englewood Cliffs, N.J.: Prentice-Hall, Inc., 1982.

McGILLEM, C. D., and G. R. COOPER, *Continuous and Discrete Signal and System Analysis*. New York: Holt, Rinehart and Winston, 1974.

MELSA, J. L., and D. G. SCHULZ, *Linear Control Systems*. New York: McGraw-Hill Book Company, 1969.

OGATA, K., *Modern Control Engineering*. Englewood Cliffs, N.J.: Prentice-Hall, Inc., 1970.

PALM, W. J., III, *Modeling, Analysis and Control of Dynamic Systems*. New York: John Wiley & Sons, Inc., 1983.

RAVEN, F. H., *Automatic Control Engineering,* 3rd ed. New York: McGraw-Hill Book Company, 1978.

SAGE, A. P., *Linear Systems Control*. Champaign, Ill.: Matrix Publishers, Inc., 1978.

SAVANT, C. J., JR., *Control System Design,* 2nd ed. New York: McGraw-Hill Book Company, 1964.

SHINNERS, S. M., *Modern Control System Theory and Application*. Reading, Mass.: Addison-Wesley Publishing Company, Inc., 1972.

TAKAHASHI, T., *Mathematics of Automatic Control*. New York: Holt, Rinehart and Winston, 1966.

FLUID POWER CONTROL

ANDERSON, B. W., *The Analysis and Design of Pneumatic Systems*. New York: John Wiley & Sons, Inc., 1967.

BELSTERLING, C. A., *Fluidic Systems Design*. New York: Wiley-Interscience, 1971.

BLACKBURN, J. F., G. REETHOF, and J. L. SHEARER, *Fluid Power Control*. Cambridge, Mass.: The MIT Press, 1960.

FOSTER, K., and G. A. PARKER, *Fluidics—Components and Circuits*. New York: Wiley-Interscience, 1970.

LEWIS, E., and H. STERN, *Design of Hydraulic Control Systems*. New York: McGraw-Hill Book Company, 1962.

MERRITT, H. E., *Hydraulic Control Systems*. New York: John Wiley & Sons, Inc., 1967.

STRINGER, J. D., *Hydraulic Systems Analysis*. London: The Macmillan Press Ltd., 1976.

NONLINEAR CONTROL SYSTEMS

ATHERTON, D. P., *Nonlinear Control Engineering*. London: Van Nostrand Reinhold Company Ltd., 1975.

GELB, A., and W. E. VANDER VELDE, *Multiple-Input Describing Functions and Nonlinear System Design*. New York: McGraw-Hill Book Company, 1968.

GIBSON, J. E., *Nonlinear Automatic Control*. New York: McGraw-Hill Book Company, 1963.

GRAHAM, D., and D. McRUER, *Analysis of Nonlinear Control Systems*. New York: John Wiley & Sons, Inc., 1961.

HSU, J. C., and A. U. MEYER, *Modern Control Principles and Applications*. New York: McGraw-Hill Book Company, 1968.

MINORSKY, N., *Theory of Nonlinear Control Systems*. New York: McGraw-Hill Book Company, 1969.

THALER, G. J., and M. P. PASTEL, *Analysis and Design of Nonlinear Feedback Control Systems*. New York: McGraw-Hill Book Company, 1962.

WEST, J. C., *Analytical Techniques for Nonlinear Control Systems*. London: English Universities Press, 1960.

DIGITAL CONTROL SYSTEMS

ÅSTRÖM, K. J., and B. WITTENMARK, *Computer-Controlled Systems—Theory and Design*. Englewood Cliffs, N.J.: Prentice-Hall, Inc., 1984.

CASSELL, D. A., *Microcomputers and Modern Control Engineering*. Reston, Va.: Reston Publishing Co., Inc., 1983.

DESHPANDE, P. B., and R. H. ASH, *Elements of Computer Process Control—With Advanced Control Applications*. Englewood Cliffs, N.J.: Prentice-Hall, Inc., 1981.

FRANKLIN, G. F., and J. D. POWELL, *Digital Control of Dynamic Systems*. Reading, Mass.: Addison-Wesley Publishing Company, Inc., 1980.

JACQUOT, R. G., *Modern Digital Control Systems*. New York: Marcel Dekker, Inc., 1981.

KUO, B. C., *Digital Control Systems*. New York: Holt, Rinehart and Winston, 1980.

LEE, T. H., G. E. ADAMS, and W. M. GAINES, *Computer Process Control—Modeling and Optimization*. New York: John Wiley & Sons, Inc., 1968.

LINDORFF, D. P., *Theory of Sampled-Data Control Systems*. New York: John Wiley & Sons, Inc., 1965.

PHILLIPS, C. L., and H. T. NAGLE, JR., *Digital Control System Analysis and Design*. Englewood Cliffs, N.J.: Prentice-Hall, Inc., 1984.

RAGAZZINI, J. R., and G. F. FRANKLIN, *Sampled-Data Control Systems*. New York: McGraw-Hill Book Company, 1958.

SAVAS, E. S., *Computer Control of Industrial Processes*. New York: McGraw-Hill Book Company, 1965.

SHINSKEY, F. G., *Process Control Systems,* 2nd ed. New York: McGraw-Hill Book Company, 1979.

ADVANCED TECHNIQUES AND MULTIVARIABLE SYSTEMS

ANDERSON, B. D. O., and J. B. MOORE, *Linear Optimal Control*. Englewood Cliffs, N.J.: Prentice-Hall, Inc., 1971.

BRYSON, A. E., JR., and Y. C. HO, *Applied Optimal Control*. Waltham, Mass.: Ginn and Company, 1969.

CHANG, S. S. L., *Synthesis of Optimum Control Systems*. New York: McGraw-Hill Book Company, 1961.

CHEN, C. T., *Introduction to Linear System Theory*. New York: Holt, Rinehart and Winston, 1970.

CROSSLEY, T. R., and B. PORTER, "Synthesis of Aircraft Modal Control Systems Having Real or Complex Eigenvalues," *The Aeronautical Journal of the Royal Aeronautical Society,* Vol. 73, pp. 138–142, Feb. 1969.

DAVISON, E. J., and H. W. SMITH, "A Note on the Design of Industrial Regulators: Integral Feedback and Feedforward Controllers," *Automatica,* Vol. 10, pp. 329–332, May 1974.

———, and S. H. WANG, "On Pole Assignment in Linear Multivariable Systems Using Output Feedback," *IEEE Transactions on Automatic Control,* Vol. AC-20, pp. 516–518, Aug. 1975.

DERUSSO, P. M., R. J. ROY, and C. M. CLOSE, *State Variables for Engineers*. New York: John Wiley & Sons, Inc., 1965.

EDMUNDS, J. M., "Control System Design and Analysis Using Closed Loop Nyquist and Bode Arrays," *International Journal of Control,* Vol. 30, pp. 773–802, Nov. 1979.

FALLSIDE, F., *Control System Design by Pole–Zero Assignment*. London: Academic Press, Inc. (London) Ltd., 1977.

HASDORFF, L., *Gradient Optimization and Nonlinear Control*. New York: John Wiley & Sons, Inc., 1976.

HAWKINS, D. J., "Pseudodiagonalization and the Inverse Nyquist Array Method," *Proceedings of the IEE,* Vol. 119, pp. 337–342, March 1972.

HUNG, Y. S., and A. G. J. MACFARLANE, *Multivariable Feedback: A Quasi-Classical Approach*. Berlin: Springer-Verlag, 1982.

JOHNSON, M. A., "Diagonal Dominance and the Method of Pseudodiagonalization," *Proceedings of the IEE,* Vol. 126, pp. 1011–1017, Oct. 1979.

KAILATH, T., *Linear Systems*. Englewood Cliffs, N.J.: Prentice-Hall, Inc., 1980.

KIRK, D. E., *Optimal Control Theory—An Introduction*. Englewood Cliffs, N.J.: Prentice-Hall, Inc., 1970.

KOSUT, R. L., "Suboptimal Control of Linear Time-Invariant Systems Subject to Control Structure Constraints," *IEEE Transactions on Automatic Control,* Vol. AC-15, pp. 557–563, Oct. 1970.

KOUVARITAKIS, B. A., "Theory and Practice of the Characteristic Locus Design Method," *Proceedings of the IEE,* Vol. 126, pp. 542–548, June 1979.

KWAKERNAAK, H., and R. SIVAN, *Linear Optimal Control Systems.* New York: Wiley-Interscience, 1972.

LAPIDUS, L., and R. LUUS, *Optimal Control of Engineering Processes.* Waltham, Mass.: Blaisdell Publishing Company, 1967.

LEININGER, G. G., "Diagonal Dominance Using Function Minimization Algorithms," *Proceedings of the IFAC Symposium on Multivariable Technological Systems,* pp. 105–112, Fredericton, Canada, July 1977.

———, "Multivariable Compensator Design Using Bode Diagrams and Nichols Charts," *Proceedings of the IFAC Symposium on Computer Aided Design,* pp. 127–132, Zürich, 1979.

LEITMAN, G., *An Introduction to Optimal Control.* New York: McGraw-Hill Book Company, 1966.

LEVINE, W. S., and M. ATHANS, "On the Determination of the Optimal Constant Output Feedback Gains for Linear Multivariable Systems," *IEEE Transactions on Automatic Control,* Vol. AC-15, pp. 44–50, Feb. 1970.

———, T. L. JOHNSON and M. ATHANS, "Optimal Limited State Variable Feedback Controllers for Linear Systems," *IEEE Transactions on Automatic Control,* Vol. AC-16, pp. 785–792, Dec. 1971.

MACFARLANE, A. G. J., "Return-Difference and Return-Ratio Matrices and Their Use in Analysis and Design of Multivariable Feedback Control Systems," *Proceedings of the IEE,* Vol. 117, pp. 2037–2049, Oct. 1970.

———, and J. J. BELLETRUTTI, "The Characteristic Locus Design Method," *Automatica,* Vol. 9, pp. 575–588, Sept. 1973.

———, and B. KOUVARITAKIS, "A Design Technique for Linear Multivariable Feedback Systems," *International Journal of Control,* Vol. 25, No. 6, pp. 837–874, 1977.

———, and I. POSTLETHWAITE, "The Generalized Nyquist Stability Criterion and Multivariable Root Loci," *International Journal of Control,* Vol. 25, pp. 81–127, Jan. 1977.

MCCAUSLAND, I., *Introduction to Optimal Control.* New York: John Wiley & Sons, Inc., 1969.

MUNRO, N., *Modern Approaches to Control System Design.* Stevenage, U.K.: Peter Peregrinus Ltd., 1979.

OGATA, K., *State Space Analysis of Control Systems.* Englewood Cliffs, N.J.: Prentice-Hall, Inc., 1967.

OLDENBURGER, R., *Optimal Control.* New York: Holt, Rinehart and Winston, 1966.

PATEL, R. V., and N. MUNRO, *Multivariable System Theory and Design.* Oxford: Pergamon Press, Ltd., 1982.

PORTER, B., and R. CROSSLEY, *Modal Control—Theory and Applications.* London: Taylor and Francis Ltd., 1972.

PRIME, H. A., *Modern Concepts in Control Theory.* London: McGraw-Hill Publishing Company Ltd., 1969.

ROSENBROCK, H. H., "Design of Multivariable Control Systems Using the Inverse Nyquist Array," *Proceedings of the IEE,* Vol. 116, pp. 1929–1936, Nov. 1969.

———, *State-Space and Multivariable Theory*. London: Thomas Nelson and Sons Ltd., 1970.

———, "Progress in the Design of Multivariable Control Systems," *Measurement and Control,* Vol. 4, pp. 9–11, Jan. 1971.

———, *Computer-Aided Control System Design*. London: Academic Press, Inc. (London) Ltd., 1974.

SAGE, A. P., *Optimum Systems Control*. Englewood Cliffs, N.J.: Prentice-Hall, Inc., 1968.

SCHULZ, D. G., and J. L. MELSA, *State Functions and Linear Control Systems*. New York: McGraw-Hill Book Company, 1967.

SMITH, H. W., and E. J. DAVISON, "Design of Industrial Regulators," *Proceedings of the IEE,* Vol. 119, pp. 1210–1215, Aug. 1972.

TAKAHASHI, Y., M. J. RABINS, and D. M. AUSLANDER, *Control and Dynamic Systems*. Reading, Mass.: Addison-Wesley Publishing Company, Inc., 1970.

VAN DE VEGTE, J., "Classical Design of Two-by-Two Systems," *International Journal of Control,* Vol. 35, No. 3, pp. 477–489, 1982.

———, "Classical Design of Three-by-Three Systems," *International Journal of Control,* Vol. 37, No. 3, pp. 503–519, 1983.

WIBERG, D. M., *State Space and Linear Systems* (Schaum's Outline Series). New York: McGraw-Hill Book Company, 1971.

WOLOVICH, W. A., *Linear Multivariable Systems*. New York: Springer-Verlag (New York), Inc., 1974.

INDEX